Mathematics
FOR BUSINESS CAREERS

Second Edition

Mathematics
FOR BUSINESS CAREERS

SECOND EDITION

Jack Cain
Oklahoma City Community College

Robert A. Carman
Santa Barbara City College

JOHN WILEY & SONS

New York • Chichester • Brisbane • Toronto • Singapore

To Patty and 'lyn

Copyright © 1985, by John Wiley & Sons, Inc.

All rights reserved. Published simultaneously in Canada.

Reproduction or translation of any part of
this work beyond that permitted by Sections
107 and 108 of the 1976 United States Copyright
Act without the permission of the copyright
owner is unlawful. Requests for permission
or further information should be addressed to
the Permissions Department, John Wiley & Sons.

Library of Congress Cataloging in Publication Data:

Cain, Jack.
  Mathematics for business careers.

  Includes index.
  1. Business mathematics. 2. Business mathematics—
Problems, exercises, etc. I. Carman, Robert A.
II. Title.
HF5691.C26 1985   513'.93   84-11949
ISBN 0-471-80856-3

Printed in the United States of America

10 9 8 7 6 5 4 3 2 1

Mathematics
FOR BUSINESS CAREERS

SECOND EDITION

Jack Cain
Oklahoma City Community College

Robert A. Carman
Santa Barbara City College

JOHN WILEY & SONS

New York • Chichester • Brisbane • Toronto • Singapore

To Patty and 'lyn

Library of Congress Cataloging in Publication Data:

Cain, Jack.
　Mathematics for business careers.

　Includes index.
　1. Business mathematics.　2. Business mathematics—
Problems, exercises, etc.　I. Carman, Robert A.
II. Title.
HF5691.C26　1985　　　513'.93　　　84-11949
ISBN 0-471-80856-3

Printed in the United States of America

10 9 8 7 6 5 4 3 2 1

PREFACE

This textbook is intended to be used in a variety of teaching situations from a traditional classroom lecture to individually paced instruction. Its primary function is to be *used* by the students.

Learning business mathematics is much like learning tennis. One can watch an instructor demonstrate a top-spin forehand or can read about it, but to master the technique, one needs practice, practice, and more practice. This textbook provides students with the necessary practice by having them participate actively in the course. A three-step procedure is followed:

Step 1: A frame is read that usually includes some explanatory material and several examples.

Step 2: Students complete practice problems.

Step 3: Answers are checked for accuracy, which provides immediate feedback.

This three-step method facilitates the students' comprehension of various math concepts by providing a means of "learning by doing."

Special features are included to make this text effective for students:

- Careful attention has been given to readability and the visual organization of the text.
- A diagnostic pretest and performance objectives keyed to the text are given at the beginning of each unit. These show students what is expected and provide a sense of direction.
- Each unit ends with a self-test covering the work of that unit.
- The format is clear and easy to follow. It respects the individual needs of each reader, providing immediate feedback at each step to assure their understanding and continued attention.
- Numerous problem sets are included so that the student has an abundant opportunity for practicing the business mathematics operations being learned. Answers to all the section problems are in the back of the text.
- Supplementary problem sets are included at the end of each unit, and additional problem sets are available in the accompanying teacher's manual. Answers to all supplementary problem sets are given in the teacher's manual.
- A light, lively conversational style of writing and a pleasant, easy-to-understand visual approach are used. The use of humor is designed to appeal to students who have in the past found business mathematics to be dry and uninteresting.

This book has been used in both classroom and individualized instruction settings and carefully field-tested with hundreds of students. Students who have used the book tell us it is helpful, interesting, even fun to work through. More important, it works.

Flexibility of use was a major criterion in the design of the book, and field-testing indicates that the book can be used successfully in a variety of course formats. It can be used as a textbook in traditional lecture-oriented courses. It is also very effective in situations where an instructor wishes to modify a traditional course by devoting a portion of class time to independent study. The book is especially useful in programs of individualized or self-paced instruction, whether in a learning lab situation, with tutors, or in totally independent study.

The second edition contains several new sections including:

1　The new ACRS depreciation method, which has been added to the traditional methods in Unit 12. Teachers may select which methods to cover.

2　A section on bank discount using bankers' interest (ordinary simple interest at exact time), which is included in Unit 8.

3　The tax tables and FICA rate, which are updated.

4　Problems and examples that are changed to reflect current interest rates, trends, and accuracy.

5　Additional problems, which are included on most problem sets.

An accompanying teacher's resource book provides:

- Information on a variety of self-paced and individualized course formats that may be used.
- Multiple forms of all unit tests, brief quizzes, and final examinations; additional problem sets.
- Answers to all problems in the unit tests, quizzes, exams, and supplementary problem sets.
- Tables and examples, from which transparency masters can be made.

We thank the many people who have contributed to the development of this book. We are indebted to the following instructors who read the preliminary edition and offered many helpful criticisms and suggestions:

Alton W. Evans,	**Tarrant County Junior College**
Clo Hampton,	**West Valley Community College**
Donald Hollin,	**Mississippi County Community College**
Jim McAnelly,	**Waubonsee Community College**
Michael R. Menaker,	**Montgomery College**
Elmer Shellenberger,	**Bethany Nazarene College**

We also thank the instructors who field-tested the preliminary edition:

Betty Coleman	Ann Martin
Ken Duncan	Jim Maisano
Joanne Forgue	Judy Mee
Juanolo Garriott	Elsie Milliron
Caroline Goad	Gus Pekara
Leon Graumann	Leodies Robinson
Tom Hutchinson	Mary Schallhorn
Deanne Ingram	Patty Staggs
Larry Irwinsky	Keith Wilson
Jim Kirk	Mary Wright

We also thank the reviewers for the second edition:

Ruth E. Branyan	**Humphreys College**
Joanne M. Collins	**Kelley Business Institute**
Donald Linner	**Essex County College**

A special note of thanks is due to Bob Wilson, Wiley Representative, who helped get this project started; to Gary DeWalt, former Business Education Editor at Wiley, whose encouragement and direction added greatly to this book; to Leonard Kruk, current Wiley Business Education Editor, who helped in the manuscript's development; to Jerry Irwinsky, who sketched all the cartoons for the chapter openings; and to Patty Cain, for working and checking the problem solutions and for her aid in the revision.

Jack Cain
Robert A. Carman

CONTENTS

vii

CONTENTS

vii

Mathematics

FOR BUSINESS CAREERS

Second Edition

Whole Numbers

SORRY BOSS! I COULDN'T FINISH THE REPORT LAST NIGHT. THE BATTERIES IN MY CALCULATOR WERE DEAD.

Whole numbers! Why this? I want to learn business mathematics, not arithmetic. I can always use my calculator to do the arithmetic but I need to know some of that heavy stuff about discounts and taxes and interest on loans.

If that sounds like your reaction to this introduction to business mathematics, relax. You're in the right place. We *are* going to explain all of those very important areas of business mathematics, and we will do it in such a way that you will be able to remember it, use it, and even go on to more advanced courses such as accounting or marketing if you wish. But to get you there we must start with the fundamentals. This book is designed so that you move quickly from concepts you understand to more difficult ones.

By the way, if you do have a calculator, warm it up, we'll be using it. A calculator is a handy helper, but it will never do the work for you or allow you to solve a problem when you do not understand the basic arithmetic concepts. Punch in the wrong numbers and you get the wrong answer—lightning fast. In this textbook you'll learn the basic mathematics ideas first, then you'll get practice in using your calculator on them.

Now let's get to work.

UNIT OBJECTIVE

When you complete this module successfully, you will be able to write whole numbers in both numerical and word form. You will also be able to add, subtract, multiply, and divide whole numbers, and use these operations in solving business problems.

This is a sample test for Unit 1. All the answers are placed immediately after the test. Work as many of the problems as you can, then check your answers and look for further directions after the answers.

		Where to Go for Help	
		Page	Frame
1. Convert the number 4,589,372 to word form.	——————	3	1
2. Convert the number: six hundred seventy-four thousand, nine hundred fifty-one to numerical form.	——————	3	1
3. Add: 857 + 26,457 + 38 + 4856.	——————	9	7
4. Subtract: 6,842,287 − 951,593.	——————	19	17
5. Multiply: 6845 × 279.	——————	25	26
6. Divide: 56,497 ÷ 356.	——————	33	35
7. Robbin's Athletic Supply Company had the following gross income for the first quarter of the year: January—$2857; February—$156 (bad month); and March—$2967. What was the total gross income for the quarter?	——————	9	7
8. Juanita's Surplus Sales purchased 398 cartons of reconditioned suspenders at $28 per carton. What was the total cost?	——————	25	26

★ If you cannot work any of these problems, start with frame 1 on page 3.
★ If you missed some of the problems, turn to the page numbers indicated.
★ If all of your answers were correct, you are probably ready to proceed to Unit 2. (If you would like more practice on Unit 1 before turning to Unit 2, try the Self-Test on page 41.)
★ Super-students—those who want to be certain they learn all of this material—will turn to frame 1 and begin work there.

Preview 1
Answers

1. Four million, five hundred eighty-nine thousand, three hundred seventy-two.

2. 674,951   3. 32,208   4. 5,890,694   5. 1,909,755   6. 158 R249   7. $5980   8. $11,144

SECTION 1: WRITING NUMBERS IN WORDS

1

> **OBJECTIVE**
>
> Convert a whole number in numerical form to word form.
> Convert a whole number in word form to numerical form.

We use *numerals* to name numbers. For example, the number of corners on a square is four, or 4, or IV in Roman numerals, or 〤 in Chinese numerals.

In our modern number system we use the ten *digits* 0, 1, 2, 3, 4, 5, 6, 7, 8, and 9 to build numerals, just as we use the twenty-six letters of the alphabet to build words.

Along with the digits, we use a *place value system* of naming numbers, where the value of any digit depends on the place in which it is located.

```
              ┌──────────────┐
              │  Tens place  │
              │ ┌────────────┤
              │ │ Ones place │
              ▼ ▼
49 ─────────▶ │ 4 │ 9 │ ◀───────────── 4 tens + 9 ones
```

The right most digit denotes the *ones* place. The next place to the left is called the *tens* place. For example, 49 represents 9 ones and 4 tens, or forty-nine. The number "49" is a *numerical form;* "forty-nine" is the *word form.*

Try writing 68 in word form.

68 = _____

Turn to **3** to check your answer.

---

2   Hi.

What are you doing in here? Lost? Window shopping? Just passing through? Nowhere in this book are you directed to frame **2**. (Notice that little **2** to the left above? That's a frame number.) Remember, in this book you move from frame to frame as directed, but not necessarily in 1-2-3 order. Follow directions and you'll never get lost.

Now, return to **1** and keep working.

---

3   68 = 8 ones and 6 tens or sixty-eight.
Easy, isn't it?

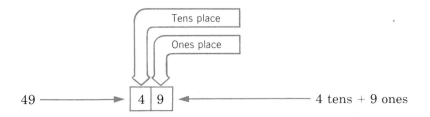

$$1000 \times 2 = 2000$$
$$100 \times 8 = 800$$
$$10 \times 4 = 40$$
$$1 \times 9 = \underline{+\ 9}$$
$$2849$$

```
2849 ─────────▶ │ 2 │ 8 │ 4 │ 9 │ ◀───────── 2 thousands + 8 hundreds +
                                              4 tens + 9 ones
```

After the tens place is the *hundreds* place and the *thousands* place. The number 5482 represents 2 ones, 8 tens, 4 hundreds, and 5 thousands or five thousand, four hundred eighty-two.

Write out the following numbers in word form.

(a)   367 = _____

(b)   9712 = _____

(c)   3045 = _____

Check your answers in **4**.

---

4   (a)   367 = three hundred sixty-seven.
    (b)   9712 = nine thousand, seven hundred twelve.
    (c)   3045 = three thousand, forty-five.

Any large number given in numerical form may be translated to words by using the following diagram.

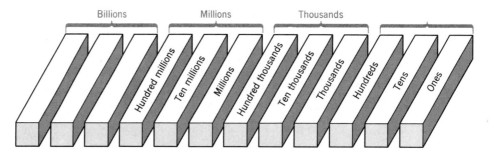

The number 14,237 can be placed in this diagram like this:

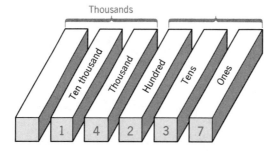

and is read "fourteen thousand, two hundred thirty-seven."

The number 47,653,290,866 becomes

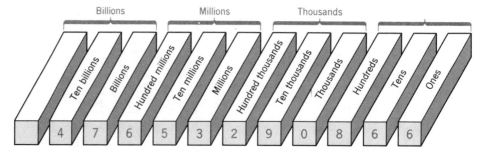

and is read "forty-seven billion, six hundred fifty-three million, two hundred ninety thousand, eight hundred sixty-six."

In each block of three digits read the digits in the normal way ("forty-seven," "six hundred fifty-three") and add the name of the block ("billion," "million"). Notice that the word "and" is not used in naming these numbers.

Write the following numbers in word form.

(a)   6,109,276 = ——————————

(b)   156,768 = ——————————

(c)   538,251,740,459 = ——————————

Turn to **5** and check your answers.

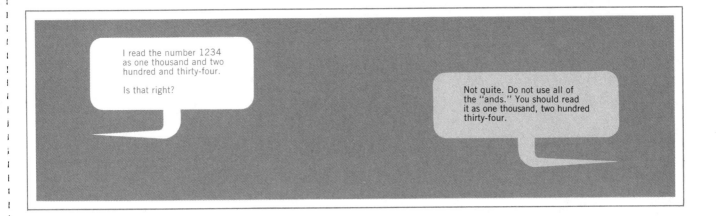

5   (a)   6,109,276 = six million, one hundred nine thousand, two hundred seventy-six.
    (b)   156,768 = one hundred fifty-six thousand, seven hundred sixty-eight.
    (c)   538,251,740,459 = five hundred thirty-eight billion, two hundred fifty-one million, seven hundred forty thousand, four hundred fifty-nine.

If you had difficulty with any of these, you should return to **1** and review. Otherwise, go to **6** for a set of practice problems on writing whole numbers.

Whole Numbers

6 The answers are on page 573.

1.1 Writing Whole Numbers

A. Write the following in word form.

1. 82 _____   2. 17 _____

3. 156 _____   4. 247 _____

5. 851 _____   6. 439 _____

7. 1825 _____   8. 7621 _____

9. 14,496 _____   10. 57,418 _____

11. 239,153 _____   12. 156,483 _____

13. 92,742,007 _____ 14. 4,785,310 _____

15. 186,014,157 _____ 16. 24,160,438 _____

17. 2,765,984 _____ 18. 5,000,701 _____

19. 23,493,631,375 _____ 20. 186,393,613,074 _____

B. Write the following in numerical form.

1. Ninety-seven _____

2. Forty-two _____

3. Six hundred fifty-two _____

4. Four hundred twenty-nine _____

5. Twenty-four thousand _____

6. Seventy-eight thousand, sixty-five _____

7. Nine thousand, six hundred five _____

8. Seven hundred forty-seven thousand, four hundred eighty-nine

   _____

9. Eighteen million, five hundred sixty-seven thousand, thirty-two

   _____

10. Eighty-three million, nine hundred three thousand, six hundred fifty-

   one _____

Date _____

Name _____

Course/Section _____

11. Twenty-six thousand, four hundred ninety-eight _____

12. Twenty-eight million, four hundred six thousand, five hundred thirty-one _____

13. Nine million, six hundred forty-one thousand, eight hundred seventeen _____

14. Ninety-eight thousand, five _____

15. Five hundred thousand, four hundred thirty-two _____

16. Thirty-nine thousand, eight hundred nine _____

17. Fifty-six billion, four hundred sixty million, five hundred thirteen thousand, six hundred twenty-seven _____

18. Fourteen billion, eight hundred twenty-one million, four hundred one thousand, fifty-two _____

19. One hundred eleven billion, two hundred ninety-three thousand, three hundred seventy-eight _____

20. Two billion, four hundred million, one hundred eighty-six.

_____

When you have had the practice you need, either return to the Preview on page 2 or continue in 7 with the study of addition of whole numbers.

SECTION 2: ADDING WHOLE NUMBERS

7

OBJECTIVE

Add a list of whole numbers and use addition to solve word problems.

Addition is the simplest arithmetic operation.

$$\begin{matrix} \$ \ \$ \\ \$ \ \$ \end{matrix} + \begin{matrix} \$ \ \$ \\ \$ \end{matrix} = \begin{matrix} \$ \ \$ \ \$ \\ \$ \ \$ \ \$ \end{matrix}$$

$$4 \ + \ 3 \ = \underline{\hspace{1.5cm}}$$

Complete the calculation and go to **9**.

---

8   Here is the completed addition square.

Add	4	2	8	7	5	6	1	3	9
2	6	4	10	9	7	8	3	5	11
4	8	6	12	11	9	10	5	7	13
7	11	9	15	14	12	13	8	10	16
5	9	7	13	12	10	11	6	8	14
1	5	3	9	8	6	7	2	4	10
9	13	11	17	16	14	15	10	12	18
6	10	8	14	13	11	12	7	9	15
8	12	10	16	15	13	14	9	11	17
3	7	5	11	10	8	9	4	6	12

Did you notice that changing the order in which you add numbers does not change their sum?

$4 + 3 = 7$ and $3 + 4 = 7$
$2 + 4 = 6$ and $4 + 2 = 6$ and so on.

If you have not already memorized the addition of one-digit numbers, it is time you did so.

Now, let's try a more difficult addition problem:

$$35 + 42 = \underline{\hspace{1.5cm}}$$

Work it out and check your answer in **10**.

---

9    $$\begin{matrix} \$ \ \$ \\ \$ \ \$ \end{matrix} + \begin{matrix} \$ \ \$ \\ \$ \end{matrix} = \begin{matrix} \$ \ \$ \ \$ \\ \$ \ \$ \ \$ \end{matrix}$$

$$4 \ + \ 3 \ = 7 \qquad \text{of course}$$

We add collections of objects by combining them into a single set and then counting and naming that new set. The numbers being added, 4 and 3 in this case, are called *addends* and 7 is the *sum* of the addition.

There are a few addition facts necessary for business you should have stored in your memory and be ready to use, whether or not you have a calculator. Complete the following table by adding the number at the top to the number at the side and placing their sum in the proper square. We have added $1 + 2 = 3$ and $4 + 3 = 7$ for you.

Add	4	2	8	7	5	6	1	3	9
2									
4								7	
7									
5									
1		3							
9									
6									
8									
3									

Check your answer in **8**.

10   You should have set it up like this:

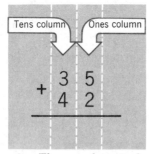

1. The numbers to be added are arranged vertically in columns.
2. The right end or ones digits are placed in the ones column, the tens digits are placed in the tens column, and so on.

Avoid the confusion of

$$\begin{array}{r} 35 \\ +\ 42 \\ \hline \end{array} \quad \text{or} \quad \begin{array}{r} 35 \\ +\ 4\ 2 \\ \hline \end{array}$$

Be Careful > The most frequent cause of errors in arithmetic is carelessness, especially in simple procedures such as lining up the digits correctly.

Carelessness in business may cost someone money and possibly a job.

Once the digits are lined up the problem is easy.

$$\begin{array}{r} 35 \\ +42 \\ \hline 77 \end{array}$$

Here is a slightly more difficult problem:

$27 + 48 =$ _____

Try it, then go to 11.

---

11   **First,** line up the addends vertically. $\begin{array}{r} 27 \\ +48 \\ \hline \end{array}$

Second, add each column, starting on the right.

Step 1

Add the ones column:

$\begin{array}{r} \overset{1}{} \\ 27 \\ +48 \\ \hline 5 \end{array}$ 7 + 8 = 15
            Write 5.
            Carry 1 ten.

Step 2

Add the tens column:

$\begin{array}{r} \overset{1}{} \\ 27 \\ +48 \\ \hline 75 \end{array}$ 1 + 2 + 4 = 7

The little 1 above the tens digit column is the "carry." The sum of the units column is 7 + 8 = 15. We write down the 5 and "carry" the 1 to the next column.

Add $429 + 758 =$ _____

For you calculator-oriented people who are in a hurry, skip to checking addition in 14.

Set it up as we did above, then turn to 12.

---

12   $429 + 758 =$ _____

First, line up the addends.

$\begin{array}{r} 429 \\ +758 \\ \hline \end{array}$

Then add column by column, carrying when necessary.

Step 1

$\begin{array}{r} \overset{1}{} \\ 429 \\ +758 \\ \hline 7 \end{array}$ 9 + 8 = 17
           Write 7.
           Carry 1 ten.

Step 2

$\begin{array}{r} \overset{1}{} \\ 429 \\ +758 \\ \hline 87 \end{array}$ 1 + 2 + 5 = 8

Step 3

$\begin{array}{r} \overset{1}{} \\ 429 \\ +758 \\ \hline 1187 \end{array}$ 4 + 7 = 11

Now, try these problems.

(a) $\$256 + 867 =$ _____

(b) $267 + 1135 + 2461 =$ _____

(c) $\$15,178 + 166 + 4415 + 27 + 13,001 =$ _____

Check your answers in **13** when you are finished.

HOW TO ADD LONG LISTS OF NUMBERS

Very often, especially in business and industry, it is necessary to add long lists of numbers. The best procedure is to break the problem down into a series of simpler additions. First add sets of two or three numbers, then add these sums to obtain the total.

For example,

```
   9
   3        12
  ─────────────
   7
   6        13        25
  ─────────────────────────
  12
   4        16
  ─────────────
  17
 + 5        22        38
  ─────────────────────────
                      63
```

You do a little more writing but carry fewer numbers in your head; the result is fewer mistakes.

Better yet, keep your eye peeled for combinations that add to 10 or 15, and work with mental addition of three addends.

```
   9
   3 ◄
   7 ◄ ──10              19
  ─────────────────────────
   6 ◄
  12 ──10
   4 ◄                   22
  ─────────────────────────
  17
   5                     22
  ─────────────────────────
                         63
```

Try these for practice:

(a)	(b)	(c)	(d)	(e)	(f)
8	7	3	11	3	13
17	6	5	7	5	17
3	8	7	2	12	11
4	5	6	5	7	14
11	9	5	6	6	15
9	3	1	7	4	8
16	7	3	13	1	9
7	12	4	6	2	16
11	8	2	5	18	12
5	16	7	14	9	7
		3	16	7	8

The answers are on page 573.

13   (a)   **Step 1**   **Step 2**   **Step 3**

^{1}		$^{1\ 1}$		$^{1\ 1}$	
$256		$256		$256	
+867	$6 + 7 = 13$	+867	$1 + 5 + 6 = 12$	+867	$1 + 2 + 8 = 11$
3	Write 3.	23	Write 2.	$1123	
	Carry 1 ten.		Carry 1 hundred.		

(b)   **Step 1**   **Step 2**

^{1}		$^{1\ 1}$	
267		267	
1135		1135	
+2461	$7 + 5 + 1 = 13$	+2461	$1 + 6 + 3 + 6 = 16$
3	Write 3.	63	Write 6.
	Carry 1 ten.		Carry 1 hundred.

      Step 3   **Step 4**

$^{1\ 1}$		$^{1\ 1}$	
267		267	
1135		1135	
+2461	$1 + 2 + 1 + 4 = 8$	+2461	$1 + 2 = 3$
863		3863	

(c)   **Step 1**   **Step 2**

^{2}		$^{1\ 2}$	
$15,178		$15,178	
166		166	
4,415		4,415	
27		27	
+13,001	$8 + 6 + 5 + 7 + 1 = 27$	+13,001	$2 + 7 + 6 + 1 + 2 + 0 = 18$
7	Write 7.	87	Write 8.
	Carry 2 tens.		Carry 1 hundred.

      Step 3   **Step 4**

$^{1\ 2}$		$^{1\ \ 1\ 2}$	
$15,178		$15,178	
166		166	
4,415		4,415	
27		27	
+13,001	$1 + 1 + 1 + 4 + 0 = 7$	+13,001	$5 + 4 + 3 = 12$
787		2,787	Write 2.
			Carry 1 ten-thousand.

      Step Last

$^{1\ \ 1\ 2}$	
$15,178	
166	
4,415	
27	
+13,001	$1 + 1 + 1 = 3$
$32,787	

Now let's try applying addition to a business problem.

Robert, salesman for the Jiffy Linoleum Company, earned the following monthly commissions: January—$576; February—$607; March—$623; April—$676; May—$246; June—$556; July—$598; August—$627; September—$653; October—$675; November—$692; December—$702. What were his total earnings for the year?

Check your answer in 14 when you are finished.

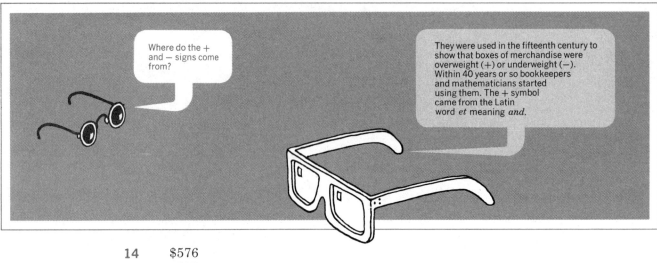

14    $576
607
623
676
246
556
598
627
653
675
692
+702

$7231

CHECKING ADDITION

Businesses use many complicated methods of "checking" addition. The easiest method is by adding in the opposite direction. In addition, normally we start at the top and add downward. In "checking," we start at the bottom and add each column upward.

Add and check 456 + 862 + 57.

Add		456	Check: ↑	456
	↓	862		862
Downward	▼	+57	Add up	+57
		1375		1375

Work and check the following problem.

Mary had a savings account balance of $478 before she made deposits of $75 and $115. What is her current balance?

Our work is in **15**.

15

Add		$478	Check: ↑	$478
	↓	75		75
Downward	▼	+115	Add up	+115
		$668		$668

You should always check your answers even if you use a calculator. Although calculators rarely make errors, calculator operators often do.

If you had any trouble with the last problem, you should return to **7** and review. Otherwise, go to **16** for a set of practice problems.

Whole Numbers

PROBLEM SET 2

16   The answers are on page 573.

1.2 Adding Whole
     Numbers

A. Add.

1. 47	2. 18	3. 27	4. 57	5. 45	6. 89
23	86	38	69	35	17

7. 73	8. 44	9. 92	10. 38	11. 88	12. 75
39	28	39	65	17	48

13. 47	14. 26	15. 76	16. 48	17. 33	18. 67
56	98	24	84	19	69

B. Add.

1. 273	2. 189	3. 726	4. 508	5. 701
142	204	387	495	829

6. 684	7. 729	8. 432	9. 708	10. 621
706	287	399	554	388

11. 386	12. 747	13. 593	14. 375	15. 906
438	59	648	486	95

C. Add.

1. 4237	2. 6489	3. 5076	4. 1684
1288	3074	4385	927

5. 7907	6. 1467	7. 3015	8. 9864
1395	2046	687	2735

9. 6872	10. 8360	11. 6009	12. 3785
493	1762	496	7643

13. 5049	14. 6709	15. 8475	16. 6008
732	9006	928	5842

D. Add.

1. 18745	2. 10674	3. 60485	4. 12008
6972	397	9766	9634

5. 9876	6. 59684	7. 40026	8. 78044
4835	29527	7085	97684

9. 94036	10. 87468	11. 34805	12. 58009
6975	92729	75297	70777

Date

Name

Course/Section

E. Arrange vertically and add.

1. $487 + 29 + 526$ = _____

2. $715 + 4293 + 184 + 19$ = _____

3. $1706 + 387 + 42 + 307$ = _____

4. $456 + 978 + 1423 + 3584$ = _____

5. $6284 + 28 + 674 + 97$ = _____

6. $6842 + 9008 + 57 + 368$ = _____

7. $322 + 46 + 5984$ = _____

8. $7268 + 209 + 178$ = _____

9. $5016 + 423 + 1075$ = _____

10. $8764 + 85 + 983 + 19$ = _____

F. Solve.

1. Lee Reed's savings of $4957 (principal) earned $294 interest for the year. What is her new principal (original principal plus interest)?

_____

2. Kermit Little's yearly automobile premium includes: bodily injury—$83, property damage—$27, medical payments—$14, uninsured motorist—$9, collision—$159, and comprehensive—$35. What is his total premium?

_____

3. For the first quarter of the year, John Boxwell earned the following wages: January—$576; February—$635; and March—$648. What were his total wages for the quarter?

_____

4. At lunch the other day, Ed the calorie counter ate the following: one slice of whole wheat bread, 55 calories; cream cheese and honey on the bread, 148 calories; yogurt, 123 calories; fresh blackberries, 45 calories. What was his total calorie count for the meal?

_____

5. During the first three months of the year, the Write Rite Typewriter Company reported the following sales:

January $3572
February $2716
March $4247

What were their sales total for this quarter of the year?

_____

6. Maynard's original balance in his checking account was $157. After he deposits checks for the following amounts, $73, $29, $152, and $8, what is his new balance?

_____

7. The Maple's Electronics Shop has two branches with the following daily sales. Calculate the total daily sales, the weekly total for each store, and the total sales for the week.

Branch	M	T	W	Th	F	S	Weekly Total
A	$567	$687	$592	$653	$751	$826	
B	497	506	519	527	632	758	
Daily Totals							

8. The French Valve Company produced the following amounts of valves for the week: Monday—157, Tuesday—163, Wednesday—159, Thursday—165, and Friday—27. What was the total for the week?

_____

9. The previous year's XKP-150 sports car cost $7859 but its price was increased this year by $457 due to inflation. What is the cost of the new year's model?

_____

10. Galyn, the rock climber, purchased the following equipment: biners—$78, slings—$14, boots—$67, hexacentrics—$27, sky hooks—$9, and a parachute—$158. What was the total cost?

_____

11. For the following sales personnel, calculate the total daily sales, weekly total for each salesperson, and the total sales for the week.

Employee	M	T	W	Th	F	Weekly Total
M. Schallhorn	$826	$921	$752	$950	$625	
B. Coleman	675	680	723	850	752	
K. Wilson	515	758	652	725	432	
J. Mee	455	872	953	862	759	
E. Milliron	520	755	659	782	526	
Total						

12. The Gadget Company produced the following number of gadgets each day: Monday—26,852, Tuesday—27,965, Wednesday—29,768, Thursday—25,263, and Friday—28,639. Calculate the weekly total.

_____

When you have completed this problem set, you may continue in 17 with the study of subtraction or return to the preview, page 2, and use it to determine the help you need next.

SECTION 3: SUBTRACTING WHOLE NUMBERS

17

> **OBJECTIVE**
>
> Find the difference between two whole numbers and use subtraction to solve word problems.

Subtraction is the reverse of the process of addition.

Addition: $3 + 4 = \square$

Subtraction: $3 + \square = 7$

Written this way, a subtraction problem asks the question, "How much must be added to a given number to produce a required amount?"

Most often, however, the numbers in a subtraction problem are written using a minus sign ($-$):

$17 - 8 = \square$ means that there is a number $\square$ such that $8 + \square = 17$.

Written this way, a subtraction problem asks you to find the difference between two numbers.

Write in the answer to this subtraction problem, then turn to **18**.

---

18   How much is 17 minus 8?

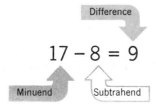

Mathematicians have given special names to numbers in subtraction problems. It will be helpful if you know them.

★ The *minuend* is the larger of the two numbers in the problem. It is the number that is being decreased.
★ The *subtrahend* is the number that is being subtracted from the minuend.
★ The *difference* is the amount that must be added to the subtrahend to produce the minuend. It is the answer to the subtraction problem.

Now try a more difficult subtraction problem:

$47 - 23 = $ _____

Work the problem and continue in **19**.

---

19   The *first* step is to write the numbers in a vertical format as you did with addition. Be careful to keep the ones digits in line in one column, the tens digits in a second column, and so on.

$\begin{array}{r} 47 \\ -23 \\ \hline \end{array}$ Notice that the minuend is written above the subtrahend—larger number on top.

Once the numbers have been arranged in this way, the difference may be written immediately.

Step 1

47
−23
‾‾‾‾
4

Ones digits: 7 − 3 = 4.

Step 2

47
−23
‾‾‾‾
24

Tens digits: 4 − 2 = 2.

The difference is 24.

With some problems, it is necessary to rewrite the minuend (larger) number using "borrowing." For example, find the difference: 74 − 18.

Step 1

6 14
$\not{7}\not{4}$
−18
‾‾‾‾
6

Borrow one ten, change the 7 in the tens place to 6, change 4 to 14, and subtract 14 − 8 = 6

Step 2

6 14
$\not{7}\not{4}$
−18
‾‾‾‾
56

Tens digits: 6 − 1 = 5.

We know you have a calculator, but to check the process, work this one by hand.

64 − 37 = _____

Check your work in **20**.

How do you say 9 − 4?

Try "nine minus four" or "nine take away four."

20   First, arrange the numbers vertically in columns.

Step 1

64
−37
‾‾‾‾

Step 2

5 14
$\not{6}\not{4}$
−37
‾‾‾‾
7

Borrow one ten, change the 6 in the tens place to 5, change 4 to 14, and subtract 14 − 7 = 7

Step 3

5 14
$\not{6}\not{4}$
−37
‾‾‾‾
27

5 − 3 = 2

Checking Subtractions   ★ Check subtraction problems by adding the answer and the subtrahend; their sum should equal the minuend.

37   Subtrahend
+27   Difference   (answer)
‾‾‾‾
64   Minuend

Try these problems for practice. Be sure to check each problem.

(a) 71	(b) $426	(c) $36,467	(d) 902
−39	−128	−19,269	−465

Solutions are in 21.

---

21 (a) **Step 1**   **Step 2**

⁶ ¹¹

71   7̸1̸   Borrow one ten,
−39   −39   change the 7 in the tens place to 6,
            ——   change the 1 in the ones place to 11,
            32          $11 - 9 = 2$, write 2,
                        $6 - 3 = 3$, write 3.

Check: 39
             +32
             ———
              71

(b) **Step 1**   **Step 2**   **Step 3**

¹ ¹⁶   ³ ¹¹ ¹⁶

$426 $4̸2̸6 $4̸2̸6̸ $16 - 8 = 8$, write 8.
−128   −128   −128 $11 - 2 = 9$, write 9.
            ————   ———— $3 - 1 = 2$, write 2.
               8        $298

Notice that in this case we must borrow twice. Borrow one ten from the tens position to make 16, then borrow one hundred from the hundreds position to make 11 in the tens position.

Check: $128
            +298
            ————
             $426

(c) **Step 1**   **Step 2**   **Step 3**   **Step 4**

                      ⁵ ¹⁷   ³ ¹⁵ ¹⁷   ² ¹⁶ ³ ¹⁵ ¹⁷

$36,467 $36,46̸7̸ $36,4̸6̸7̸ $3̸6̸,4̸6̸7̸ $17 - 9 = 8$, write 8.
−19,269   −19,269   −19,269   −19,269 $15 - 6 = 9$, write 9.
————————   ————————   ————————   ———————— $3 - 2 = 1$, write 1.
                     8   198 $17,198 $16 - 9 = 7$, write 7.
                                                          $2 - 1 = 1$, write 1.

Check: $19,269
            +17,198
            ————————
             $36,467

(d) **Step 1**   **Step 2**   **Step 3**   **Check:**   465
                                                                                    +437
                   ⁸ ¹⁰   ⁸ ⁹ ¹²   ————
902   9̸0̸2   9̸0̸2̸ $12 - 5 = 7$, write 7.   902
−465   −465   −465 $9 - 6 = 3$, write 3.
————   ————   ———— $8 - 4 = 4$, write 4.
                          437

In problem (d) we first borrow 1 hundred from the hundreds position to get a 10 in the tens place, then we borrow 1 ten from the tens place to get a 12 in the ones place.

Now, do you want more worked examples of problems containing a lot of borrowing similar to this last one? If so, go to 22. Otherwise, go to 23.

---

22   Let's work through a few examples.

(a) **Step 1**   **Step 2**   **Step 3**

$$
\begin{array}{r} \$400 \\ -167 \\ \hline \end{array}
\qquad
\begin{array}{r} {}^{3\ 10} \\ \$4\cancel{0}0 \\ -167 \\ \hline \end{array}
\qquad
\begin{array}{r} {}^{3\ 9\ 10} \\ \$\cancel{4}\cancel{0}\cancel{0} \\ -167 \\ \hline \$233 \end{array}
$$

Check: $\begin{array}{r} \$167 \\ +233 \\ \hline \$400 \end{array}$

(b) **Step 1**   **Step 2**   **Step 3**   **Step 4**

$$
\begin{array}{r} 5006 \\ -2487 \\ \hline \end{array}
\qquad
\begin{array}{r} {}^{4\ 10} \\ 5\cancel{0}06 \\ -2487 \\ \hline \end{array}
\qquad
\begin{array}{r} {}^{4\ 9\ 10} \\ 5\cancel{0}\cancel{0}6 \\ -2487 \\ \hline \end{array}
\qquad
\begin{array}{r} {}^{4\ 9\ 9\ 16} \\ 5\cancel{0}\cancel{0}\cancel{6} \\ -2487 \\ \hline 2519 \end{array}
$$

Check: $\begin{array}{r} 2487 \\ +2519 \\ \hline 5006 \end{array}$

(c) **Step 1**   **Step 2**   **Step 3**   **Step 4**

$$
\begin{array}{r} \$24{,}632 \\ -5{,}718 \\ \hline \end{array}
\qquad
\begin{array}{r} {}^{2\ 12} \\ \$24{,}6\cancel{3}\cancel{2} \\ -5{,}718 \\ \hline 14 \end{array}
\qquad
\begin{array}{r} {}^{3\ 16\ 2\ 12} \\ \$24{,}\cancel{6}\cancel{3}\cancel{2} \\ -5{,}718 \\ \hline 914 \end{array}
\qquad
\begin{array}{r} {}^{1\ 13\ 16\ 2\ 12} \\ \$2\cancel{4}{,}\cancel{6}\cancel{3}\cancel{2} \\ -5{,}718 \\ \hline \$18{,}914 \end{array}
$$

Check: $\begin{array}{r} \$5{,}718 \\ +18{,}914 \\ \hline \$24{,}632 \end{array}$

Any subtraction problem that involves borrowing should always be checked in this way. It is very easy to make a mistake in this process.

Next, go to 23.

---

23   Let's apply subtraction to a practical business situation.

A new machine for Craig's Cleaning Company originally cost $13,843. It depreciated $4937 during the past year. What is the machine's book value (current value)?

$\begin{array}{r} \$13{,}843 \\ -4{,}937 \\ \hline \$\ 8{,}906 \end{array}$ **Check:** $\begin{array}{r} \$\ 4{,}937 \\ 8{,}906 \\ \hline \$13{,}843 \end{array}$

Now it's your turn. Try the following problem.

Jim originally had a bank balance of $1235 but he withdrew $477. What is his new balance?
Check your answer in 24.

---

24     $\begin{array}{r} \$1235 \\ -477 \\ \hline \$\ 758 \end{array}$ **Check:** $\begin{array}{r} \$\ 477 \\ +758 \\ \hline \$1235 \end{array}$

Easy, isn't it? Be careful; check all subtraction problems, even if you are using a calculator.

Turn to 25 for a set of practice problems on subtraction.

Whole Numbers

PROBLEM SET 3

25 The answers are on page 574.

1.3 Subtracting Whole
Numbers

A. Subtract.

1. 13	2. 9	3. 12	4. 15	5. 8	6. 13	7. 8	8. 7	9. 11
7	4	5	9	6	8	0	7	7

10. 6	11. 16	12. 16	13. 10	14. 13	15. 5	16. 18	17. 12	18. 10
5	7	8	7	6	5	9	9	3

19. 11	20. 5	21. 10	22. 8	23. 14	24. 13	25. 12	26. 9	27. 15
8	1	2	7	6	9	3	2	6

28. 11	29. 9	30. 12	31. 14	32. 9	33. 9	34. 1	35. 11	36. 14
4	0	8	5	3	6	0	5	8

37. 15	38. 11	39. 12	40. 17	41. 13	42. 15	43. 18	44. 14	45. 16
7	6	7	9	6	8	0	7	9

46. 7	47. 12	48. 17	49. 0	50. 9
4	4	8	0	7

B. Subtract.

1. 40	2. 63	3. 78	4. 33	5. 51	6. 85	7. 36	8. 60	9. 42
27	19	49	17	39	28	17	43	27

10. 91	11. 52	12. 47	13. 70	14. 94	15. 34	16. 55	17. 56	18. 93
63	16	29	48	57	9	29	18	8

C. Subtract.

1. 546	2. 640	3. 409	4. 914	5. 476	6. 219	7. 747	8. 564
357	182	324	37	195	43	593	298

9. 400	10. 316	11. 803	12. 327	13. 632	14. 525	15. 438	16. 701
127	118	88	276	58	480	409	556

D. Subtract.

1. 6218	2. 8704	3. 6084	4. 30209	5. 13042	6. 8000
3409	923	386	1367	524	321

7. 57022	8. 46804	9. 5007	10. 10785	11. 10000	12. 31072
980	9476	266	888	386	4265

13. 48093	14. 384000	15. 27004	16. 60754	17. 42003	18. 9005
500	67360	4582	5295	17064	1767

Date

Name

Course/Section

E. **Solve.**

1. On March 1, Mrs. Pennywatcher had a balance of $635 in her checking account. She deposited checks of $352 and $114, and wrote checks for $37, $216, $147, and $106 during March. How much did she have left in her account at the end of the month?

   _____

2. A Pulsar TV set costs $348. A Zoomar TV set costs $259. What is the difference in costs for the two sets?

   _____

3. Don services his car every 7500 miles. If he has driven 5954 miles since his last service, how much farther will he drive before he services his car again?

   _____

4. Cindy's monthly salary is $684. Every month she has the following deductions: Federal income tax—$107, state tax—$22, FICA—$40, and insurance—$28. What is Cindy's take-home pay?

   _____

5. An automobile originally cost $7476. During the first year it depreciated $3948. What is the value of the car at the end of the year?

   _____

6. Mary's Get-Rich-Quick Gold Mine stock originally cost $5035. It is now worth $987. How much has the value of the stock decreased?

   _____

7. Al had a balance of $452 on his bank charge card. If he paid $195 on his account, what is his new balance?

   _____

8. A new deluxe calculator has been reduced from $72 to $49. How much was it reduced?

   _____

9. Don has paid $3792 on his new car loan of $5688. What is the current balance of his loan?

   _____

10. The Wright Sports Shop sold 137 pairs of shoes last week. If they originally had 539 pairs in stock, calculate the number of pairs left in stock.

    _____

When you have had the practice you need, go to 26 to study multiplication of whole numbers or return to the Preview for this unit on page 2.

SECTION 4: MULTIPLYING WHOLE NUMBERS

26

> **OBJECTIVE**
>
> Find the product of two whole numbers and use multiplication to solve word problems.

Sam purchased six records at $5 each. What was the total cost? We can answer the question several ways:

1. Count dollars:

   $$$$$ $$$$$ $$$$$ $$$$$ $$$$$ $$$$$

2. Add $5's:
   $5 + $5 + $5 + $5 + $5 + $5 = ?

3. Multiply:
   $6 \times 5 = ?$

We're not sure about Sam's mathematical ability, but most people would multiply. Multiplication is a short-cut method of counting or repeated addition.

What was the total cost?

Complete this problem in the manner you choose, then go to **27**.

27

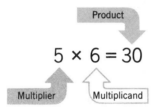

In a multiplication statement the *multiplicand* is the number to be multiplied, the *multiplier* is the number multiplying the multiplicand, and the *product* is the result of the multiplication. The multiplier and the multiplicand are called the *factors* of the product.

In order to become skillful at multiplication, you must know the one-digit multiplication table from memory.

Complete the table below by multiplying the number at the top by the number at the side and placing their product in the proper square. We have multiplied $3 \times 4 = 12$ and $2 \times 5 = 10$ for you.

Multiply	2	5	8	1	3	6	9	7	4
1									
7									
5	10								
4					12				
9									
2									
6									
3									
8									

Check your answers in **28**.

28   Here is the completed multiplication table.

Multiply	2	5	8	1	3	6	9	7	4
1	2	5	8	1	3	6	9	7	4
7	14	35	56	7	21	42	63	49	28
5	10	25	40	5	15	30	45	35	20
4	8	20	32	4	12	24	36	28	16
9	18	45	72	9	27	54	81	63	36
2	4	10	16	2	6	12	18	14	8
6	12	30	48	6	18	36	54	42	24
3	6	15	24	3	9	18	27	21	12
8	16	40	64	8	24	48	72	56	32

Not even the best of calculators can help you if you can't do one-digit multiplication. Practice until you are able to quickly perform these from memory.

Notice that the product of any number and 1 is that same number. For example,

$1 \times 2 = 2$
$1 \times 6 = 6$

or even

$1 \times 753 = 753$

Zero has been omitted from the table because the product of any number and zero is zero. For example,

$0 \times 2 = 0$
$0 \times 7 = 0$
$0 \times 395 = 0$

The multiplication of larger numbers is based on the one-digit number multiplication table. Find the product of

$34 \times 2 =$ _____

Try it, then go to 29.

___

29   First, arrange the factors to be multiplied vertically, with the ones digits in a single column, tens digits in a second column, and so on.

  34
$\times 2$

Step 1

  34
$\times 2$ Multiply ones:
   8     $4 \times 2 = 8$; write 8.

Step 2

  34
$\times 2$ Multiply $2 \times 3 = 6$,
 68   write 6.

Try the following multiplication problem.

  28
$\times 3$

Be careful, this one has a "carry." Check your work in 30.

___

30   **Step 1**   **Step 2**

$$\begin{array}{r} \scriptstyle 2 \\ 28 \\ \times 3 \\ \hline 4 \end{array}$$
Multiply ones:
$3 \times 8 = 24$.
Write 4, carry 2.

$$\begin{array}{r} \scriptstyle 2 \\ 28 \\ \times 3 \\ \hline 84 \end{array}$$
Multiply $3 \times 2 = 6$.
Add the carry, $6 + 2 = 8$.
Write 8.

Try this technique on the following three problems.

(a) $\begin{array}{r} 43 \\ \times 5 \\ \hline \end{array}$ (b) $\begin{array}{r} \$258 \\ \times 7 \\ \hline \end{array}$ (c) $\begin{array}{r} \$26{,}495 \\ \times 9 \\ \hline \end{array}$

Check your work in **31**.

---

31   (a)   **Step 1**   **Step 2**

$$\begin{array}{r} \scriptstyle 1 \\ 43 \\ \times 5 \\ \hline 5 \end{array}$$
Multiply $5 \times 3 = 15$.
Write 5, carry 1.

$$\begin{array}{r} \scriptstyle 1 \\ 43 \\ \times 5 \\ \hline 215 \end{array}$$
Multiply $5 \times 4 = 20$.
Add the carry,
$20 + 1 = 21$. Write 21.

(b)   **Step 1**   **Step 2**

$$\begin{array}{r} \scriptstyle 5 \\ \$258 \\ \times 7 \\ \hline 6 \end{array}$$
$7 \times 8 = 56$
Write 6, carry 5.

$$\begin{array}{r} \scriptstyle 4\ 5 \\ \$258 \\ \times 7 \\ \hline 06 \end{array}$$
$7 \times 5 = 35$. Add the carry,
$35 + 5 = 40$.
Write 0, carry 4.

Step 3

$$\begin{array}{r} \scriptstyle 4\ 5 \\ \$258 \\ \times 7 \\ \hline \$1806 \end{array}$$
$7 \times 2 = 14$. Add the carry,
$14 + 4 = 18$, write 18.

(c)   **Step 1**   **Step 2**

$$\begin{array}{r} \scriptstyle 4 \\ \$26495 \\ \times 9 \\ \hline 5 \end{array}$$
$9 \times 5 = 45$
Write 5, carry 4.

$$\begin{array}{r} \scriptstyle 8\ 4 \\ \$26495 \\ \times 9 \\ \hline 55 \end{array}$$
$9 \times 9 = 81$
Add the carry, $81 + 4 = 85$,
write 5, carry 8.

Step 3   **Step 4**

$$\begin{array}{r} \scriptstyle 4\ 8\ 4 \\ \$26495 \\ \times 9 \\ \hline 455 \end{array}$$
$9 \times 4 = 36$
Add the carry,
$36 + 8 = 44$,
write 4, carry 4.

$$\begin{array}{r} \scriptstyle 5\ 4\ 8\ 4 \\ \$26495 \\ \times 9 \\ \hline 8455 \end{array}$$
$9 \times 6 = 54$
Add the carry, $54 + 4 = 58$,
write 8, carry 5.

Step Last

$$\begin{array}{r} \scriptstyle 5\ 4\ 8\ 4 \\ \$26495 \\ \times 9 \\ \hline \$238455 \end{array}$$
$9 \times 2 = 18$
Add the carry, $18 + 5 = 23$,
write 23.

Next, let's work a calculation involving a two-digit multiplier, 24 × 57.

Step 1 Multiply 4 × 57.

$$\begin{array}{r} {\scriptstyle 2} \\ 57 \\ \times 24 \\ \hline 228 \end{array}$$

4 × 7 = 28. Write 8, carry 2.
4 × 5 = 20. Add the carry.
20 + 2 = 22. Write 22.

Step 2 Multiply 2 × 57 and indent one place to the left. This allows for multiplying 57 by 2 tens.

$$\begin{array}{r} {\scriptstyle 1} \\ {\scriptstyle 2} \\ 57 \\ \times 24 \\ \hline 228 \\ 114 \end{array}$$

2 × 7 = 14. Write 4, carry 1.
2 × 5 = 10. Add the carry.
10 + 1 = 11. Write 11.

Step 3 Add

$$\begin{array}{r} {\scriptstyle 1} \\ {\scriptstyle 2} \\ 57 \\ \times 24 \\ \hline 228 \\ 114 \\ \hline 1368 \end{array}$$

Add.

Try these:

(a) $\begin{array}{r} \$64 \\ \times 37 \\ \hline \end{array}$ (b) $\begin{array}{r} 327 \\ \times 145 \\ \hline \end{array}$ (c) $\begin{array}{r} 342 \\ \times 102 \\ \hline \end{array}$ (d) $\begin{array}{r} \$5847 \\ \times 3256 \\ \hline \end{array}$

Check your answers in **32**.

MULTIPLICATION SHORT-CUTS

There are hundreds of quick ways to multiply various numbers. Most of them are only quick if you are already a math whiz and will confuse you more than help you. Here are a few that are easy to do and easy to remember.

1. To multiply by 10, annex a zero on the right end of the multiplicand. For example,

    34 × 10 = 340
    256 × 10 = 2560

    Multiplying by 100 or 1000 is similar.

    34 × 100 = 3400
    256 × 1000 = 256000

2. To multiply by a number ending in zeros, carry the zeros forward to the answer. For example,

    $$\begin{array}{r} 26 \\ \times 20 \\ \hline \end{array} \longrightarrow \begin{array}{r} 26 \\ \times 2\,0 \\ \hline 52\,0 \end{array}$$

Multiply 26 × 2 and attach the zero on the right. The product is 520.

```
 34          34
×2100  →   ×2100
            34 ▏▏
            68 ▼▼
          71400
```

3. If both multiplier and multiplicand end in zeros, bring all zeros forward to the answer.

```
 230        23|0
×200  →    ×2|00
           46|000
```

Attach three zeros to the product of 23 × 2.

Another example:

```
  1000
  ×100
100,000
```

This sort of multiplication is mostly a matter of counting zeros.

32 (a)
```
   $64
   ×37
   448     7 × 64 = 448
   192     3 × 64 = 192
 $2368
```

(b)
```
   327
  ×145
  1635     5 × 327 = 1635
  1308     4 × 327 = 1308
   327     1 × 327 = 327
 47415
```

(c)
```
   342
  ×102
   684     2 × 342 = 684
   000     0 × 342 = 000
   342     1 × 342 = 342
 34884
```

(d)
```
    $5847
   ×3256
   35082    6 × 5847 = 35082
   29235    5 × 5847 = 29235
   11694    2 × 5847 = 11694
   17541    3 × 5847 = 17541
$19037832
```

CHECKING
MULTIPLICATION

The easiest method for checking multiplication is by interchanging the multiplier and the multiplicand, then remultiplying.

Work the following business application problem and check your answer.

Larry's Photo Shop sold 47 cameras for $198 each. What was the total income from these camera sales?

Check your work in **33**.

33
```
  $198         Check:      47
  ×47                   ×$198
  1386                    376
   792                    423
 $9306                    47
                        $9306
```

Always remember to check your work, even if you are using a calculator. It's extremely easy to hit the wrong keys.

Go to **34** for a set of practice problems on multiplication.

THE SEXY SIX

Here are the six most often missed one-digit multiplications:

"Inside"

$9 \times 8 = 72$
$9 \times 7 = 63$
$9 \times 6 = 54$
$8 \times 7 = 56$
$8 \times 6 = 48$
$7 \times 6 = 42$

It may help you to notice that in these multiplications the "inside" digits, such as 8 and 7, are consecutive and the digits of the answer add to nine: $7 + 2 = 9$. This is true for *all* one-digit numbers multiplied by 9.

Be certain you have these memorized.

(There is nothing very sexy about them, but we did get your attention, didn't we?)

Whole Numbers

PROBLEM SET 4

The answers are on page 575.

1.4 Multiplying Whole Numbers

A. Multiply.

1. 7	2. 9	3. 7	4. 7	5. 6	6. 9
6	8	8	9	8	6

7. 8	8. 6	9. 6	10. 9	11. 8	12. 8
9	7	9	7	7	6

B. Multiply.

1. 29	2. 67	3. 72	4. 27	5. 47	6. 88	7. 64	8. 37
3	6	8	9	6	9	5	7

9. 39	10. 42	11. 58	12. 87	13. 94	14. 49	15. 17	16. 23
4	7	5	3	6	8	9	7

17. 47	18. 53	19. 77	20. 36	21. 48	22. 35	23. 64	24. 72
6	8	4	9	15	43	27	38

25. 90	26. 41	27. 86	28. 18	29. 34	30. 28	31. 66	32. 71
56	72	83	65	57	91	25	19

33. 59	34. 18	35. 29	36. 82	37. 78	38. 35	39. 94	40. 43
76	81	32	76	49	58	95	64

C. Multiply.

1. 305	2. 145	3. 3006	4. 481	5. 8043	6. 765	7. 809
123	516	125	203	37	502	47

8. 1107	9. 3706	10. 4210	11. 708	12. 6401	13. 684	14. 319
98	102	304	58	773	45	708

15. 2043	16. 354	17. 2008	18. 923	19. 563	20. 8745
670	88	198	47	107	583

D. Solve.

1. A portable TV can be bought on credit for $25 down and 12 monthly payments of $15 each. What is the total cost?

   _____

2. A room has 26 square yards of floor space. If carpeting costs $13 per square yard, what would it cost to carpet the room?

   _____

Date _____

Name _____

Course/Section _____

3. If you manage to save $23 per week, how much money will you have in a year (52 weeks)?

_____

4. A used car is advertised for sale at $950 or the installment price of 30 monthly payments of $42. How much would you save by paying cash?

_____

5. Janice's $18,000 insurance policy has a cash value of $258 per $1000. Calculate the total cash value.

_____

6. Mary purchased 265 shares of Get-Rich-Quick Gold Mine stock for $19 per share. What was the total cost not including brokerage fee?

_____

7. Alice enrolled in 14 credit hours this semester. If tuition is $17 per credit hour, what is the total cost of her tuition?

_____

8. The largest flag in the world is 104 feet by 235 feet. What is its total area? (Area = length × width)

_____

9. What is the cost of six dozen widgets at $17 each?

_____

10. An assembly line at the Ace Widget Company turns out 2457 widgets per hour. How many widgets are produced in a 40-hour week?

_____

11. If you worked 39 hours at $17 per hour, what would be your total pay?

_____

12. What is the total cost of six dozen gadgets at $157 each?

_____

When you have had the practice you need, turn to **35** to study the division of whole numbers, or return to the Preview on page 2 and continue.

SECTION 5: DIVIDING WHOLE NUMBERS

35

OBJECTIVE

Divide whole numbers and use division to solve word problems.

Division is the reverse process for multiplication. It enables us to separate a given quantity into equal parts. The mathematical phrase $12 \div 3$ is read "twelve divided by three," and it asks us to separate a collection of 12 objects into 3 equal parts. The mathematical phrases

$$12 \div 3 \qquad 3\overline{)12} \qquad \frac{12}{3} \qquad \text{and} \qquad 12/3$$

all represent division and are all read "twelve divided by three."

Perform this division: $12 \div 3 = $ _____

Turn to **36** to continue.

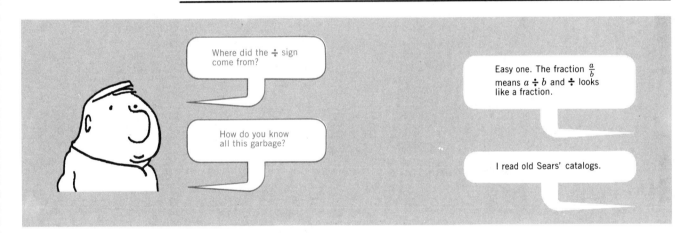

36

$$12 \div 3 = 4$$

Quotient

Dividend

Divisor

In this division problem, 12, the number being divided, is called the *dividend;* 3, the number used to divide, is called the *divisor;* and 4, the result of the division is called the *quotient,* from a Latin word meaning "how many times."
One way to perform division is to reverse the multiplication process.

$$24 \div 4 = \square \qquad \text{means that } 4 \times \square = 24$$

If the one-digit multiplication tables are firmly in your memory, you will recognize immediately that $\square = 6$. By placing a 6 in $\square$ you make the statement $4 \times \square = 24$ true.

Try these:

$35 \div 7 =$ _____ $42 \div 6 =$ _____

$28 \div 4 =$ _____ $56 \div 7 =$ _____

$45 \div 5 =$ _____ $18 \div 3 =$ _____

$70 \div 10 =$ _____ $63 \div 9 =$ _____

$30 \div 5 =$ _____ $72 \div 8 =$ _____

Check your answers in **37**.

37
$35 \div 7 = 5$ $42 \div 6 = 7$
$28 \div 4 = 7$ $56 \div 7 = 8$
$45 \div 5 = 9$ $18 \div 3 = 6$
$70 \div 10 = 7$ $63 \div 9 = 7$
$30 \div 5 = 6$ $72 \div 8 = 9$

Business people use division every day.

You should be able to do all of these quickly by working backward from the one-digit multiplication tables.

How do we divide dividends that are larger than 9×9 and therefore not in the multiplication table? Obviously, we need a better procedure.

For example, in the problem $96 \div 8$, start by arranging the divisor and dividend horizontally.

Step 1

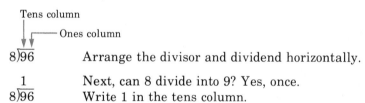

$8\overline{)96}$ Arrange the divisor and dividend horizontally.

$\dfrac{1}{8\overline{)96}}$ Next, can 8 divide into 9? Yes, once.
                  Write 1 in the tens column.

Step 2

$\begin{array}{r} 1 \\ 8\overline{)\ 96} \\ -8 \\ \hline 16 \end{array}$ ← Multiply $8 \times 1 = 8$.
           Subtract $9 - 8 = 1$ and bring down the
           next digit, 6.

Step 3

$\begin{array}{r} 12 \\ 8\overline{)\ 96} \\ -8 \\ \hline 16 \end{array}$ Next $16 \div 8 = 2$. Write 2 in
           the ones column.

Step 4

$\begin{array}{r} 12 \\ 8\overline{)\ 96} \\ -8 \\ \hline 16 \\ -16 \\ \hline 0 \end{array}$ ← Multiply $8 \times 2 = 16$ and subtract $16 - 16 = 0$.

34   1 WHOLE NUMBERS

$96 \div 8 = 12$. We can check our answer by multiplying the divisor by the quotient, which should yield the dividend.

Check: $8 \times 12 = 96$.

Now, try the following problems. We know you can work it on your calculator, but try it longhand.

(a) $976 \div 8$ (b) $3179 \div 6$

Check your answers, then turn to **38** to see our work.

38   (a)

```
      122
 8) 976
    −8
    ──
     17
    −16
    ───
     16
    −16
    ───
      0
```

Step 1. Line up divisor and dividend.
Divide 8 into 9? Once, write 1.

Step 2. $8 \times 1 = 8$, subtract $9 - 8 = 1$, and bring down 7.

Step 3. Divide 8 into 17 twice. Write 2.

Step 4. Multiply $8 \times 2 = 16$. Subtract $17 - 16 = 1$ and bring down 6.

Step 5. Divide 8 into 16 twice. Write 2.

Step 6. $8 \times 2 = 16$. Subtract $16 - 16 = 0$.

Check: $8 \times 122 = 976$

(b)

```
      529
 6) 3179
    −30
    ───
     17
    −12
    ───
      59
     −54
     ───
       5  ←Remainder
```

Step 1. Line up divisor and dividend.
Divide 6 into 3? No.
Divide 6 into 31? 5 times, write 5.

Step 2. $6 \times 5 = 30$. Subtract $31 - 30 = 1$, and bring down 7.

Step 3. Divide 6 into 17 twice. Write 2.

Step 4. $6 \times 2 = 12$. Subtract $17 - 12 = 5$, and bring down 9.

Step 5. Divide 6 into 59, 9 times.
Write 9.

Step 6. $6 \times 9 = 54$. Subtract $59 - 54 = 5$.
5 is the remainder.

Checking Division With A Remainder

Notice that $3179 \div 6$ doesn't "come out even." There is a remainder of 5. When checking problems with a remainder, first multiply the divisor by the quotient, then add the remainder. The result should equal the dividend.

Check:
$$6 \times 529 = 3174$$
$$\underline{+ \quad 5} \quad \text{Remainder}$$
$$3179 \quad \text{Dividend}$$

Now try a problem using a two-digit divisor.

$5096 \div 31 = \underline{\qquad}$

The procedure is the same as above. Check your answer in **39**.

39

$$
\begin{array}{r}
164 \\
31{\overline{)\,5096}} \\
-31 \\
\hline
199 \\
-186 \\
\hline
136 \\
-124 \\
\hline
12 \\
\end{array}
$$
← Remainder

Step 1. Divide 31 into 5? No.
Divide 31 into 50? Once, write 1.

Step 2. $31 \times 1 = 31$. Subtract $50 - 31 = 19$, and bring down 9.

Step 3. Divide 31 into 199, 6 times.
Write 6.

Step 4. $31 \times 6 = 186$. Subtract $199 - 186 = 13$, and bring down 6.

Step 5. Divide 31 into 136, 4 times.
Write 4.

Step 6. $31 \times 4 = 124$. Subtract $136 - 124 = 12$. Remainder $= 12$.

Check:
$$31 \times 164 = 5084$$
$$\underline{+ \quad 12}$$
$$5096$$

Try a business problem. Check your answer by multiplying.

The Giant Corporation purchased 849 Giant Makers for $646,938. What was the cost per Giant Maker?

Check your answer in **40**.

40

$$
\begin{array}{r}
\$ 762 \\
849\overline{)\$646938} \\
\underline{5943} \\
5263 \\
\underline{5094} \\
1698 \\
\underline{1698}
\end{array}
$$

Check: $\quad 762 \leftarrow$ Quotient
$\quad\quad \times\ 849 \leftarrow$ Divisor

$$
\begin{array}{r}
6858 \\
3048 \\
\underline{6096} \\
646938 \leftarrow \text{Dividend}
\end{array}
$$

Always remember to check your work. In business you are working with real-world situations where accuracy is crucial.

Turn to 41 for a set of practice problems on division.

Whole Numbers

PROBLEM SET 5

41   The answers are on page 575.

1.5   Division of Whole Numbers

A.   Divide.

1. $63 \div 7 = $ _____
2. $84 \div 7 = $ _____
3. $92 \div 8 = $ _____

4. $56 \div 8 = $ _____
5. $72 \div 0 = $ _____
6. $65 \div 5 = $ _____

7. $37 \div 5 = $ _____
8. $45 \div 9 = $ _____
9. $71 \div 7 = $ _____

10. $7 \div 1 = $ _____
11. $6 \div 6 = $ _____
12. $13 \div 0 = $ _____

13. $\dfrac{32}{4} = $ _____
14. $\dfrac{18}{3} = $ _____
15. $\dfrac{28}{7} = $ _____

16. $\dfrac{42}{6} = $ _____
17. $\dfrac{54}{9} = $ _____
18. $\dfrac{63}{7} = $ _____

B.   Divide.

1. $245 \div 7 = $ _____
2. $369 \div 9 = $ _____
3. $167 \div 7 = $ _____

4. $126 \div 3 = $ _____
5. $228 \div 4 = $ _____
6. $232 \div 5 = $ _____

7. $310 \div 6 = $ _____
8. $360 \div 8 = $ _____
9. $337 \div 3 = $ _____

10. $\dfrac{132}{3} = $ _____
11. $\dfrac{147}{7} = $ _____
12. $\dfrac{216}{8} = $ _____

13. $7\overline{)364}$
14. $6\overline{)222}$
15. $8\overline{)704}$
16. $5\overline{)625}$
17. $4\overline{)201}$
18. $9\overline{)603}$

C.   Divide.

1. $322 \div 14 = $ _____
2. $357 \div 17 = $ _____
3. $382 \div 19 = $ _____

4. $407 \div 13 = $ _____
5. $936 \div 24 = $ _____
6. $502 \div 10 = $ _____

7. $700 \div 28 = $ _____
8. $701 \div 36 = $ _____
9. $730 \div 81 = $ _____

10. $\dfrac{451}{11} = $ _____
11. $\dfrac{901}{17} = $ _____
12. $\dfrac{989}{23} = $ _____

13. $31\overline{)682} = $ _____
14. $43\overline{)507} = $ _____
15. $61\overline{)732} = $ _____

16. $12\overline{)408} = $ _____
17. $33\overline{)303} = $ _____
18. $13\overline{)928} = $ _____

D.   Divide.

1. $2001 \div 21 = $ _____
2. $3328 \div 32 = $ _____
3. $2016 \div 21 = $ _____

4. $3536 \div 17 = $ _____
5. $1000 \div 7 = $ _____
6. $5029 \div 47 = $ _____

7. $2000 \div 9 = $ _____
8. $1880 \div 11 = $ _____
9. $2400 \div 75 = $ _____

Date

Name

Course/Section

10. $7\overline{)7000}$ = _____   11. $14\overline{)4275}$ = _____   12. $27\overline{)8405}$ = _____

13. $71\overline{)6005}$ = _____   14. $31\overline{)3105}$ = _____   15. $53\overline{)6307}$ = _____

16. $231\overline{)14091}$ = _____   17. $411\overline{)42020}$ = _____   18. $603\overline{)48843}$ = _____

E. Solve.

1. Typing at the rate of 72 words per minute, how long would it take to type 1800 words?

   _____

2. If your yearly income is $13,780, what is your average weekly income?

   _____

3. Roberta drove 938 miles in 14 hours. What was her average speed? Was she exceeding the 55 mph speed limit?

   _____

4. Mary's annual insurance premium is $348. If she makes 12 equal monthly payments, what is the amount of each payment?

   _____

5. Sam financed his new car through Money Magician Finance Company. The total amount financed is $4272, in equal monthly payments of $178. What is the number of payments?

   _____

6. Galyn purchased a new super deluxe stereo system on credit. The total finance charge was $4536 with 24 equal monthly payments. What is Galyn's monthly payment?

   _____

7. Storm's Steam Cleaning had an annual profit of $8625. If the profit is split equally between the three partners, how much does each partner receive?

   _____

8. The Ace Widget Company sold 257 deluxe widgets for $4883. What was the selling price per widget?

   _____

9. The French Valve Company produced 12,240 valves. If the valves are packaged eight per carton, how many cartons were needed?

   _____

10. Platt's Speed Shop has set aside $9724 for employee bonuses. Calculate the bonus of each of the 17 employees if they all share equally.

    _____

When you have had the practice you need, turn to 42 for the Self-Test covering whole numbers.

Whole Numbers

42   Self-Test

The answers are on page 576.

1.   Write the number 2,847,391 in word form.

   _____

2.   Write the number, forty-eight million, five hundred sixty-one thousand, seventy-two, in numerical form.

   _____

3.   3847 + 278 + 12,936 + 24 =   _____

4. $2856 − 1947 =   _____

5. $278 × 47 =   _____

6. $16,055 ÷ 65 =   _____

7.   2,871,114 − 792,205 =   _____

8.   660,528 ÷ 278 =   _____

9.   A new stereo system can be purchased in installments for $35 down and 12 monthly payments of $27. What is the total cost?

   _____

10.   If your annual salary is $14,352, what is your weekly salary?

   _____

Date _____

Name _____

Course/Section _____

2 Fractions

When most people think of fractions, they think of amounts less than one, a proper fraction, such as $\frac{1}{2}$, $\frac{3}{4}$, or $\frac{2}{3}$.

In our cartoon, the man in charge of production is thinking of an improper fraction, a fraction larger than one, namely $\frac{4}{3}$. Hence, the new cost is larger than the original cost.

Although using improper fractions isn't improper, you must be careful. It's important that all business employers and employees have a thorough knowledge of fractions.

Are you ready? This "fraction" of the text is easy and fun. The "proper" way to begin is to start on the Preview.

UNIT OBJECTIVE

After you complete this unit successfully, you will be able to add, subtract, multiply, and divide fractions. You will also be able to write fractions as mixed numbers and improper fractions.

PREVIEW 2   This is a sample test for Unit 2. All the answers are placed immediately after the test. Work as many of the problems as you can, then check your answers and look after the answers for further directions.

1. $\dfrac{3}{4} = \dfrac{?}{12}$

    _____   45   1

2. Reduce to lowest terms $\dfrac{18}{39} =$

    _____   45   1

3. Convert to improper fraction
$4\dfrac{2}{3} =$

    _____   45   1

4. Convert to a mixed number
$\dfrac{37}{5} =$

    _____   45   1

5. $1\dfrac{1}{3} \times 2\dfrac{3}{5}$

    _____   55   16

6. $3\dfrac{1}{2} \div 1\dfrac{3}{4}$

    _____   61   23

7. $5\dfrac{1}{8} - 3\dfrac{1}{3}$

    _____   69   32

8. Ruth purchased the following lengths of fabric: $1\frac{7}{8}$ yd, $2\frac{1}{3}$ yd, $3\frac{1}{4}$ yd, and $2\frac{1}{2}$ yd. What was the total yardage purchased?

    _____   69   32

* If you cannot work any of these problems, start with frame 1 on page 45.
* If you missed some of the problems, turn to the page numbers indicated.
* If all of your answers were correct, you are probably ready to proceed to Unit 3. (If you would like more practice on Unit 2 before turning to Unit 3, try the Self-Test on page 83.)
* Super-students—those who want to be certain they learn all of this material— will turn to frame 1 and begin work there.

Answers

1. $\dfrac{9}{12}$ 2. $\dfrac{6}{13}$

3. $\dfrac{14}{3}$ 4. $7\dfrac{2}{5}$

5. $3\dfrac{7}{15}$ 6. 2

7. $1\dfrac{19}{24}$ 8. $9\dfrac{23}{24}$

Preview 2
Answers

SECTION 1: RENAMING FRACTIONS

---

1 | **OBJECTIVE**

Convert an improper fraction to a mixed number.
Convert a mixed number to an improper fraction.
Reduce or expand a fraction to higher terms.

Consider the rectangular area below. What happens when we break it into equal parts?

2 parts, each one-half of the whole

3 parts, each one-third of the whole

4 parts, each one-fourth of the whole

Divide this area [ ] into fifths by drawing vertical lines.

Try it, then go to **2**.

---

2 Notice that the five parts of "fifths" are equal in area.

A fraction is normally written as the division of two whole numbers:

$$\frac{2}{3}, \quad \frac{3}{4}, \quad \text{or} \quad \frac{26}{12}$$

One of the five equal areas above would be "one-fifth" or $\frac{1}{5}$ of the entire area.

$\frac{1}{5} = \frac{1 \text{ shaded part}}{5 \text{ parts total}}$

How would you label this portion of the area? $= ?$

Continue in **3**.

---

3 $\frac{3}{5} = \frac{3 \text{ shaded total}}{5 \text{ parts total}}$

The fraction $\frac{3}{5}$ implies an area equal to three of the five original portions.

$$\frac{3}{5} = 3 \times \left(\frac{1}{5}\right)$$

The two numbers that form a fraction are given special names to simplify talking about them. In the fraction $\frac{3}{5}$ the upper number, 3, is called the *numerator* from the Latin *numero* meaning number. It is a count of the number of parts. The lower number is called the *denominator* from the Latin *nomen* or *name*. It tells us the name of the part being counted.

$$\frac{3}{5}$$

3 ◄ Numerator, the number of parts

5 ◄ Denominator, the name of the part, "fifths"

A textbook costs $6 and I have $5. What fraction of its cost do I have? Write the answer as a fraction.

_____, numerator = _____, denominator = _____

Check your answer in 4.

---

4    $$$$$ $ $5 is $\frac{5}{6}$ of the total cost.

$$\frac{5}{6}$$

numerator = 5, denominator = 6

Complete these sentences by writing in the correct fraction.

(a)   If we divide a length into eight equal parts, each part will be _____ of the total length.

(b)   Then three of these parts will represent _____ of the total length.

(c)   Eight of these parts will be _____ of the total length.

(d)   Ten of these parts will be _____ of the total length.

Check your answers in 5.

---

5   (a) $\frac{1}{8}$ (b) $\frac{3}{8}$ (c) $\frac{8}{8}$ (d) $\frac{10}{8}$

The original length is used as a standard and any other length—smaller or larger—can be expressed as a fraction of it. A *proper fraction* is a number less than 1, as you would suppose a fraction should be. It represents a quantity less than the standard. For example, $\frac{1}{2}$, $\frac{2}{3}$, and $\frac{17}{20}$ are all proper fractions. Notice that for a proper fraction, the numerator is less than the denominator—the top number is less than the bottom number in the fraction.

An *improper fraction* is a number greater than 1 and represents a quantity greater than the standard. If a standard length is 8 inches, a length of 11 inches would be $\frac{11}{8}$ of the standard. Notice that for an improper fraction the numerator is greater than the denominator—top number greater than the bottom number in the fraction.

Circle the proper fractions in the following list.

$$\frac{3}{2}, \quad \frac{3}{4}, \quad \frac{7}{8}, \quad \frac{5}{4}, \quad \frac{15}{12}, \quad \frac{1}{16}, \quad \frac{35}{32}, \quad \frac{7}{50}, \quad \frac{65}{64}, \quad \frac{105}{100}$$

Go to 6 when you have finished.

6   You should have circled the following proper fractions:

$\frac{3}{4}, \frac{7}{8}, \frac{1}{16}$, and $\frac{7}{50}$. All are numbers less than 1. In each the numerator is less than the denominator.

The improper fraction $\frac{7}{3}$ can be shown graphically as follows:

Unit standard = [diagram]

   $\frac{1}{3}$ = [diagram]

then,

   $\frac{7}{3}$ = [diagram] (seven, count 'em)

We can rename this number by regrouping.

[diagram]   2 + $\frac{1}{3}$ or 2 $\frac{1}{3}$ standard units

A *mixed number* is an improper fraction written as the sum of a whole number and a proper fraction.

$$\frac{7}{3} = 2 + \frac{1}{3} \qquad \text{or} \qquad 2\frac{1}{3}$$

We usually omit the + sign and write $2 + \frac{1}{3}$ as $2\frac{1}{3}$, and read it as "two and one-third." The numbers $1\frac{1}{2}$, $2\frac{3}{5}$, and $16\frac{2}{3}$ are all written as mixed numbers.

Changing An
Improper Fraction
To A Mixed Number

To write an improper fraction as a mixed number, divide numerator by denominator and form a new fraction as shown:

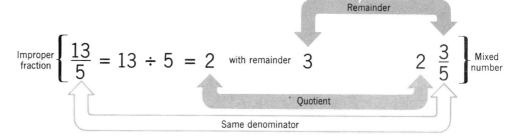

Now you try it. Rename $\frac{23}{4}$ as a mixed number. $\frac{23}{4}$ = _____

Follow the procedure shown above, then turn to 8.

---

7   (a) $\frac{9}{5} = 1\frac{4}{5}$ (b) $\frac{13}{4} = 3\frac{1}{4}$ (c) $\frac{27}{8} = 3\frac{3}{8}$

     (d) $\frac{31}{5} = 6\frac{1}{5}$ (e) $\frac{41}{12} = 3\frac{5}{12}$ (f) $\frac{17}{2} = 8\frac{1}{2}$

Changing A Mixed Number To An Improper Fraction

The reverse process, rewriting a mixed number as an improper fraction is equally simple.

Work in a clockwise direction,
first multiply, 5 x 2 = 10
then add the numerator 10 + 3 = 13

10 + 3 = 13

$$2\frac{3}{5} = \frac{13}{5}$$

Same denominator

5 x 2 = 10

STEPS
1. **Multiply denominator times whole number.**
2. **Add numerator.**
3. **Sum over original denominator.**

Graphically, $2\frac{3}{5}$ = $\frac{5}{5}$

$\frac{5}{5}$

or $\frac{13}{5}$, count them. $\frac{3}{5}$

Now you try it. Rewrite these mixed numbers as improper fractions.

(a) $3\frac{1}{6}$ (b) $4\frac{3}{5}$ (c) $1\frac{1}{2}$ (d) $8\frac{2}{3}$ (e) $15\frac{3}{8}$ (f) $9\frac{3}{4}$

Check your answers in 9.

---

8   $\frac{23}{4} = 23 \div 4 = 5$ with remainder 3 $\longrightarrow 5\frac{3}{4}$

Now try these for practice. Write each improper fraction as a mixed number.

(a) $\frac{9}{5}$ (b) $\frac{13}{4}$ (c) $\frac{27}{8}$ (d) $\frac{31}{5}$ (e) $\frac{41}{12}$ (f) $\frac{17}{2}$

The answers are in 7.

---

9   (a) $3\frac{1}{6} = \frac{19}{6}$ (b) $4\frac{3}{5} = \frac{23}{5}$ (c) $1\frac{1}{2} = \frac{3}{2}$

     (d) $8\frac{2}{3} = \frac{26}{3}$ (e) $15\frac{3}{8} = \frac{123}{8}$ (f) $9\frac{3}{4} = \frac{39}{4}$

Two fractions are said to be *equivalent* if they are numerals or names for the same number. For example, $\frac{1}{2} = \frac{2}{4}$ since both fractions represent the same portion of some standard amount.

$\frac{1}{2}$

$\frac{2}{4}$

There is a very large set of fractions equivalent to $\frac{1}{2}$.

$$\frac{1}{2} = \frac{2}{4} = \frac{3}{6} = \frac{4}{8} = \frac{5}{10} = \cdots = \frac{48}{96} = \frac{61}{122} = \frac{1437}{2874} \text{ and so on.}$$

Each fraction is a name for the same number, and we can use these fractions interchangeably.

Raising A Fraction To Higher Terms

To obtain a fraction equivalent to any given fraction multiply the original, numerator and denominator, by the same nonzero number. For example,

$$\frac{1}{2} = \frac{1 \times 3}{2 \times 3} = \frac{3}{6}$$

or

$$\frac{2}{3} = \frac{2 \times 5}{3 \times 5} = \frac{10}{15}$$

Multiplying by $\frac{3}{3}$ or $\frac{5}{5}$ is the same as multiplying by 1, and one times any number is equal to the number.

Rename as shown: $\quad \dfrac{3}{4} = \dfrac{?}{20}$

Check your work in **10**.

---

10 $\quad \dfrac{3}{4} = \dfrac{3 \times ?}{4 \times ?} = \dfrac{3 \times 5}{4 \times 5} = \dfrac{15}{20}$

$\qquad\qquad 4 \times ? = 20 \qquad\qquad\qquad\qquad$ or $? = 20 \div 4$
$\qquad\qquad\quad ?$ must be 5 $\qquad\qquad\qquad\qquad\qquad ? = \;\; 5$

The number value of the fraction has not changed, we have simply renamed it.

Practice with these.

(a) $\dfrac{5}{6} = \dfrac{?}{42}$ $\qquad$ (b) $\dfrac{7}{16} = \dfrac{?}{48}$ $\qquad$ (c) $\dfrac{3}{7} = \dfrac{?}{56}$ $\qquad$ (d) $1\dfrac{2}{3} = \dfrac{?}{12}$

Look in **12** for the answers.

---

11 $\quad \dfrac{90}{105} = \dfrac{90 \div 5}{105 \div 5} = \dfrac{18}{21} = \dfrac{18 \div 3}{21 \div 3} = \dfrac{6}{7}$

This process of dividing by, or eliminating common factors is called *cancelling*. Note that the order of cancelling the factors doesn't affect the final answer.

Usually we don't write out the division, but instead we write the results above the numerator and below the denominator.

$$\frac{\overset{18}{\overset{6}{\cancel{\cancel{90}}}}}{\underset{21}{\underset{7}{\cancel{\cancel{105}}}}} = \frac{18}{21} = \frac{6}{7} \qquad \begin{array}{l}\textbf{First, } \text{divide by 5.} \\ \textbf{Second, } \text{divide by 3.}\end{array}$$

Reduce the following fractions to lowest terms:

(a) $\dfrac{15}{84}$ $\quad$ (b) $\dfrac{21}{35}$ $\quad$ (c) $\dfrac{4}{12}$ $\quad$ (d) $\dfrac{154}{1078}$ $\quad$ (e) $\dfrac{256}{208}$ $\quad$ (f) $\dfrac{378}{405}$

The answers are in **13**.

---

12 (a) $\dfrac{5}{6} = \dfrac{5 \times 7}{6 \times 7} = \dfrac{35}{42}$ $\qquad\qquad$ (b) $\dfrac{7}{16} = \dfrac{7 \times 3}{16 \times 3} = \dfrac{21}{48}$

$\qquad$ (c) $\dfrac{3}{7} = \dfrac{3 \times 8}{7 \times 8} = \dfrac{24}{56}$ $\qquad\qquad$ (d) $1\dfrac{2}{3} = \dfrac{5}{3} = \dfrac{5 \times 4}{3 \times 4} = \dfrac{20}{12}$

Very often in working with fractions you will be asked to *reduce a fraction to lowest terms*. This means to replace it with the most simple fraction in its set of equivalent fractions. To reduce $\frac{15}{30}$ to its lowest terms means to replace it by $\frac{1}{2}$.

$$\frac{15}{30} = \frac{1 \times 15}{2 \times 15} = \frac{1}{2}$$

and $\frac{1}{2}$ is the simplest equivalent fraction to $\frac{15}{30}$ because its numerator and denominator are the smallest whole numbers of any in the set $\frac{1}{2}, \frac{2}{4}, \frac{3}{6}, \frac{4}{8}, \ldots \frac{15}{30} \ldots$

How can you find the simplest equivalent fraction? For example, how would you reduce $\frac{30}{42}$ to lowest terms?

First, identify a common factor of the numerator and denominator. A common factor is a number that evenly divides the numerator and denominator. **Then divide the numerator and denominator by the same factor.**

Two is a factor of 30 and 42.

$$\frac{30}{42} = \frac{30 \div 2}{42 \div 2} = \frac{15}{21}$$

Next, continue dividing by common factors until the fraction is reduced as far as possible.

Three is a factor of 15 and 21.

$$\frac{15}{21} = \frac{15 \div 3}{21 \div 3} = \frac{5}{7}$$

$$\frac{30}{42} = \frac{15}{21} = \frac{5}{7}$$

If we had noticed that 6 is a factor of 30 and 42, we could have reduced $\frac{30}{42}$ in one step.

$$\frac{30}{42} = \frac{30 \div 6}{42 \div 6} = \frac{5}{7}$$

The important point is to continue reducing until there are no more common factors.

Your turn. Reduce $\dfrac{90}{105}$ to lowest terms.

Look in 11 for the answer.

---

13 (a) $\dfrac{\overset{5}{\cancel{15}}}{\underset{28}{\cancel{84}}} = \dfrac{5}{28}$ Divide by 3.

(b) $\dfrac{\overset{3}{\cancel{21}}}{\underset{5}{\cancel{35}}} = \dfrac{3}{5}$ Divide by 7.

(c) $\dfrac{\overset{1}{\cancel{4}}}{\underset{3}{\cancel{12}}} = \dfrac{1}{3}$ Divide by 4.

(d) $\dfrac{\overset{77}{\cancel{154}}}{\underset{539}{\cancel{1078}}} = \dfrac{\overset{11}{\cancel{77}}}{\underset{77}{\cancel{539}}} = \dfrac{\overset{1}{\cancel{11}}}{\underset{7}{\cancel{77}}} = \dfrac{1}{7}$ First, divide by 2.
Divide by 7.
Divide by 11.

(e) $\dfrac{\overset{128}{\cancel{256}}}{\underset{104}{\cancel{208}}} = \dfrac{\overset{64}{\cancel{128}}}{\underset{52}{\cancel{104}}} = \dfrac{\overset{32}{\cancel{64}}}{\underset{26}{\cancel{52}}} = \dfrac{\overset{16}{\cancel{32}}}{\underset{13}{\cancel{26}}} = \dfrac{16}{13}$ First, divide by 2.
Divide by 2.
Divide by 2.
Divide by 2.

(f) $\dfrac{\overset{126}{\cancel{378}}}{\underset{135}{\cancel{405}}} = \dfrac{\overset{42}{\cancel{126}}}{\underset{45}{\cancel{135}}} = \dfrac{\overset{14}{\cancel{42}}}{\underset{15}{\cancel{45}}} = \dfrac{14}{15}$
    First, divide by 3.
    Divide by 3.
    Divide by 3.

Remember, if you cancelled the common factors in a different order, your final answer will be the same although the middle results may be different.

Reduce the fraction $\dfrac{6}{3}$ to lowest terms.

Check your answer in **14**.

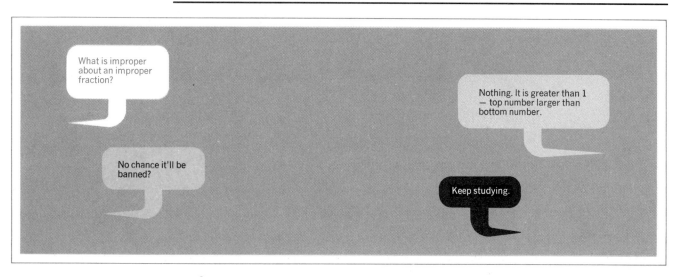

14   $\dfrac{6}{3} = \dfrac{\overset{2}{\cancel{6}}}{\underset{1}{\cancel{3}}} = \dfrac{2}{1}$ or   simply 2.

Any whole number may be written as a fraction by using a denominator equal to 1.

$3 = \dfrac{3}{1}$ $4 = \dfrac{4}{1}$ and so on

Of course, the number 1 can be written as any fraction whose numerator and denominator are equal.

$1 = \dfrac{2}{2} = \dfrac{3}{3} = \dfrac{4}{4} = \ \ldots \ = \dfrac{72}{72} = \dfrac{1257}{1257} \ \ldots$ and so on

Now turn to **15** for some practice working with fractions.

2 Fractions

PROBLEM SET 1

15   The answers are on page 576.

2.1 Renaming
     Fractions

A. Write as an improper fraction.

1. $2\frac{1}{3} =$ ___

2. $4\frac{2}{5} =$ ___

3. $7\frac{1}{2} =$ ___

4. $13\frac{3}{7} =$ ___

5. $8\frac{3}{4} =$ ___

6. $4 =$ ___

7. $1\frac{2}{3} =$ ___

8. $5\frac{5}{6} =$ ___

9. $3\frac{7}{8} =$ ___

10. $2\frac{3}{5} =$ ___

11. $16\frac{1}{10} =$ ___

12. $70\frac{5}{9} =$ ___

13. $12\frac{1}{40} =$ ___

14. $15\frac{5}{11} =$ ___

15. $37\frac{2}{3} =$ ___

B. Write as a mixed number.

1. $\frac{17}{2} =$ _____

2. $\frac{23}{3} =$ _____

3. $\frac{8}{5} =$ _____

4. $\frac{19}{4} =$ _____

5. $\frac{37}{6} =$ _____

6. $\frac{28}{3} =$ _____

7. $\frac{37}{8} =$ _____

8. $\frac{29}{7} =$ _____

9. $\frac{34}{25} =$ _____

10. $\frac{47}{9} =$ _____

11. $\frac{211}{4} =$ _____

12. $\frac{170}{23} =$ _____

13. $\frac{43}{10} =$ _____

14. $\frac{125}{6} =$ _____

15. $\frac{139}{15} =$ _____

C. Reduce to lowest terms.

1. $\frac{26}{30} =$ _____

2. $\frac{12}{15} =$ _____

3. $\frac{8}{10} =$ _____

4. $\frac{27}{54} =$ _____

5. $\frac{5}{40} =$ _____

6. $\frac{18}{45} =$ _____

7. $\frac{7}{42} =$ _____

8. $\frac{16}{18} =$ _____

9. $\frac{9}{27} =$ _____

10. $\frac{21}{56} =$ _____

11. $\frac{42}{120} =$ _____

12. $\frac{54}{144} =$ _____

13. $\frac{36}{216} =$ _____

14. $\frac{280}{490} =$ _____

15. $\frac{115}{207} =$ _____

D. Complete these fractions.

1. $\frac{7}{8} = \frac{}{16}$

2. $\frac{3}{5} = \frac{}{45}$

3. $\frac{3}{4} = \frac{}{12}$

4. $2\frac{5}{12} = \frac{}{60}$

5. $\frac{1}{9} = \frac{}{63}$

6. $1\frac{2}{7} = \frac{}{35}$

7. $\frac{5}{8} = \frac{}{32}$

8. $5\frac{3}{5} = \frac{}{25}$

9. $\frac{1}{2} = \frac{}{78}$

10. $\frac{2}{3} = \frac{}{51}$

11. $8\frac{1}{4} = \frac{}{44}$

12. $5\frac{6}{7} = \frac{}{14}$

13. $\frac{11}{12} = \frac{}{72}$

14. $3\frac{7}{10} = \frac{}{50}$

15. $9\frac{5}{9} = \frac{}{54}$

When you have had the practice you need, either return to the Preview on page 44, or continue in 16 with the study of multiplication of fractions.

Date

Name

Course/Section

SECTION 2: MULTIPLICATION OF FRACTIONS

16

> **OBJECTIVE**
>
> Multiply two or more given fractions.

The simplest arithmetic operation with fractions is multiplication and, happily, it is easy to show graphically. The multiplication of a whole number and a fraction may be illustrated this way:

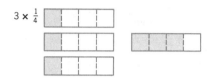

$$3 \times \frac{1}{4} = \frac{1}{4} + \frac{1}{4} + \frac{1}{4} = \frac{3}{4}$$

which is three segments each $\frac{1}{4}$ unit long.

Any fraction such as $\frac{3}{4}$ can be thought of as a product

$$3 \times \frac{1}{4}$$

The product of two fractions can also be shown graphically.

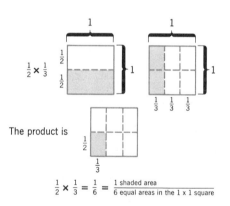

$\frac{1}{2} \times \frac{1}{3} = \frac{1}{6} = \frac{1 \text{ shaded area}}{6 \text{ equal areas in the } 1 \times 1 \text{ square}}$

Another way to solve this is

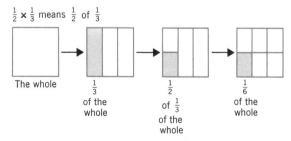

In general, we calculate this product as $\frac{1}{2} \times \frac{1}{3} = \frac{1 \times 1}{2 \times 3} = \frac{1}{6}$.

The product of two fractions is a fraction whose numerator is the product of their numerators and whose denominator is the product of their denominators.

Multiply $\dfrac{5}{6} \times \dfrac{2}{3}$.

(a) $\dfrac{5}{6} \times \dfrac{2}{3} = \dfrac{10}{18}$ Go to **17**.

(b) $\dfrac{5}{6} \times \dfrac{2}{3} = \dfrac{5}{9}$ Go to **18**.

(c) I don't know how to do it and I can't   Go to **19**.
figure out how to draw the little boxes

17   Right.

$$\frac{5}{6} \times \frac{2}{3} = \frac{5 \times 2}{6 \times 3} = \frac{10}{18}$$

Now reduce this answer to lowest terms and turn to **18**.

18   Excellent.

$$\frac{5}{6} \times \frac{2}{3} = \frac{\overset{5}{\cancel{10}}}{\underset{9}{\cancel{18}}} = \frac{5}{9}$$

Always reduce your answer to lowest terms. In this problem you probably recognized that 6 was evenly divisible by 2 and did it like this:

$$\frac{5}{\underset{3}{\cancel{6}}} \times \frac{\overset{1}{\cancel{2}}}{3} = \frac{5}{9}$$

Divide top and bottom of the fraction by 2.

It will save you time and effort if you eliminate common factors, such as the 2 above, *before* you multiply.

Do you see that $3 \times \frac{1}{4}$ is really the same sort of problem?

$$3 = \frac{3}{1}, \quad \text{so} \quad 3 \times \frac{1}{4} = \frac{3}{1} \times \frac{1}{4} = \frac{3 \times 1}{1 \times 4} = \frac{3}{4}$$

Test your understanding with these problems. Multiply as shown:

(a) $\dfrac{7}{8} \times \dfrac{2}{3} =$ _____   (b) $\dfrac{8}{12} \times \dfrac{3}{16} =$ _____   (c) $\dfrac{15}{4} \times \dfrac{9}{10} =$ _____

(d) $\dfrac{3}{2} \times \dfrac{2}{3} =$ _____   (e) $1\dfrac{1}{2} \times \dfrac{2}{5} =$ _____   (f) $4 \times \dfrac{7}{8} =$ _____

(g) $3\frac{5}{6} \times \frac{3}{10} =$ _____   (h) $1\frac{4}{5} \times \frac{2}{3} \times \frac{1}{4} =$ _____

(i) $2\frac{1}{4} \times \frac{1}{24} \times 2\frac{2}{3} =$ _____

Hint: Change mixed numbers such as $1\frac{1}{2}$ and $3\frac{5}{6}$ to improper fractions, then multiply as usual. Multiply three fractions like two.

Have you reduced all answers to lowest terms?

The correct answers are in **20**.

___

19 Now, now, don't panic. You needn't draw the little boxes to do the calculation. Try it this way:

$$\frac{5}{6} \times \frac{2}{3} = \frac{5 \times 2}{6 \times 3}$$ ⬅ Multiply the numerators

◁ Multiply the denominators

Finish the calculation and then return to **16** and choose an answer.

___

20 (a) $\dfrac{7}{\overset{}{\underset{4}{\cancel{8}}}} \times \dfrac{\overset{1}{\cancel{2}}}{3} = \dfrac{7}{12}$ (b) $\dfrac{\overset{1}{\cancel{8}}}{\underset{4}{\cancel{12}}} \times \dfrac{\overset{1}{\cancel{3}}}{\underset{2}{\cancel{16}}} = \dfrac{1}{8}$

(c) $\dfrac{\overset{3}{\cancel{15}}}{4} \times \dfrac{9}{\underset{2}{\cancel{10}}} = \dfrac{27}{8} = 3\dfrac{3}{8}$ (d) $\dfrac{\overset{1}{\cancel{3}}}{\underset{1}{\cancel{2}}} \times \dfrac{\overset{1}{\cancel{2}}}{\underset{1}{\cancel{3}}} = 1$

(e) $1\dfrac{1}{2} \times \dfrac{2}{5} = \dfrac{3}{\underset{1}{\cancel{2}}} \times \dfrac{\overset{1}{\cancel{2}}}{5} = \dfrac{3}{5}$ (f) $4 \times \dfrac{7}{8} = \dfrac{4}{1} \times \dfrac{7}{\underset{2}{\cancel{8}}} = \dfrac{7}{2} = 3\dfrac{1}{2}$

(g) $3\dfrac{5}{6} \times \dfrac{3}{10} = \dfrac{23}{\underset{2}{\cancel{6}}} \times \dfrac{\overset{1}{\cancel{3}}}{10} = \dfrac{23}{20} = 1\dfrac{3}{20}$

(h) $1\dfrac{4}{5} \times \dfrac{2}{3} \times \dfrac{1}{4} = \dfrac{\overset{3}{\cancel{9}}}{5} \times \dfrac{\overset{1}{\cancel{2}}}{\underset{1}{\cancel{3}}} \times \dfrac{1}{\underset{2}{\cancel{4}}} = \dfrac{3}{10}$

(i) $2\dfrac{1}{4} \times \dfrac{1}{24} \times 2\dfrac{2}{3} = \dfrac{\overset{}{\cancel{9}}}{4} \times \dfrac{1}{\underset{\cancel{8}}{\underset{1}{\cancel{24}}}}^{\overset{1}{\cancel{3}}} \times \dfrac{\overset{1}{\cancel{8}}}{\cancel{3}} = \dfrac{1}{4}$

Now try a practical business problem.

Leroy's office measures $14\frac{2}{3}$ feet by $20\frac{3}{4}$ feet. What is the area? (Area is length times width.)

Check your work in **21**.

___

21     $14\frac{2}{3} \times 20\frac{3}{4} = \frac{44}{3} \times \frac{83}{4} = \frac{913}{3} = 304\frac{1}{3}$ square feet.

In multiplying fractions, change the mixed numbers to improper fractions, then cancel to reduce the answer to lowest terms.

Easy, isn't it!

Now turn to 22 for a set of practice problems on multiplication.

2 Fractions

PROBLEM SET 2

22 The answers are on page 576.

2.2 Multiplication of Fractions

A. Multiply and reduce the answer to lowest terms.

1. $\dfrac{1}{2} \times \dfrac{1}{4} =$ _____

2. $\dfrac{2}{3} \times \dfrac{1}{6} =$ _____

3. $\dfrac{2}{5} \times \dfrac{2}{3} =$ _____

4. $\dfrac{3}{8} \times \dfrac{1}{3} =$ _____

5. $\dfrac{4}{5} \times \dfrac{1}{6} =$ _____

6. $\dfrac{5}{3} \times \dfrac{1}{2} =$ _____

7. $6 \times \dfrac{1}{2} =$ _____

8. $\dfrac{5}{6} \times \dfrac{3}{5} =$ _____

9. $\dfrac{8}{9} \times 3 =$ _____

10. $\dfrac{5}{16} \times \dfrac{8}{3} =$ _____

11. $\dfrac{11}{12} \times \dfrac{4}{15} =$ _____

12. $\dfrac{3}{7} \times \dfrac{3}{8} =$ _____

13. $\dfrac{8}{3} \times \dfrac{5}{12} =$ _____

14. $14 \times \dfrac{3}{4} =$ _____

15. $\dfrac{7}{8} \times \dfrac{13}{14} =$ _____

16. $\dfrac{5}{9} \times \dfrac{36}{25} =$ _____

17. $\dfrac{12}{8} \times \dfrac{15}{9} =$ _____

18. $\dfrac{32}{5} \times \dfrac{15}{16} =$ _____

19. $\dfrac{4}{7} \times \dfrac{49}{2} =$ _____

20. $\dfrac{16}{6} \times \dfrac{15}{28} =$ _____

B. Multiply and reduce the answer to lowest terms.

1. $4\dfrac{1}{2} \times \dfrac{2}{3} =$ _____

2. $3\dfrac{1}{5} \times 1\dfrac{1}{4} =$ _____

3. $6 \times 1\dfrac{1}{3} =$ _____

4. $\dfrac{3}{8} \times 3\dfrac{1}{2} =$ _____

5. $2\dfrac{1}{6} \times 1\dfrac{1}{2} =$ _____

6. $7\dfrac{3}{4} \times 8 =$ _____

7. $\dfrac{5}{7} \times 1\dfrac{7}{15} =$ _____

8. $1\dfrac{2}{9} \times \dfrac{3}{11} =$ _____

9. $4\dfrac{3}{5} \times 15 =$ _____

10. $3\dfrac{3}{8} \times 1\dfrac{7}{9} =$ _____

11. $10\dfrac{5}{6} \times 3\dfrac{3}{10} =$ _____

12. $4\dfrac{5}{11} \times \dfrac{2}{7} =$ _____

13. $34 \times 2\dfrac{3}{17} =$ _____

14. $9\dfrac{7}{8} \times \dfrac{4}{5} =$ _____

15. $7\dfrac{9}{10} \times 1\dfrac{1}{4} =$ _____

16. $14 \times 3\dfrac{1}{3} =$ _____

17. $11\dfrac{6}{7} \times \dfrac{7}{8} =$ _____

18. $5\dfrac{1}{6} \times 2\dfrac{3}{5} =$ _____

19. $18 \times 1\dfrac{5}{27} =$ _____

20. $3\dfrac{1}{5} \times 1\dfrac{7}{8} =$ _____

C. Solve.

1. A jet cruises at 450 mph for $3\frac{2}{5}$ hours. How far did it travel?

   _____

2. How many feet of wood are required to make 16 shelves that are each $5\frac{3}{4}$ feet long?

   _____

3. What is the area in square miles of a farm $1\frac{5}{16}$ miles long by $\frac{2}{3}$ mile wide?

   _____

Date

Name

Course/Section

4. Sam's base pay rate is $8 per hour. If he gets paid $1\frac{1}{2}$ times his base pay rate for overtime, what is his overtime rate?

_____

5. The scale on a map is 1 cm equals $12\frac{1}{2}$ kilometers. What actual distance is represented by a map distance of $8\frac{4}{5}$ cm?

_____

6. How many miles can you travel on $8\frac{1}{10}$ gallons of gas if your car gets $27\frac{1}{2}$ miles per gallon?

_____

7. What is the value of 16 shares of stock priced at $24\frac{5}{8}$ per share?

_____

8. What is the total cost of $5\frac{3}{4}$ yards of fabric selling for $8 per yard?

_____

9. Calculate the value of 60 shares of stock valued at $5\frac{3}{8}$ per share.

_____

10. If the annual depreciation is $\frac{10}{12}$ times $\frac{4}{15}$ times the total depreciation, and the total depreciation is $900, calculate the annual depreciation.

_____

When you have had the practice you need, either return to the Preview on page 44 or continue in **23** with the study of division of fractions.

_____

SECTION 3: DIVISION OF FRACTIONS

23

> **OBJECTIVE**
>
> Divide two given fractions.

Addition and multiplication are both reversible arithmetic operations. For example,

2×3 and 3×2 Both equal 6.
$4 + 5$ and $5 + 4$ Both equal 9.

In subtraction and division this kind of exchange is not allowed and because it is not allowed, many people find these operations very troublesome.

$7 - 5 = 2$

but $5 - 7$ is not equal to 2 and is not even a counting number.

$8 \div 4 = 2$

but $4 \div 8$ is not equal to 2 and is a very different kind of number.

In the division of fractions it is particularly important that you set up the process correctly.

Are these numbers equal?

"8 divided by 4" $4\overline{)8}$ $8 \div 4$ $\dfrac{4}{8}$

Choose an answer: (a) Yes   Go to **25**
                      (b) No   Go to **26**

---

24   The divisor is $\dfrac{1}{2}$.

The division $5 \div \dfrac{1}{2}$ is read "5 divided by $\dfrac{1}{2}$," and it asks how many $\dfrac{1}{2}$ unit lengths are included in a length of 5 units.

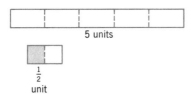

Division is defined in terms of multiplication.

$8 \div 4 = \square$ asks that you find a number $\square$ such that $8 = \square \times 4$. It is easy to see that $\square = 2$.

$5 \div \dfrac{1}{2} = \square$ asks that you find a number $\square$ such that $5 = \square \times \dfrac{1}{2}$. The number $5 \div \dfrac{1}{2}$ answers the question "How many $\dfrac{1}{2}$s are there in 5?" Working backwards from this multiplication and the diagram above, find

$5 \div \dfrac{1}{2} = $ _____

Hop ahead to **27** to continue.

---

25   Your answer is incorrect. Be very careful about this.

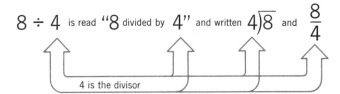

In the above problem you are being asked to divide a set of 8 objects into sets of 4 objects. The divisor 4 is the denominator or bottom number of the fraction.

$8 \div 4$ or "8 divided by 4" are *not* equal to $\frac{4}{8}$ or $8\overline{)4}$.

In the division $5 \div \frac{1}{2}$, which number is the divisor?

Check your answer in **24**.

---

26   Right you are.

The phrase "8 divided by 4" is written $8 \div 4$. We can also write this as $4\overline{)8}$ or $\frac{8}{4}$.

In all of these the divisor is 4, and you are being asked to divide a set of 8 objects into sets of 4 objects.

In the division $5 \div \frac{1}{2}$ which number is the divisor?

Check your answer in **24**.

---

27 $5 \div \frac{1}{2} = 10$

There are ten $\frac{1}{2}$ unit lengths contained in the 5-unit length.

Using a drawing of this sort to solve a division problem is difficult and clumsy. We need a simple rule. Here it is.

> **To divide by a fraction, invert the divisor and multiply.**

For example,

$$5 \div \frac{1}{2} = 5 \times \frac{2}{1} = 5 \times 2 = 10$$

The divisor $\frac{1}{2}$ has been inverted

2 Fractions

PROBLEM SET 3

31 The answers are on page 577.

2.3 Division of Fractions

A. Divide and reduce the answer to lowest terms.

1. $\dfrac{5}{6} \div \dfrac{1}{2} =$ _____

2. $\dfrac{3}{4} \div \dfrac{3}{7} =$ _____

3. $6 \div \dfrac{2}{3} =$ _____

4. $\dfrac{1}{2} \div 6 =$ _____

5. $\dfrac{5}{12} \div \dfrac{4}{3} =$ _____

6. $\dfrac{4}{18} \div \dfrac{1}{2} =$ _____

7. $8 \div \dfrac{1}{3} =$ _____

8. $\dfrac{7}{20} \div \dfrac{4}{5} =$ _____

9. $\dfrac{6}{13} \div \dfrac{3}{4} =$ _____

10. $3 \div \dfrac{2}{5} =$ _____

11. $\dfrac{1}{2} \div \dfrac{1}{2} =$ _____

12. $\dfrac{1}{2} \div \dfrac{1}{3} =$ _____

13. $\dfrac{3}{14} \div \dfrac{6}{5} =$ _____

14. $\dfrac{3}{5} \div \dfrac{1}{3} =$ _____

15. $\dfrac{3}{4} \div \dfrac{5}{16} =$ _____

16. $\dfrac{2}{9} \div \dfrac{8}{3} =$ _____

B. Divide and reduce the answer to lowest terms.

1. $1\dfrac{1}{2} \div \dfrac{1}{6} =$ _____

2. $2\dfrac{3}{4} \div \dfrac{3}{8} =$ _____

3. $6 \div 1\dfrac{1}{2} =$ _____

4. $2\dfrac{1}{4} \div 3 =$ _____

5. $3\dfrac{1}{7} \div 2\dfrac{5}{14} =$ _____

6. $8\dfrac{2}{5} \div 1\dfrac{2}{5} =$ _____

7. $3\dfrac{1}{2} \div 2 =$ _____

8. $4\dfrac{1}{2} \div 1\dfrac{3}{4} =$ _____

9. $6\dfrac{2}{5} \div 5\dfrac{1}{3} =$ _____

10. $10 \div 1\dfrac{1}{5} =$ _____

11. $4\dfrac{1}{6} \div 3\dfrac{1}{3} =$ _____

12. $15\dfrac{5}{6} \div 9\dfrac{1}{2} =$ _____

13. $7\dfrac{1}{7} \div 8\dfrac{1}{3} =$ _____

14. $11\dfrac{2}{3} \div 2\dfrac{2}{9} =$ _____

15. $1\dfrac{1}{5} \div 1\dfrac{1}{2} =$ _____

16. $13\dfrac{1}{3} \div 2\dfrac{3}{6} =$ _____

C. Divide.

1. $\dfrac{\frac{8}{1}}{\frac{1}{2}} =$ _____

2. $\dfrac{\frac{3}{4}}{2} =$ _____

3. $\dfrac{\frac{2}{3}}{6} =$ _____

4. $\dfrac{2\frac{1}{3}}{\frac{3}{4}} =$ _____

5. $\dfrac{12}{\frac{2}{3}} =$ _____

6. $\dfrac{15}{\frac{3}{5}} =$ _____

7. Divide $\dfrac{3}{4}$ by $\dfrac{7}{8}$ _____

8. Divide 2 by $\dfrac{1}{3}$ _____

9. Divide $1\dfrac{7}{8}$ by $\dfrac{3}{2}$ _____

10. Divide $1\dfrac{1}{2}$ by $7\dfrac{1}{4}$ _____

11. Divide 5 by $\dfrac{2}{7}$ _____

12. Divide $11\dfrac{1}{3}$ by $\dfrac{2}{3}$ _____

Date _____

Name _____

Course/Section _____

D. Solve.

1. If you drive $59\frac{1}{2}$ miles in $1\frac{3}{4}$ hours, what is your average speed?

   _____

2. A length of fabric $6\frac{7}{8}$ yards long is divided into 5 equal pieces. What is the length of each piece?

   _____

3. How many pieces of wood each $2\frac{1}{2}$ inches long can be cut from a strip $142\frac{1}{2}$ inches long?

   _____

4. How many shares of stock selling for $\$12\frac{3}{8}$ each can be purchased for \$495?

   _____

5. If your car used $3\frac{1}{4}$ gallons of gasoline in $34\frac{1}{2}$ miles, what is your average miles per gallon?

   _____

6. A large size can of fruit weighs $3\frac{1}{4}$ pounds. Into how many $\frac{1}{8}$-lb servings can this be divided?

   _____

7. A box of modeling clay weighs $45\frac{1}{2}$ pounds. How many $1\frac{3}{4}$ pound packages can be obtained from one such box?

   _____

8. How many shares of stock selling for $\$4\frac{5}{8}$ per share can be purchased for \$185?

   _____

When you have had the practice you need, either return to the Preview on page 44 or continue in **32** with the addition and subtraction of fractions.

32

OBJECTIVE

Add two or more fractions.

Subtract two fractions.

Adding Fractions

At heart, adding fractions is a matter of counting.

$$\frac{1}{5} + \frac{3}{5} = \frac{1+3}{5} = \frac{4}{5}$$

$\frac{1}{5}$ [bar] 1 fifth

$+$

$\frac{3}{5}$ [bar] $+$ 3 fifths

$=$

$\frac{4}{5}$ [bar] $=$ 4 fifths, count them.

Add $\dfrac{2}{7} + \dfrac{3}{7} =$ _____

Check your answer in **33**.

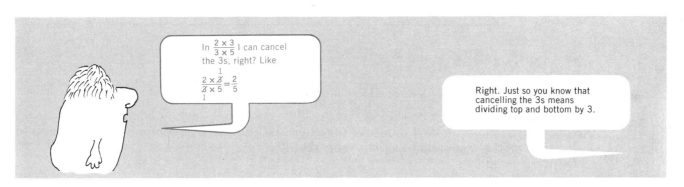

In $\frac{2 \times 3}{3 \times 5}$ I can cancel the 3s, right? Like

$$\frac{2 \times \cancel{3}^{1}}{\cancel{3}_{1} \times 5} = \frac{2}{5}$$

Right. Just so you know that cancelling the 3s means dividing top and bottom by 3.

33    $\dfrac{2}{7} + \dfrac{3}{7} = \dfrac{2+3}{7} = \dfrac{5}{7}$

$\frac{2}{7}$ [bar] 2 sevenths

$+$

$\frac{3}{7}$ [bar] $+$ 3 sevenths

$=$

$\frac{5}{7}$ [bar] 5 sevenths or $\frac{5}{7}$

Fractions having the same denominator are called *like* fractions. In the problem above, $\frac{2}{7}$ and $\frac{3}{7}$ both have denominator 7 and are like fractions. **Adding like fractions is easy: *first,* add the numerators to find the numerator of the sum and, *second,* use the denominator the fractions have in common as the denominator of the sum.**

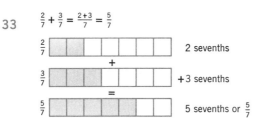

$$\frac{2}{9} + \frac{5}{9} = \frac{2+5}{9} = \frac{7}{9}$$ Add numerators.
Same denominator.

Adding three or more like fractions presents no special problems:

$$\frac{3}{12} + \frac{1}{12} + \frac{5}{12} = \underline{\hspace{1.5cm}}$$

Add the fractions as shown above, then turn to **34**.

---

34 $\qquad \dfrac{3}{12} + \dfrac{1}{12} + \dfrac{5}{12} = \dfrac{3+1+5}{12} = \dfrac{9}{12} = \dfrac{3}{4}$

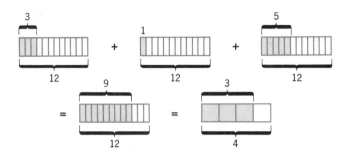

Notice that we reduce the sum to lowest terms.

Try these problems for exercise:

(a) $\quad \dfrac{1}{8} + \dfrac{3}{8} = \underline{\hspace{1.5cm}}$ $\qquad$ (b) $\quad \dfrac{7}{9} + \dfrac{5}{9} = \underline{\hspace{1.5cm}}$

Go to **35** to check your work.

---

35 $\quad$ (a) $\quad \dfrac{1}{8} + \dfrac{3}{8} = \dfrac{1+3}{8} = \dfrac{4}{8} = \dfrac{1}{2}$, reduced to lowest terms.

$\qquad$ (b) $\quad \dfrac{7}{9} + \dfrac{5}{9} = \dfrac{7+5}{9} = \dfrac{12}{9} = \dfrac{4}{3} = 1\dfrac{1}{3}$

Adding Mixed Numbers $\qquad$ **To add mixed numbers together, we add the whole numbers together, add the fractions together, then simplify.**

$$2\frac{2}{5} + 3\frac{4}{5} + 7\frac{1}{5} = 2 + 3 + 7 + \frac{2}{5} + \frac{4}{5} + \frac{1}{5}$$

$$= 12 + \frac{2+4+1}{5} = 12 + \frac{7}{5}$$

$$= 12 + 1\frac{2}{5} = 12 + 1 + \frac{2}{5} = 13\frac{2}{5}$$

Lining up the mixed number vertically usually simplifies the process.

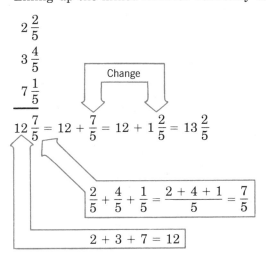

$$2\frac{2}{5}$$

$$3\frac{4}{5}$$

$$7\frac{1}{5}$$

$$12\frac{7}{5} = 12 + \frac{7}{5} = 12 + 1\frac{2}{5} = 13\frac{2}{5}$$

$$\frac{2}{5} + \frac{4}{5} + \frac{1}{5} = \frac{2 + 4 + 1}{5} = \frac{7}{5}$$

$$2 + 3 + 7 = 12$$

Try the following problems. Be sure to express your answer as a mixed number and reduce to lowest terms.

(a) $2\frac{2}{7} + 5\frac{4}{7} + 3\frac{5}{7}$ (b) $2 + 3\frac{1}{2}$

(c) $2\frac{1}{2} + 6 + 4\frac{1}{2}$ (d) $3\frac{5}{8} + 4\frac{7}{8} + 1\frac{1}{8} + 12\frac{7}{8}$

Check your work in **37**.

---

36 $\frac{3}{4} + \frac{2}{3} = \frac{9}{12} + \frac{8}{12} = \frac{17}{12} = 1\frac{5}{12}$

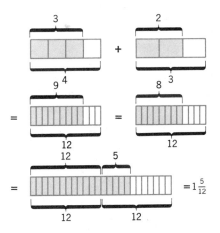

Least Common Denominator   We change the original fractions to equivalent fractions with the same denominator and then add as before. The new denominator should be the smallest number evenly divisible by the other denominators. The new denominator is called the *Least Common Denominator* or *LCD*.

How do you know what number to use as the LCD? Sometimes it can be guessed by trial and error. A better method is the *LCD FINDER*.

The LCD Finder uses prime numbers. Numbers divisible only by themselves and one are called *prime numbers*. Prime numbers include 2, 3, 5, 7, 11, 13, and so on. Numbers such as $4 = 2 \times 2$, $6 = 2 \times 3$, $9 = 3 \times 3$ are not prime numbers.

Step 1. Arrange the denominators in a row.

Step 2. Divide by any prime number that can be divided into two or more denominators. Bring down any denominators that are not divided.

Step 3. Continue until there are no more divisors.

Step 4. Multiply all the divisors and the numbers in the last row to find the LCD.

This method is really much easier than it sounds. An example will help clarify it.

Example: Find the LCD of the fractions $\frac{1}{4}, \frac{3}{7}$, and $\frac{5}{14}$.

Step 1. Arrange the denominators   4   7   14

Step 2. Both 4 and 14 are divisible ⟶ $2\,|\,4$ 7   14
by 2. $4 \div 2 = 2$, $14 \div 2 = 7$ 2   7   7
Bring down the 7 to make
a new row.

Step 3. There are now 2 sevens; | Multiply. |
therefore, divide by 7.
Bring down the 2 for a new row   2 | 4   7   14
The LCD is $2 \times 7 \times 2 \times 1 \times 1$ or   7 | 2   7   7
$2 \times 7 \times 2 = 28$.   2   1   1

Using the LCD FINDER may seem like a long procedure, but with a little practice, it's easy.

Now it's your turn. Try finding the LCD of $\frac{3}{10}, \frac{1}{4}, \frac{7}{20}$, and $\frac{5}{8}$.

Check your answer in **38**.

---

37   (a) $2\frac{2}{7}$

           $5\frac{4}{7}$

        $+3\frac{5}{7}$

     $10\frac{11}{7} = 10 + \frac{11}{7} = 10 + 1\frac{4}{7} = 11\frac{4}{7}$

               $\frac{2}{7} + \frac{4}{7} + \frac{5}{7} = \frac{2+4+5}{7} = \frac{11}{7}$

               $2 + 5 + 3 = 10$

   (b)   2

        $+3\frac{1}{2}$

       $5\frac{1}{2}$

(c)

$$2\frac{1}{2}$$

$$6$$

$$+4\frac{1}{2}$$

$$12\frac{2}{2} = 12 + \frac{2}{2} = 12 + 1 = 13$$

(d)

$$3\frac{5}{8}$$

$$4\frac{7}{8}$$

$$1\frac{1}{8}$$

$$+12\frac{7}{8}$$

$$20\frac{20}{8} = 20 + \frac{20}{8} = 20 + 2\frac{4}{8} = 20 + 2 + \frac{4}{8} = 22 + \frac{4}{8} = 22 + \frac{1}{2} = 22\frac{1}{2}$$

How do we add fractions whose denominators are not the same?

$$\frac{2}{3} + \frac{3}{4} = ?$$

The problem is to find a simple numeral that names this new number. One way to find it is to change these fractions to equivalent fractions with the same denominator.

$$\frac{3}{4} = \frac{3 \times 3}{4 \times 3} = \frac{9}{12}$$

(Equivalent fractions are discussed on page 48 in frame **9**.)

$$\frac{2}{3} = \frac{2 \times 4}{3 \times 4} = \frac{8}{12}$$

Now add these fractions.

Continue in **36**.

The original denominators

38

		10	4	20	8
Divide by 2 ⟶	2	10	4	20	8
Divide by 2 ⟶	2	5	2	10	4
Divide by 5 ⟶	5	5	1	5	2
		1	1	1	2

LCD = 2 × 2 × 5 × 1 × 1 × 1 × 2 or 2 × 2 × 5 × 2 = 40

It's easy once you get the hang of it.

Try a few more.

Find the LCD of the following fractions.

(a) $\dfrac{2}{3}, \dfrac{1}{8}, \dfrac{5}{12}, \dfrac{3}{4}$ (b) $\dfrac{7}{12}, \dfrac{5}{18}, \dfrac{1}{20}$

(c) $\dfrac{1}{10}, \dfrac{7}{8}, \dfrac{4}{15}, \dfrac{3}{4}, \dfrac{5}{6}$ (d) $\dfrac{2}{3}, \dfrac{3}{4}, \dfrac{5}{6}, \dfrac{7}{9}$

Check your answers in **39**.

---

39   (a)

2	3	8	12	4
2	3	4	6	2
3	3	2	3	1
	1	2	1	1

(b)

2	12	18	20
2	6	9	10
3	3	9	5
	1	3	5

$\text{LCD} = 2 \times 2 \times 3 \times 1 \times 2 \times 1 \times 1 = 24$ $\text{LCD} = 2 \times 2 \times 3 \times 1 \times 3 \times 5 = 180$

(c)

2	10	8	15	4	6
2	5	4	15	2	3
3	5	2	15	1	3
5	5	2	5	1	1
	1	2	1	1	1

(d)

2	3	4	6	9
3	3	2	3	9
	1	2	1	3

$\text{LCD} = 2 \times 3 \times 1 \times 2 \times 1 \times 3 = 36$

$\begin{aligned}\text{LCD} &= 2 \times 2 \times 3 \times 5 \times 1 \times 2 \times 1 \\ &\quad \times 1 \times 1 = 120\end{aligned}$

Now let's use the LCD to solve an addition problem.

$$\dfrac{7}{15} + \dfrac{5}{6} + \dfrac{3}{10}$$

First, we must find the LCD.

2	15	6	10
3	15	3	5
5	5	1	5
	1	1	1

$\text{LCD} = 2 \times 3 \times 5 \times 1 \times 1 \times 1 = 30$

Next, change all fractions to equivalent fractions with 30 in the denominator.

$$\dfrac{7}{15} = \dfrac{?}{30} = \dfrac{7 \times 2}{15 \times 2} = \dfrac{14}{30}$$

$$\dfrac{5}{6} = \dfrac{?}{30} = \dfrac{5 \times 5}{6 \times 5} = \dfrac{25}{30}$$

$$\dfrac{3}{10} = \dfrac{?}{30} = \dfrac{3 \times 3}{10 \times 3} = \dfrac{9}{30}$$

Finally, add and reduce.

$$\dfrac{7}{15} + \dfrac{5}{6} + \dfrac{3}{10} = \dfrac{14}{30} + \dfrac{25}{30} + \dfrac{9}{30} = \dfrac{14 + 25 + 9}{30} = \dfrac{48}{30} = 1\dfrac{18}{30} = 1\dfrac{3}{5}$$

Remember, to add fractions:

1. **Find the LCD of the denominators of the fractions to be added.**
2. **Write the fractions so that they are equivalent fractions with the LCD.**
3. **The answer numerator is the sum of the numerators. The denominator of the answer is the LCD.**
4. **Reduce to lowest terms.**

(c)
$$2\frac{1}{2}$$
$$6$$
$$+4\frac{1}{2}$$

$$12\frac{2}{2} = 12 + \frac{2}{2} = 12 + 1 = 13$$

(d)
$$3\frac{5}{8}$$
$$4\frac{7}{8}$$
$$1\frac{1}{8}$$
$$+12\frac{7}{8}$$

$$20\frac{20}{8} = 20 + \frac{20}{8} = 20 + 2\frac{4}{8} = 20 + 2 + \frac{4}{8} = 22 + \frac{4}{8} = 22 + \frac{1}{2} = 22\frac{1}{2}$$

How do we add fractions whose denominators are not the same?

$$\frac{2}{3} + \frac{3}{4} = ?$$

The problem is to find a simple numeral that names this new number. One way to find it is to change these fractions to equivalent fractions with the same denominator.

$$\frac{3}{4} = \frac{3 \times 3}{4 \times 3} = \frac{9}{12}$$

(Equivalent fractions are discussed on page 48 in frame **9**.)

$$\frac{2}{3} = \frac{2 \times 4}{3 \times 4} = \frac{8}{12}$$

Now add these fractions.

Continue in **36**.

The original denominators

38

Divide by 2 ⟶	2	10	4	20	8
Divide by 2 ⟶	2	5	2	10	4
Divide by 5 ⟶	5	5	1	5	2
		1	1	1	2

LCD $= 2 \times 2 \times 5 \times 1 \times 1 \times 1 \times 2$ or $2 \times 2 \times 5 \times 2 = 40$

It's easy once you get the hang of it.

Try a few more.

Find the LCD of the following fractions.

(a) $\dfrac{2}{3}, \dfrac{1}{8}, \dfrac{5}{12}, \dfrac{3}{4}$ (b) $\dfrac{7}{12}, \dfrac{5}{18}, \dfrac{1}{20}$

(c) $\dfrac{1}{10}, \dfrac{7}{8}, \dfrac{4}{15}, \dfrac{3}{4}, \dfrac{5}{6}$ (d) $\dfrac{2}{3}, \dfrac{3}{4}, \dfrac{5}{6}, \dfrac{7}{9}$

Check your answers in **39**.

---

39 (a)

2	3	8	12	4
2	3	4	6	2
3	3	2	3	1
	1	2	1	1

(b)

2	12	18	20
2	6	9	10
3	3	9	5
	1	3	5

$\text{LCD} = 2 \times 2 \times 3 \times 1 \times 2 \times 1 \times 1 = 24$ $\text{LCD} = 2 \times 2 \times 3 \times 1 \times 3 \times 5 = 180$

(c)

2	10	8	15	4	6
2	5	4	15	2	3
3	5	2	15	1	3
5	5	2	5	1	1
	1	2	1	1	1

(d)

2	3	4	6	9
3	3	2	3	9
	1	2	1	3

$\text{LCD} = 2 \times 3 \times 1 \times 2 \times 1 \times 3 = 36$

$$\text{LCD} = 2 \times 2 \times 3 \times 5 \times 1 \times 2 \times 1 \times 1 \times 1 = 120$$

Now let's use the LCD to solve an addition problem.

$$\dfrac{7}{15} + \dfrac{5}{6} + \dfrac{3}{10}$$

First, we must find the LCD.

2	15	6	10
3	15	3	5
5	5	1	5
	1	1	1

$\text{LCD} = 2 \times 3 \times 5 \times 1 \times 1 \times 1 = 30$

Next, change all fractions to equivalent fractions with 30 in the denominator.

$$\dfrac{7}{15} = \dfrac{?}{30} = \dfrac{7 \times 2}{15 \times 2} = \dfrac{14}{30}$$

$$\dfrac{5}{6} = \dfrac{?}{30} = \dfrac{5 \times 5}{6 \times 5} = \dfrac{25}{30}$$

$$\dfrac{3}{10} = \dfrac{?}{30} = \dfrac{3 \times 3}{10 \times 3} = \dfrac{9}{30}$$

Finally, add and reduce.

$$\dfrac{7}{15} + \dfrac{5}{6} + \dfrac{3}{10} = \dfrac{14}{30} + \dfrac{25}{30} + \dfrac{9}{30} = \dfrac{14 + 25 + 9}{30} = \dfrac{48}{30} = 1\dfrac{18}{30} = 1\dfrac{3}{5}$$

Remember, to add fractions:

1. **Find the LCD of the denominators of the fractions to be added.**
2. **Write the fractions so that they are equivalent fractions with the LCD.**
3. **The answer numerator is the sum of the numerators. The denominator of the answer is the LCD.**
4. **Reduce to lowest terms.**

If this way of adding fractions seems rather long and involved, here is why:

1. It *is* involved, but it is the only sure way to arrive at the answer unless you own a fancy calculator that works with fractions.
2. It is new to you and you will need lots of practice at it before it comes quickly. Take each problem step by step, work slowly at first, and gradually you will become very quick at adding fractions.

Practice by adding the following:

(a) $\dfrac{2}{15} + \dfrac{4}{9}$ (b) $\dfrac{5}{16} + \dfrac{7}{12} + \dfrac{1}{6}$ (c) $\dfrac{5}{12} + \dfrac{2}{9} + \dfrac{3}{8} + \dfrac{1}{6}$

Check your answers in **40**.

---

40   (a) $3\ \underline{|\ 15 \qquad 9}$ LCD $= 3 \times 5 \times 3 = 45$
                    $5 \qquad 3$

$$\frac{2}{15} = \frac{?}{45} = \frac{2 \times 3}{15 \times 3} = \frac{6}{45}$$

$$\frac{4}{9} = \frac{?}{45} = \frac{4 \times 5}{9 \times 5} = \frac{20}{45}$$

$$\frac{2}{15} + \frac{4}{9} = \frac{6}{45} + \frac{20}{45} = \frac{6 + 20}{45} = \frac{26}{45}$$

(b) $2\ \underline{|\ 16 \qquad 12 \qquad 6}$ LCD $= 2 \times 2 \times 3 \times 4 \times 1 \times 1 = 48$
     $2\ \underline{|\ \ 8 \qquad\ \ 6 \qquad 3}$
     $3\ \underline{|\ \ 4 \qquad\ \ 3 \qquad 3}$
          $4 \qquad\ \ 1 \qquad 1$

$$\frac{5}{16} = \frac{?}{48} = \frac{5 \times 3}{16 \times 3} = \frac{15}{48}$$

$$\frac{7}{12} = \frac{?}{48} = \frac{7 \times 4}{12 \times 4} = \frac{28}{48}$$

$$\frac{1}{6} = \frac{?}{48} = \frac{1 \times 8}{6 \times 8} = \frac{8}{48}$$

$$\frac{5}{16} + \frac{7}{12} + \frac{1}{6} = \frac{15}{48} + \frac{28}{48} + \frac{8}{48} = \frac{15 + 28 + 8}{48} = \frac{51}{48} = 1\frac{3}{48} = 1\frac{1}{16}$$

(c) $2\ \underline{|\ 12 \qquad 9 \qquad 8 \qquad 6}$ LCD $= 2 \times 2 \times 3 \times 1 \times 3 \times 2 \times 1 = 72$
     $2\ \underline{|\ \ 6 \qquad 9 \qquad 4 \qquad 3}$
     $3\ \underline{|\ \ 3 \qquad 9 \qquad 2 \qquad 3}$
          $1 \qquad 3 \qquad 2 \qquad 1$

$$\frac{5}{12} = \frac{?}{72} = \frac{5 \times 6}{12 \times 6} = \frac{30}{72}$$

$$\frac{2}{9} = \frac{?}{72} = \frac{2 \times 8}{9 \times 8} = \frac{16}{72}$$

$$\frac{3}{8} = \frac{?}{72} = \frac{3 \times 9}{8 \times 9} = \frac{27}{72}$$

$$\frac{1}{6} = \frac{?}{72} = \frac{1 \times 12}{6 \times 12} = \frac{12}{72}$$

$$\frac{5}{12} + \frac{2}{9} + \frac{3}{8} + \frac{1}{6} = \frac{30}{72} + \frac{16}{72} + \frac{27}{72} + \frac{12}{72} = \frac{30 + 16 + 27 + 12}{72} = \frac{85}{72} = 1\frac{13}{72}$$

When adding mixed numbers, remember to add the whole numbers together and add the fractions together.

Add $3\frac{5}{6} + 17\frac{7}{12} + 9\frac{3}{10}$

$$\begin{array}{r|ccc} 2 & 6 & 12 & 10 \\ \hline 3 & 3 & 6 & 5 \\ \hline & 1 & 2 & 5 \end{array} \qquad \text{LCD} = 2 \times 3 \times 1 \times 2 \times 5 = 60$$

$$3\frac{5}{6} = 3\frac{50}{60} \qquad\qquad \frac{5}{6} = \frac{?}{60} = \frac{5 \times 10}{6 \times 10} = \frac{50}{60}$$

$$17\frac{7}{12} = 17\frac{35}{60} \qquad\qquad \frac{7}{12} = \frac{?}{60} = \frac{7 \times 5}{12 \times 5} = \frac{35}{60}$$

$$9\frac{3}{10} = 9\frac{18}{60} \qquad\qquad \frac{3}{10} = \frac{?}{60} = \frac{3 \times 6}{10 \times 6} = \frac{18}{60}$$

$$29\frac{103}{60} = 29 + 1\frac{43}{60} = 30\frac{43}{60}$$

Try adding some mixed numbers.

(a) $4\frac{5}{6} + 12\frac{3}{7} + 8\frac{9}{14}$ 　　　　(b) $13\frac{5}{12} + 2\frac{7}{8} + 5\frac{9}{16}$

Turn to **43** to check your answers.

---

41 $\dfrac{3}{8} - \dfrac{1}{8} = \dfrac{3-1}{8} = \dfrac{2}{8} = \dfrac{1}{4}$

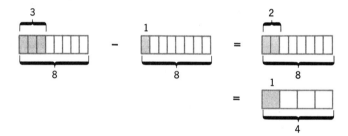

Easy enough? If the denominators are the same, we subtract numerators and write this difference over the common denominator.

Subtract $\dfrac{3}{4} - \dfrac{1}{5} = $ _____

Our solution is in **42**.

---

42 The LCD of 4 and 5 is 20.

$$\frac{3}{4} = \frac{?}{20} = \frac{3 \times 5}{4 \times 5} = \frac{15}{20}$$

$$\frac{1}{5} = \frac{?}{20} = \frac{1 \times 4}{5 \times 4} = \frac{4}{20}$$

Then $\dfrac{3}{4} - \dfrac{1}{5} = \dfrac{15}{20} - \dfrac{4}{20} = \dfrac{11}{20}$

When the two fractions have different denominators, we must change them to equivalent fractions with the LCD as denominator before subtracting.

If you have not yet learned how to find the Least Common Denominator (LCD) turn to **36**. Otherwise, continue by finding the following differences.

(a) $\dfrac{5}{6} - \dfrac{1}{4}$ (b) $\dfrac{11}{15} - \dfrac{2}{9}$

Check your answers in **44**.

---

43   (a)

$$
\begin{array}{r|ccc}
2 & 6 & 7 & 14 \\
7 & 3 & 7 & 7 \\
\hline
  & 3 & 1 & 1
\end{array}
$$

LCD $= 2 \times 7 \times 3 \times 1 \times 1 = 42$

$$4\dfrac{5}{6} = 4\dfrac{35}{42}$$
$$12\dfrac{3}{7} = 12\dfrac{18}{42}$$
$$+8\dfrac{9}{14} = 8\dfrac{27}{42}$$
$$\overline{\phantom{+8\dfrac{9}{14}}}$$
$$24\dfrac{80}{42} = 24 + 1\dfrac{38}{42} = 25\dfrac{38}{42} = 25\dfrac{19}{21}$$

(b)

$$
\begin{array}{r|ccc}
2 & 12 & 8 & 16 \\
2 & 6 & 4 & 8 \\
2 & 3 & 2 & 4 \\
\hline
  & 3 & 1 & 2
\end{array}
$$

LCD $= 2 \times 2 \times 2 \times 3 \times 1 \times 2 = 48$

$$13\dfrac{5}{12} = 13\dfrac{20}{48}$$
$$2\dfrac{7}{8} = 2\dfrac{42}{48}$$
$$+5\dfrac{9}{16} = 5\dfrac{27}{48}$$
$$\overline{\phantom{+5\dfrac{9}{16}}}$$
$$20\dfrac{89}{48} = 20 + \dfrac{89}{48} = 20 + 1\dfrac{41}{48} = 21\dfrac{41}{48}$$

Subtracting Fractions   Once you have mastered the process of adding fractions, subtraction is very simple indeed.

Calculate $\dfrac{3}{8} - \dfrac{1}{8} =$ _____

Turn to **41**.

---

44   (a)   The LCD is 12.

$$\dfrac{5}{6} = \dfrac{5 \times 2}{6 \times 2} = \dfrac{10}{12} \qquad \dfrac{1}{4} = \dfrac{1 \times 3}{4 \times 3} = \dfrac{3}{12}$$

$$\dfrac{5}{6} - \dfrac{1}{4} = \dfrac{10}{12} - \dfrac{3}{12} = \dfrac{10 - 3}{12} = \dfrac{7}{12}$$

(b)   The LCD is 45.

$$\dfrac{11}{15} = \dfrac{11 \times 3}{15 \times 3} = \dfrac{33}{45} \qquad \dfrac{2}{9} = \dfrac{2 \times 5}{9 \times 5} = \dfrac{10}{45}$$

$$\dfrac{11}{15} - \dfrac{2}{9} = \dfrac{33}{45} - \dfrac{10}{45} = \dfrac{33 - 10}{45} = \dfrac{23}{45}$$

The easiest method of subtracting mixed numbers is to subtract the whole number parts and subtract the fractions.

Subtract $17\frac{5}{6} - 12\frac{2}{9}$.   LCD = 18

$$
\begin{array}{ll}
17\frac{5}{6} = & 17\frac{15}{18} \\
-12\frac{2}{9} = & -12\frac{4}{18} \\
\hline
& 5\frac{11}{18}
\end{array}
\qquad
\frac{5}{6} = \frac{5\times 3}{6\times 3} = \frac{15}{18} \\
\qquad
\frac{2}{9} = \frac{2\times 2}{9\times 2} = \frac{4}{18}
$$

Oftentimes we must use borrowing when subtracting mixed numbers.

Subtract $28\frac{1}{6} - 15\frac{5}{8}$.   LCD = 24

$$
\begin{array}{lll}
28\frac{1}{6} = & 28\frac{4}{24} = & 27\frac{28}{24} \\
-15\frac{5}{8} = & -15\frac{15}{24} = & -15\frac{15}{24} \\
\hline
& & 12\frac{13}{24}
\end{array}
\qquad
\left(28\frac{4}{24} = 27 + 1\frac{4}{24} = 27 + \frac{28}{24}\right)
$$

When we borrow 1, we change it to $\frac{24}{24}$ and add it to the $\frac{4}{24}$,

$$1 + \frac{4}{24} = \frac{24}{24} + \frac{4}{24} = \frac{28}{24}.$$

Try the following problems.

(a) $14\frac{2}{3} - 9\frac{1}{4}$ (b) $18\frac{1}{4} - 12\frac{9}{10}$

(c) $53\frac{5}{14} - 27\frac{3}{4}$ (d) $243\frac{1}{6} - 195\frac{5}{9}$

Check your answers in **45**.

---

45 (a) The LCD is 12.

$$
\begin{array}{ll}
14\frac{2}{3} = & 14\frac{8}{12} \\
-9\frac{1}{4} = & -9\frac{3}{12} \\
\hline
& 5\frac{5}{12}
\end{array}
$$

(b) The LCD is 20.

$$
\begin{array}{lll}
18\frac{1}{4} = & 18\frac{5}{20} = & 17\frac{25}{20} \\
-12\frac{9}{10} = & -12\frac{18}{20} = & -12\frac{18}{20} \\
\hline
& & 5\frac{7}{20}
\end{array}
\qquad
18\frac{5}{20} = 17 + 1\frac{5}{20} = 17 + \frac{25}{20}
$$

(c) The LCD is 28.

$$53 \frac{5}{14} = \quad 53 \frac{10}{28} = \quad 52 \frac{38}{28}$$

$$-27 \frac{3}{4} = -27 \frac{21}{28} = -27 \frac{21}{28}$$

$$25 \frac{17}{28}$$

$$53 \frac{10}{28} = 52 + 1 \frac{10}{28} = 52 + \frac{38}{28}$$

(d) The LCD is 18.

$$243 \frac{1}{6} = \quad 243 \frac{3}{18} = \quad 242 \frac{21}{18}$$

$$-195 \frac{5}{9} = -195 \frac{10}{18} = -195 \frac{10}{18}$$

$$47 \frac{11}{18}$$

$$243 \frac{3}{18} = 242 + 1 \frac{3}{18} = 242 + \frac{21}{18}$$

Work the following word problem.

Stock in the Up & Down Company opened at $27\frac{3}{8}$ yesterday and closed at $24\frac{3}{4}$. What was its net loss in price?

Check your work in 46.

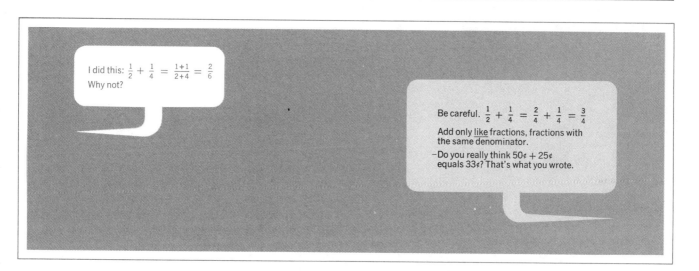

46   The LCD is 8.

$$27 \frac{3}{8} = \quad 27 \frac{3}{8} = \quad 26 \frac{11}{8}$$

$$27 \frac{3}{8} = 26 + 1 \frac{3}{8} = 26 + \frac{11}{8} = 26 \frac{11}{8}$$

$$-24 \frac{3}{4} = -24 \frac{6}{8} = -24 \frac{6}{8}$$

$$2 \frac{5}{8}$$

Now turn to 47 for a set of practice problems on addition and subtraction of fractions.

2 Fractions

2.4 Addition and
Subtraction of Fractions

47 The answers are on page 577.

A. Add or subtract as shown and reduce to lowest terms.

1. $\dfrac{3}{5} + \dfrac{4}{5} = $ ___

2. $\dfrac{5}{12} + \dfrac{11}{12} = $ ___

3. $\dfrac{1}{8} + \dfrac{7}{8} = $ ___

4. $\dfrac{7}{15} + \dfrac{3}{15} = $ ___

5. $\dfrac{7}{8} - \dfrac{5}{8} = $ ___

6. $\dfrac{3}{4} - \dfrac{1}{4} = $ ___

7. $\dfrac{7}{12} - \dfrac{2}{12} = $ ___

8. $\dfrac{15}{16} - \dfrac{9}{16} = $ ___

9. $\dfrac{53}{64} + \dfrac{19}{64} = $ ___

10. $\dfrac{17}{32} + \dfrac{19}{32} = $ ___

11. $\dfrac{53}{64} - \dfrac{5}{64} = $ ___

12. $\dfrac{15}{32} - \dfrac{7}{32} = $ ___

13. $\dfrac{2}{16} + \dfrac{7}{16} + \dfrac{13}{16} = $ ___

14. $\dfrac{7}{18} + \dfrac{10}{18} + \dfrac{13}{18} = $ ___

15. $\dfrac{1}{7} + \dfrac{6}{7} + \dfrac{4}{7} = $ ___

16. $\dfrac{7}{8} + \dfrac{1}{8} - \dfrac{3}{8} = $ ___

17. $\dfrac{5}{12} + \dfrac{11}{12} - \dfrac{1}{12} = $ ___

18. $\dfrac{7}{20} + \dfrac{13}{20} - \dfrac{11}{20} = $ ___

B. Add or subtract as shown and reduce to lowest terms.

1. $\dfrac{7}{8} + \dfrac{3}{4} = $ ___

2. $\dfrac{7}{8} - \dfrac{3}{4} = $ ___

3. $\dfrac{11}{12} + \dfrac{7}{18} = $ ___

4. $\dfrac{11}{12} - \dfrac{7}{18} = $ ___

5. $\dfrac{5}{12} + \dfrac{3}{16} = $ ___

6. $\dfrac{3}{7} + \dfrac{2}{5} = $ ___

7. $\dfrac{11}{48} + \dfrac{3}{64} = $ ___

8. $\dfrac{10}{27} + \dfrac{15}{24} = $ ___

9. $\dfrac{7}{12} - \dfrac{5}{16} = $ ___

10. $\dfrac{5}{7} - \dfrac{3}{5} = $ ___

11. $\dfrac{18}{64} - \dfrac{11}{48} = $ ___

12. $\dfrac{16}{27} - \dfrac{5}{24} = $ ___

13. $1\dfrac{1}{4} + \dfrac{3}{8} = $ ___

14. $2\dfrac{3}{4} + 1\dfrac{5}{18} = $ ___

15. $3\dfrac{5}{12} + 1\dfrac{15}{16} = $ ___

16. $3\dfrac{5}{8} + 1\dfrac{2}{7} = $ ___

17. $3\dfrac{5}{8} - 2\dfrac{7}{8} = $ ___

18. $1\dfrac{1}{8} - \dfrac{3}{4} = $ ___

19. $3\dfrac{5}{12} - 1\dfrac{15}{16} = $ ___

20. $2\dfrac{3}{35} - 1\dfrac{5}{14} = $ ___

C. Calculate and reduce.

1. $2 - \dfrac{1}{3} = $ ___

2. $6 - 4\dfrac{3}{16} = $ ___

3. $8 - \dfrac{11}{4} = $ ___

4. $3 - 1\dfrac{1}{5} = $ ___

5. $6\dfrac{1}{2} + 5\dfrac{3}{4} + 8\dfrac{1}{2} = $ ___

6. $\dfrac{7}{8} + 2\dfrac{1}{5} - 1\dfrac{1}{3} = $ ___

7. $\dfrac{1}{2} + \dfrac{1}{5} + \dfrac{1}{8} = $ ___

8. $\dfrac{1}{2} + \dfrac{1}{3} + \dfrac{1}{4} + \dfrac{1}{5} = $ ___

9. $1\dfrac{2}{3}$ subtracted from $4\dfrac{3}{4} = $ ___

10. $2\dfrac{3}{8}$ less than $4\dfrac{7}{10} = $ ___

Date

Name

Course/Section

11. $6\frac{5}{12}$ reduced by $1\frac{3}{16}$ = _____   12. $4\frac{3}{5}$ less than $6\frac{1}{2}$ = _____

D. Solve.

1. Sara's time card showed that she worked the following hours last week: Monday $7\frac{1}{4}$ hours, Tuesday $6\frac{1}{2}$ hours, Wednesday $5\frac{3}{4}$ hours, Thursday 8 hours, and Friday $9\frac{1}{6}$ hours. What total time did she work? What was her total pay at $6 per hour?

   _____

2. Bob owned $173\frac{1}{4}$ acres of land. He sold $95\frac{1}{2}$ acres. What is his current total acreage after the sale?

   _____

3. Helen purchased the following lengths of material: $3\frac{5}{8}$ yards, $2\frac{1}{2}$ yards, $1\frac{3}{4}$ yards. What was the total yardage Helen purchased?

   _____

4. On a recent trip we traveled $560\frac{9}{10}$ miles and stopped for gas three times using $8\frac{1}{2}$ gallons, $10\frac{3}{5}$ gallons, and $9\frac{3}{10}$ gallons.

   (a) How much gas was used? _____

   (b) What was our average mileage per gallon? _____

5. The four sides of a plot of land are $120\frac{3}{4}$ ft, $85\frac{5}{8}$ ft, $116\frac{2}{3}$ ft, and $91\frac{5}{8}$ ft. What is the total distance around the edge of this lot?

   _____

6. If $5\frac{1}{4}$ yards of cloth are cut from a roll containing $17\frac{5}{8}$ yards, how much remains on the roll?

   _____

7. Stock in the Acme Celery Company opened at $47\frac{3}{8}$ yesterday on the New York Stalk Exchange and closed at $45\frac{3}{16}$. What was the decrease?

   _____

8. Mary's weight dropped from $123\frac{1}{2}$ to $114\frac{7}{8}$ lb while she was on a diet. How much weight did she lose?

   _____

9. When $2\frac{3}{4}$ lb of ground meat is cooked, the finished product weighs $2\frac{1}{3}$ lb. How much weight is lost in cooking?

   _____

10. Joel originally owned $37\frac{1}{2}$ acres of land. After he purchases the adjoining property of $15\frac{5}{8}$ acres, what will be his total acreage?

    _____

When you have had the practice you need, turn to 48 for the Self-Test on fractions.

2 Fractions

The answers are on page 577.

1. Write $7\frac{3}{16}$ as an improper fraction. _____

2. Write $\frac{37}{11}$ as a mixed number. _____

3. $\frac{3}{8} = \frac{?}{40}$ _____

4. Reduce to lowest terms $\frac{195}{255}$. _____

5. $\frac{1}{4} + \frac{2}{3} + \frac{2}{5} =$ _____

6. $11\frac{3}{8} + 12\frac{1}{4} + 18\frac{2}{3} =$ _____

7. $6\frac{1}{4} - 3\frac{2}{3} =$ _____

8. $1\frac{3}{5} \times \frac{5}{12} \times 7\frac{1}{2} =$ _____

9. $1\frac{3}{10} \div 4\frac{3}{4} =$ _____

10. On a recent trip we traveled $577\frac{2}{10}$ miles and stopped for gas three times, using $11\frac{2}{5}$ gallons, $9\frac{3}{10}$ gallons, and $10\frac{1}{2}$ gallons.

    (a) How much gas was used? _____

    (b) What was our average mileage per gallon? _____

Date _____

Name _____

Course/Section _____

3 Decimal Numbers

WAIT! THE DECIMAL POINT WAS IN THE WRONG PLACE ON THAT FINANCIAL REPORT.

How many times have you accidentally gotten the decimal point in the wrong place? Embarrassing, isn't it? Usually it isn't a matter of life or death. In the above case, it would have been. You can be sure it cost the young accountant his job.

This unit is designed to help individuals use decimal numbers properly. Accuracy is essential in all business transactions! In Unit 3, emphasis has been placed on the need for accuracy when using decimals.

Let's get started.

Sharpen your pencil and we'll get right to the "point"—the decimal point, that is.

UNIT OBJECTIVE

When you complete this unit successfully, you will be able to round, add, subtract, multiply, and divide decimal numbers and use these operations in solving word problems. You will also be able to change fractions to decimal numbers and decimal numbers to fractions.

PREVIEW 3   This is a sample test for Unit 3. All of the answers are placed immediately after the preview. Work as many problems as you can, then check your answers and look for further directions after the answers.

<table>
<tr><td></td><td></td><td colspan="2">Where to Go for Help</td></tr>
<tr><td></td><td></td><td>Page</td><td>Frame</td></tr>
<tr><td>1.</td><td>$4.1672 + 17.009 + 2.9$</td><td>87</td><td>1</td></tr>
<tr><td>2.</td><td>$81.62 - 79.627$</td><td>87</td><td>1</td></tr>
</table>

3. 13.05×4.6 _____ 97 10

4. 210.84×3.4 _____ 97 10

5. $4.32 \div 0.064$ _____ 97 10

6. Round 4.1563 to two decimal places. _____ 97 10

7. Change $\frac{13}{35}$ to a decimal and round to three decimal digits. _____ 111 26

8. Change 7.325 to a fraction in lowest terms. _____ 111 26

9. Mary originally had a savings account balance of $347.86. After deposits of $119.95, $210.28, and $97.12, what is her new balance? _____ 87 1

★ If you cannot work any of these problems, start with frame 1 one page 87.
★ If you missed some of the problems, turn to page numbers indicated.
★ If all of your answers were correct, you are probably ready to proceed to Unit 4. (If you would like more practice on Unit 3 before turning to Unit 4, try the Self-Test on page 117).
★ Super-students—those who want to be certain they learn all of this material—will turn to frame 1 and begin work there.

Preview 3
Answers

1. 24.0762

2. 1.993

3. 60.03

4. 716.856

5. 67.5

6. 4.16

7. 0.371

8. $7\frac{13}{40}$

9. $775.21

SECTION 1: ADDITION AND SUBTRACTION OF DECIMALS

1

OBJECTIVE
Add and subtract decimal numbers.

Reading Decimal Numbers

In Unit 1 you learned that whole numbers are written in a place value system. A number such as

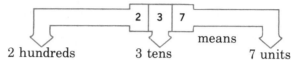

2 hundreds   3 tens   7 units

This way of writing numbers can be extended to fractions. A *decimal* number is a fraction whose denominator is 10 or 100, 1,000, 10,000, etc.

For example,

$0.6 \quad = 6 \text{ tenths} \qquad = \dfrac{6}{10}$

$0.05 \quad = 5 \text{ hundredths} \qquad = \dfrac{5}{100}$

$0.32 \quad = 32 \text{ hundredths} \qquad = \dfrac{32}{100}$

$0.004 = 4 \text{ thousandths} \qquad = \dfrac{4}{1000}$

$0.267 = 267 \text{ thousandths} = \dfrac{267}{1000}$

Decimal form   Fraction form

Write the decimal number 0.526 in fraction form.

Check your answer in 2

---

2     $0.526 = 526 \text{ thousandths} = \dfrac{526}{1000}$

Decimal notation enables us to extend the idea of place value to numbers less than one. A decimal number often has both a whole number part and a fraction part. For example,

$$324.576 = 324 + 0.576 = 324 + \frac{576}{1000} = 324\frac{576}{1000}$$

You are already familiar with this way of interpreting decimal numbers from working with money.

$$\$243.78 = \$243\frac{78}{100} = 243\frac{78}{100} \text{ dollars}$$

Of course this is similar to our method of writing checks. We will discuss this further in Unit 9.

Write the following in fraction form.

(a) $86.42   (b)   43.607
(c)   14.5060   (d)   235.22267

Compare your answers with ours in 3.

---

3   (a) $86.42 = $86 $\frac{42}{100}$ = 86 $\frac{42}{100}$ dollars

(b)   43.607 = 43 $\frac{607}{1000}$

(c)   14.5060 = 14 $\frac{5060}{10000}$

(d)   235.22267 = 235 $\frac{22267}{100000}$

The decimal number 3,254,935.4728 should be interpreted as

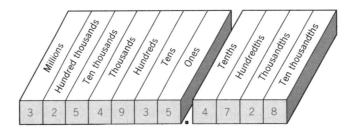

It may be read "three million, two hundred fifty-four thousand, nine hundred thirty-five, and four thousand seven hundred twenty-eight ten thousandths."

Notice that the decimal point is read "and."

Most often, however, this number is read more simply as "three million, two hundred fifty-four thousand, nine hundred thirty-five, point four, seven, two, eight." This way of reading the number is easiest to write, to say, and to understand.

In the decimal number 86.423 the digits 4, 2, and 3 are called *decimal digits*. The number 43.6708 has four decimal digits. All digits to the right of the decimal point, those that name the fractional part of the number, are decimal digits.

Lou Price was calculating compound interest and found the value 2.6915881 in the table. How many decimal digits are included in the number?

Count, then turn to 4.

---

4   The number 2.6915881 has seven decimal digits, 6, 9, 1, 5, 8, 8, and 1. These are the digits to the right of the decimal point.

Notice that the decimal point is simply a way of separating the whole number part from the fraction part; it is a place marker. In whole numbers the decimal point usually is not written, but its location should be clear to you.

$$2 = \quad 2.$$

The decimal point

or $$324 = 324.$$

The decimal point

Very often additional zeros are annexed to the decimal number without changing its value. For example,

8.5 = 8.50 = 8.5000 and so on

6 = 6. = 6.0 = 6.00 and so on

The value of the number is not changed but the additional zeros may be useful, as we shall see.

The decimal number .6 is often written 0.6. The zero added on the left is used to call attention to the decimal point. It is easy to mistake .6 for 6, but the decimal point in 0.6 cannot be overlooked.

Adding Decimal Numbers

Add the following decimal numbers:

$$0.2 + 0.5 = \frac{2}{10} + \frac{5}{10} = \underline{\hspace{2cm}}$$

Try it, using what you know about adding fractions, then turn to 5.

5   $0.2 + 0.5 = \frac{2}{10} + \frac{5}{10} = \frac{2+5}{10} = \frac{7}{10} = 0.7$

Of course we need not use the fraction form in order to add decimal numbers. As with whole numbers, we may **arrange the digits in vertical columns and add directly. Digits of the same place value are put in the same vertical column. Decimal points are always lined up vertically.**

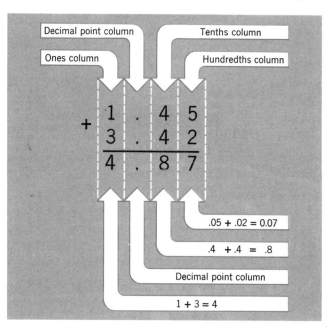

If one of the addends is written with fewer decimal digits than the other, annex as many zeros as needed to write both addends with the same number of decimal digits.

$$\begin{array}{r} 2.345 \\ +1.5 \\ \hline \end{array} \quad \text{becomes} \quad \begin{array}{r} 2.345 \\ +1.500 \\ \hline \end{array}$$

Except for the preliminary step of lining up decimal points, addition of decimal numbers is exactly the same process as addition of whole numbers.

Add the following decimal numbers by hand. Hold off on using your calculator here.

(a) $4.02 + 3.67 = \underline{\hspace{2cm}}$ (d) $14.6 + 1.2 + 3.15 = \underline{\hspace{2cm}}$

(b) $13.2 + 1.57 = \underline{\hspace{2cm}}$ (e) $5.7 + 3.4 = \underline{\hspace{2cm}}$

(c) $23.007 + 1.12 = \underline{\hspace{2cm}}$ (f) $42.768 + 9.37 = \underline{\hspace{2cm}}$

Arrange each addend vertically, placing the decimal points in the same vertical column, then add as with whole numbers.

Check your work in 6.

---

6   (a)   Decimal points in line vertically.

$4.02
$3.67
$7.69 ←— 2 + 7 = 9   Add cents (hundredths).
     0 + 6 = 6   Add 10¢ units (tenths).
     4 + 3 = 7   Add dollars (units).

(b)   Decimal points in line.

13.20
+1.57   Annex a zero to provide the same number of decimal
——   digits as in the other addend.
14.77

     Place answer decimal point in the same vertical line.

(c)   23.007   (d)   14.60
     +1.120   1.20
     ——   +3.15
     24.127   ——
                                     18.95

            1
(e)   5.7   1 1   1
     +3.4   (f)   42.768
     ——   +9.370
     9.1   ——
       —— 7 + 4 = 11   Write 1.   52.138
                     Carry 1.
       —— 1 + 5 + 3 = 9.

Beware!   You must line up the decimal points carefully to be certain of getting a correct answer.

Subtracting Decimal Numbers

Subtraction is equally simple if you are careful to line up decimal points and attach any needed zeros before you begin work.

For example, $437.56 − $41

           3 13
is $437.56   Decimal points in a vertical line.
     −$ 41.00 ←—Attach zeros. (Remember that $41 is $41. or $41.00.)
     ————
     $396.56

        —— Answer decimal point in the same vertical line.

or again, 19.452 − 7.3615

      3 15 1 10
  19.4520   —— Decimal points in a vertical line.
  −7.3615   Attach zero.
  ————
  12.0905

     —— Answer decimal point in the same vertical line.

Try these problems in longhand to test yourself on subtraction of decimal numbers. Check the answers on your calculator, if you like.

(a) $37.66 − 14.57 = _____

(b)   248.3 − 135.921 = _____

(c)   6.4701 − 3.2 = _____

(d)   7.304 − 2.59 = _____

Work carefully. The answers are in **7**.

---

7   (a) $37.66   ┐   ── Line up decimal points.

5 16

7   (a) $37.66
    −14.57
    $23.09   **Check:** 14.57 + 23.09 = 37.66

       ── Line up decimal points.

(b)   248.300
  −135.921 ── Attach zeros.
    112.379   **Check:** 135.921 + 112.379 = 248.300

      ── Answer decimal point in the same vertical line.

(c)   6.4701
 −3.2000
   3.2701   **Check:** 3.2000 + 3.2701 = 6.4701

(d)   7.304
 −2.590
   4.714   **Check:** 2.590 + 4.714 = 7.304

Notice that each problem is checked by comparing the sum of the difference (answer) and subtrahend (number subtracted) with the minuend. Avoid careless mistakes by always checking your answer.

Try the following business problem.

Jim had a savings account balance of $753.47 before withdrawing $165.19. What is his current balance?

Check your answer in **8**.

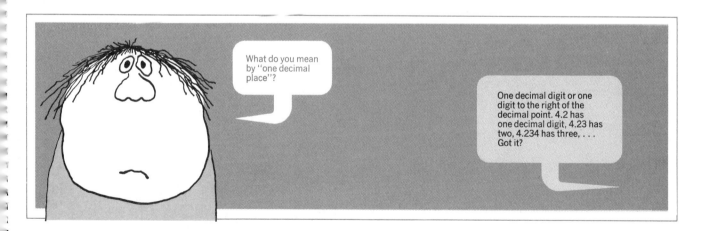

8   $753.47
 −165.19
 $588.28   **Check:** $165.19 + 588.28 = $753.47

Always remember to check your answers.

Now for a set of practice problems on addition and subtraction of decimal numbers turn to **9**.

---

3 Decimal Numbers

PROBLEM SET 1

9   The answers are on page 577.

3.1   Addition and Subtraction of Decimals

A. Add or subtract as shown.

1. $0.5 + 0.3 =$ _____
2. $0.7 + 0.9 =$ _____
3. $0.9 + 0.6 =$ _____
4. $0.4 + 0.2 =$ _____
5. $0.1 + 0.8 =$ _____
6. $0.5 + 0.7 =$ _____
7. $0.8 + 0.8 =$ _____
8. $0.3 + 0.4 =$ _____
9. $0.8 + 0.9 =$ _____
10. $0.8 - 0.7 =$ _____
11. $0.9 - 0.2 =$ _____
12. $5.6 - 2.3 =$ _____
13. $0.9 - 0.4 =$ _____
14. $4.9 - 2.6 =$ _____
15. $2.9 - 1.1 =$ _____
16. $1.7 - 0.3 =$ _____
17. $3.7 - 0.4 =$ _____
18. $8.7 - 3.5 =$ _____
19. $0.7 + 0.9 + 0.3 =$ _____
20. $5.2 + 1.7 + 3.0 =$ _____
21. $2.8 + 0.9 + 1.1 =$ _____
22. $0.5 + 0.8 + 0.1 =$ _____
23. $2.6 + 4.5 + 1.9 =$ _____
24. $8.3 + 2.6 + 7.2 =$ _____
25. $1.4 + 3.6 + 0.5 =$ _____
26. $0.3 + 0.6 + 0.5 =$ _____
27. $8.8 + 3.4 + 5.3 =$ _____
28. $3.3 - 1.7 =$ _____
29. $9.3 - 2.6 =$ _____
30. $7.2 - 6.6 =$ _____
31. $4.2 - 0.6 =$ _____
32. $7.1 - 5.8 =$ _____
33. $5.7 - 3.9 =$ _____
34. $8.5 - 5.9 =$ _____
35. $3.0 - 0.4 =$ _____
36. $1.1 - 0.7 =$ _____

B. Add or subtract as shown.

1. $14.21 + 6.8 =$ _____
2. $\$2.83 + \$12.19 =$ _____
3. $.687 + .93 =$ _____
4. $3.76 + 23.43 =$ _____
5. $\$7.04 + \$23.56 =$ _____
6. $5.702 + .784 =$ _____
7. $75.6 + 2.57 =$ _____
8. $\$52.37 + \$98.74 =$ _____
9. $.096 + 5.82 =$ _____
10. $507.18 + 321.42 =$ _____
11. $4.0983 + 12.1036 =$ _____
12. $623.09 + 408.19 =$ _____

Date

Name

Course/Section

13. 45.6725 + 18.0588 = _____   14. 212.78 + 25.46 = _____

15. 70.3042 + 58.0643 = _____   16. 37 + .09 + 3.5 + 4.605 = _____

17. $14.75 + $9.08 + $3.76 = _____   18. .721 + 48.06 + 22 + .09 = _____

19. $52.19 + $17.43 + $38.75 = _____   20. 24.17 − 4.8 = _____

21. $33.40 − $18.04 = _____   22. 54.5 − 3.16 = _____

23. 7.83 − 6.79 = _____   24. $11.36 − $7.50 = _____

25. 75.08 − 32.75 = _____   26. 10.05 − 3.42 = _____

27. $20.00 − $13.48 = _____   28. 14.22 − 7.8 = _____

29. $40 − $3.82 = _____   30. 30 − 7.984 = _____

31. 3.892 − .995 = _____   32. $65 − $47.35 = _____

33. 13 − 6.04 = _____   34. 1.0487 − .6728 = _____

C. Calculate.

1. 148.002 + 3.459 = _____   2. 632.9 − 30.246 = _____

3. 68.708 + 27.18 = _____   4. 517.03 − 425.88 = _____

5. 7.865 + 308.9 = _____   6. 23.745 − 9.06745 = _____

7. .9437 + 15.0988 = _____   8. 68.4 − 32.9067 = _____

9. 8.939 + 10.072 = _____   10. 9.77803 − 6.42829 = _____

D. Solve.

1. A salesperson at Rita's Boutique made the following sales. Shown is the amount of money given the salesperson. What is the change?

	Amount of Sale	Salesperson Received	Change Returned
a	$11.42	$20	_____
b	2.85	5	_____
c	5.98	10	_____
d	21.47	25	_____
e	12.98	50	_____
f	59.60	70	_____

2. Can you balance a checkbook? At the start of a shopping spree your balance was $472.33. While shopping you wrote checks for $12.57, $8.95, $4, $7.48, and $23.98. What is your new balance?

_____

3. At the start of the semester Joe College bought a math textbook for $11.95, a history book for $12.99, a biology book for $14.50, and an English workbook for $7.40. What was the total cost of these four books?

_____

4. On a recent shopping spree at the Happy Peanut Health Food Store you bought the following:

1 quart celery juice	$0.75
1 jar honey	$1.46
Granny's Granola	$1.38
Sunflower seeds	$0.59
Wheat germ oil	$3.98

How much change should you receive from a $10 bill?

_____

5. Jack's annual auto insurance premium includes bodily injury—$84.50, property damage—$27.75, medical payments—$18.50, collision—$55.25, comprehensive—$32, and uninsured motorist—$7. Calculate his total premium.

_____

6. A new XKP-350 has a base price of $5952.55 with the following options:

Custom deluxe seat and shoulder belts	$ 29.78
Adjustable seat backs	37.19
Tinted glass	57.86
Air conditioning	406.72
Sport mirrors	35.86
Power brakes	25.86
5-Speed transmission	257.36
Power steering	146.83
Steel belted radial whitewall tires	63.89
AM-FM stereo radio	157.21
Shipping	298.26

Calculate the total price.

_____

7. The Radio Hut has four stores. Complete the following table to summarize the total monthly sales for all stores, the quarterly totals for each store, and the total sales for the quarter.

Month	A	B	C	D	Monthly Totals
January	$13,846.29	$12,465.23	$23,156.18	$ 9,476.18	_____
February	13,756.82	13,567.30	25,203.25	10,386.22	_____
March	14,050.26	13,406.27	24,860.29	9,956.28	_____
Quarter Total	_____	_____	_____	_____	_____

8. Silvia had the following deductions from her pay: federal income tax—$29.80, FICA—$14.50, and medical insurance—$14.50. Calculate her "take-home" pay for a week with gross pay being $247.86.

_____

9. At the start of a long trip your mileage odometer read 18327.4, and at the end of the trip it read 23015.2. How far did you travel?

_____

10. Pauline originally had $567.86 in her savings account. Calculate her balance after deposits of $57.82, $129.50, and $26.27 and a withdrawal of $57.83.

----

11. A certain machine part is 1.0076 in. thick. What is its thickness after 0.099 in. is ground off?

----

12. A new calculator originally cost $18.74, but is now reduced $6.95. What is the reduced price?

When you have had the practice you need, either return to the Preview on page 85 or continue in frame 10 with the study of multiplication and division of decimal numbers.

SECTION 2: MULTIPLICATION AND DIVISION OF DECIMALS

10

> **OBJECTIVE**
>
> Multiply and divide decimal numbers.

Multiplying Decimal Numbers

A decimal number is only a fraction. Multiplication of decimals should therefore be no more difficult than the multiplication of fractions.

Find $0.5 \times 0.3 =$ _____

Write out the two numbers as fractions and multiply, then choose an answer.

(a)   15   Go to **11**.
(b)   1.5   Go to **12**.
(c)   0.15   Go to **13**.

---

11   You answered that $0.5 \times 0.3 = 15$ and that is incorrect. Both 0.5 and 0.3 are less than 1; therefore their product is also less than 1, and 15 is not a reasonable answer.

Try calculating the product this way:

$$0.5 = \frac{5}{10} \qquad 0.3 = \frac{3}{10}$$

$$0.5 \times 0.3 = \frac{5}{10} \times \frac{3}{10}$$

Complete this multiplication, then return to **10** and choose a better answer.

---

12   Your answer is incorrect. Don't get discouraged; we'll never tell.

Convert the decimals to fractions with 10 as denominator.

$$0.5 = \frac{5}{10} \qquad 0.3 = \frac{3}{10}$$

Then multiply:

$$0.5 \times 0.3 = \frac{5}{10} \times \frac{3}{10} = \underline{\qquad\qquad}$$

Complete this multiplication, then return to **10** and choose a better answer.

---

13   Excellent.

$$0.5 = \frac{5}{10} \qquad 0.3 = \frac{3}{10}$$

$$0.5 \times 0.3 = \frac{5}{10} \times \frac{3}{10} = \frac{5 \times 3}{10 \times 10} = \frac{15}{100} = 0.15$$

Of course it would be very, very clumsy and time-consuming to calculate every decimal multiplication in this way. We need a simpler method. Here is the procedure most often used.

Step 1.   Multiply the two decimal numbers as if they were whole numbers. Pay no attention to the decimal points.

Step 2. **The sum of the decimal digits in both of the factors will give you the number of decimal digits in the product.**

For example,

3.2 × .41 = _____

Step 1. Multiply without regard to the decimal points.

$$\begin{array}{r} 32 \\ \times 41 \\ \hline 1312 \end{array}$$

Step 2. Count decimal digits in the factors:

3.2 has *one* decimal digit (2)
0.41 has *two* decimal digits (4 and 1)

The total number of decimal digits in the two factors is 3. The product will have *three* decimal digits. Count over *three* digits to the left in the product.

1.312

Three decimal digits.

Try these simple decimal multiplications:

(a)   0.5 × 0.5 = _____   (b)   0.1 × 0.1 = _____

(c)   10 × 0.6 = _____   (d)   2 × 0.003 = _____

(e)   0.01 × 0.02 = _____   (f)   0.04 × 0.005 = _____

Follow the steps outlined above. Count decimal digits carefully. Check your answers in **14**.

_____

14   (a)   0.5 × 0.5 = _____

First, multiply 5 × 5 = 25. **Second,** count decimal digits.

0.5 × 0.5

One   One   = a total of *two* decimal digits.
decimal   decimal
digit   digit

Count over *two* decimal digits from the right

25   The product is 0.25.

Two decimal digits

(b)   0.1 × 0.1   1 × 1 = 1

Count over *two* decimal digits from the right. Since there are not two decimal digits in the product, attach a few on the left.

1 ⟶ 0.01

Two decimal digits

So 0.1 × 0.1 = 0.01

(c)   10 × 0.6   10 × 6 = 60

Count over *one* decimal digit from the right so that 10 × 0.6 = 6.0.

Notice that multiplication by 10 simply shifts the decimal place one digit to the right.

$10 \times 6.2 = 62$

$10 \times 0.075 = 0.75$

$10 \times 8.123 = 81.23$ and so on.

(d)　$2 \times 0.\underbrace{003}$　　　　$2 \times 3 = 6$

　　　　Three　　　　Count over *three* decimal digits.
　　　　decimal
　　　　digits　　　　$0.\underbrace{006}$　　$2 \times 0.003 = 0.006$

　　　　　　　　　Three
　　　　　　　　　decimal
　　　　　　　　　digits

(e)　$0.\underbrace{01} \times 0.\underbrace{02}$　　　　$1 \times 2 = 2$

　　　Two　　　Two　　　Count over *four* decimal digits.
　　　decimal　decimal
　　　digits　　digits　　$0.\underbrace{0002}$　　　$0.01 \times 0.02 = 0.0002$

　　　Total of four　　　Four decimal
　　　decimal digits　　digits

(f)　$0.\underbrace{04} \times 0.\underbrace{005}$　　　　$4 \times 5 = 20$

　　　Two　　　Three　　Count over *five* decimal digits.
　　　decimal　decimal
　　　digits　　digits　　$0.\underbrace{00020}$　　$0.04 \times 0.005 = 0.00020$
　　　　　　　　　　　　　　　　　　　or 0.0002
　　　Total of five　　　Five decimal
　　　decimal digits　　digits

Do not try to do this entire process mentally until you are certain you will not misplace zeros.

Multiplication of larger decimal numbers is performed in exactly the same manner. Try these:

(a)　$4.302 \times 12.05 = $ _____

(b)　$6.715 \times 2.002 = $ _____

(c)　$3.144 \times 0.00125 = $ _____

Look in 15 for the answers.

15　(a)　Multiply　　4302
　　　　　　　　　$\times 1205$
　　　　　　　　　5183910　(If you cannot do this multiplication correctly turn to page 25 in Unit 1 for help with multiplication of whole numbers.)

The factors contain a total of *five* decimal digits (*three* in 4.302 and *two* in 12.05). Count over five decimal digits from the right in the product.

$51.\underbrace{83910}$

so that

$4.302 \times 12.05 = 51.8391$

(b)　Multiply　　6715　　　6.715 has *three* decimal digits.
　　　　　　　　$\times 2002$　　2.002 has *three* decimal digits.
　　　　　　　13.443430　　A total of *six* decimal digits.

　　　　　　Six decimal
　　　　　　digits

$6.715 \times 2.002 = 13.44343$

(c)   Multiply   3144   3.144 has *three* decimal digits.
$$\underline{\times 125}$$ 0.00125 has *five* decimal digits.

.00393000   A total of *eight* decimal digits.

Eight decimal
digits

$3.144 \times 0.00125 = 0.00393$

Now go to 16 for a look at division of decimal numbers.

Dividing Decimal Numbers

Division of decimal numbers is very similar to division of whole numbers. For example,

16    $6.8 \div 1.7$ can be written $\dfrac{6.8}{1.7}$

and, if we multiply top and bottom of the fraction by 10,

$$\frac{6.8}{1.7} = \frac{6.8 \times 10}{1.7 \times 10} = \frac{68}{17}$$

$\dfrac{68}{17}$ is a normal whole number division.

$$\frac{68}{17} = 68 \div 17 = 4$$

Therefore $6.8 \div 1.7 = 4$.   **Check:** $1.7 \times 4 = 6.8$

Rather than take the trouble to write the division as a fraction, we may use a short cut.

Example

Step 1.   Write the divisor and dividend in standard long division form. $6.8 \div 1.7$

$1.7\overline{)6.8}$

Step 2.   Shift the decimal point in the divisor to the right so as to make the divisor a whole number. $1.7.\overline{)}$

Step 3.   Shift the decimal point in the dividend *the same number of digits.* (Add zeros if necessary.) $1.7\overline{)6.8}$

Step 4.   Place the decimal point in the answer space directly above the new decimal position in the dividend. $17.\overline{)68.}$

Step 5.   Complete the division exactly as you would with whole numbers. The decimal points in divisor and dividend may now be ignored.

$$17.\overline{)68.}$$
$$\underline{68.}$$

with $4.$ above.

$6.8 \div 1.7 = 4$

Notice in Steps 2 and 3 that we have simply multiplied both divisor and dividend by 10.

Repeat the process above with this division:

$1.38 \div 2.3$

Work carefully, then compare your work with ours in 17.

---

17   Let's do it step by step.

$2.3\overline{)1.38}$ Shift the decimal point in the divisor 2.3 one digit to the right so that the divisor becomes a whole number. Then shift the decimal point in the dividend 1.38 the *same* number of digits. 2.3 becomes 23. and 1.38 becomes 13.8. This is the same as multiplying both numbers by 10.

$$2.3. \overline{)1.3.8}$$

$$23. \overline{)13.8}$$ Place the answer decimal point directly above the decimal point in the dividend.

$$23. \overline{)\begin{array}{r} .6 \\ 13.8 \\ \underline{13\ 8} \end{array}}$$ Divide as you would with whole numbers.

$6 \times 23 = 138$

$1.38 \div 2.3 = 0.6$ **Check:** $2.3 \times 0.6 = 1.38$

Never forget to check your answer.

How would you do this one?

$2.6 \div 0.052 = \underline{\quad ? \quad}$

Look in **18** for the solution after you have tried it.

---

18 $0.052. \overline{)2.6}$

To shift the decimal place three digits in the dividend, we must attach several zeros to its right.

$0.052. \overline{)2.600.}$ Now place the decimal point in the answer space above that in the dividend.

$$52. \overline{)\begin{array}{r} 50. \\ 2600. \\ \underline{260} \\ 0 \\ \underline{0} \end{array}}$$

$5 \times 52 = 260$

$2.6 \div 0.052 = 50$ **Check:** $0.052 \times 50 = 2.6$

Shifting the decimal point three digits and attaching zeros to the right of the decimal point in this way is equivalent to multiplying both divisor and dividend by 1000.

Try this problem set.

(a) $3.5 \div 0.001 = \underline{\hspace{2in}}$

(b) $\$9 \div 0.02 = \underline{\hspace{2in}}$

(c) $0.365 \div 18.25 = \underline{\hspace{2in}}$

(d) $\$8.80 \div 3.2 = \underline{\hspace{2in}}$

The answers are in **19**.

---

19 (a) $0.001. \overline{)3.500.}$

$$1. \overline{)\begin{array}{r} 3500. \\ 3500. \end{array}}$$ $3.5 \div 0.001 = 3500$

Check: $0.001 \times 3500 = 3.5$

(b) 0.02.)$9.00.

$$\begin{array}{r} \$450. \\ 2.\overline{)\$900.} \end{array} \qquad \$9 \div 0.02 = \$450$$

Check: $0.02 \times \$450 = \$9$

(c) 18.25.)0.36.5

$$\begin{array}{r} .02 \\ 1825\overline{)36.50} \\ \underline{36\ 50} \longleftarrow 2 \times 1825 = 3650 \end{array}$$

$0.365 \div 18.25 = 0.02$

Check: $18.25 \times 0.02 = 0.365$

(d) 3.2.)$8.8.0

$$\begin{array}{r} \$2.75 \\ 32.\overline{)88.00} \\ \underline{64} \quad\longleftarrow 2 \times 32 = 64 \\ 24\ 0 \\ \underline{22\ 4} \quad\longleftarrow 7 \times 32 = 224 \\ 1\ 60 \\ \underline{1\ 60} \quad\longleftarrow 5 \times 32 = 160 \end{array}$$

$\$8.80 \div 3.2 = \$2.75$

Check: $3.2 \times \$2.75 = \$8.80$

If the dividend is not exactly divisible by the divisor, we must either stop the process after some preset number of decimal places in the answer or we must round the answer. We do not generally indicate a remainder in decimal division.

Turn to **20** for some rules for rounding.

Rounding

Rounding is a process of approximating a number. To round a number means to find another number roughly equal to the given number but expressed less precisely.

20   If you have a calculator with an eight-digit display, frequently your answer must be rounded. In business mathematics, you are often calculating money and the final answer must be rounded to the nearest cent or hundredths position.

For example,

$\$24,847.847 = \$24,847.85$ rounded to the nearest
                                    cent or hundreths

$\$432.57 = \$400$ rounded to the nearest hundred dollars
          $= \$430$ rounded to the nearest ten dollars
          $= \$433$ rounded to the nearest dollar
          and so on.

1.376521 is equal to 1.377 rounded to three decimal digits
                     or 1.4   rounded to the nearest tenth
                     or 1   rounded to the nearest whole number

When computing interest, Judy's calculator displayed 45.746753 as the final answer. Round her answer to two decimal places (cents).

Check your answer in **21**.

---

21 $45.746753 = $45.75 rounded to the nearest cents position.

For most rounding follow this simple rule:

Example

Step 1.   Determine the number of digits or   Round 2.832 to one decimal place
the place to which the number is to
be rounded. Mark it with a ∧.   2.8∧32

Step 2.   If the digit to the right of the mark   2.8∧32 becomes
is less than 5, replace all digits to   2.800
the right of the mark by zeros. If   or
the zeros are decimal digits, you   2.8
may discard them.

Step 3.   If the digit to the right of the mark   3.4∧62 becomes
is equal to or larger than 5, in-   3.5
crease the digit to the left by 1.

Judy got the following results from her calculator. Round all the results to the nearest cent (two decimal places).

(a)   137.85602
(b)   86315.174
(c)   1456.2171
(d)   2.3749956

Follow the rules, round these numbers, then check your work in **22**.

---

22   (a)   137.85602 = 137.86 rounded to two decimal places.
(Write 137.85∧602; since 6 is larger than 5, increase this to 6.)
(b)   86315.174 = 86,315.17 rounded to the nearest hundredth. (Write 86,315.17∧4; since 4 is less than 5, drop the 4.)
(c)   1456.2171 = 1456.22 rounded to two decimal places. (Write 1456.21∧71; since 7 is larger than 5, increase the 1 to 2.)
(d)   2.3749956 = 2.37 rounded to the nearest cent. (Write 2.37∧49956; since 4 is less than 5, drop the 99956.)

There are a few very specialized situations where this rounding rule is *not* used.

1.   Engineers use a more complex rule when rounding a number that ends in 5.
2.   In most retail stores, fractions of a cent are rounded up in determining selling price. Three items for 25¢ or 8⅓¢ each are rounded to 9¢ each.

Our rule will be quite satisfactory for most of your work in business mathematics.

Divide as shown and round your answer to two decimal places.

$6.84 ÷ 32.7 = ⎯⎯⎯⎯⎯⎯⎯⎯

Careful now.

Check your work in **23**.

How do I do a problem like 90 −25.4?

Easy. Write 90 as 90.0, then line up the decimal points

```
 90.0
−25.4 and subtract
```

```
 90.0
−25.4
 64.6 Then check your
         answer:
```

25.4 + 64.6 = 90.0

AVERAGES

The average of a set of numbers is the single number that best represents the whole set. One simple kind of average is the arithmetic average or arithmetic mean defined as

$$\text{Arithmetic average} = \frac{\text{sum of measurements}}{\text{number of measurements}}$$

For example, the average weight of the five middle linemen on our college football team is

$$\text{Average weight} = \frac{216 \text{ lb} + 235 \text{ lb} + 224 \text{ lb} + 213 \text{ lb} + 239 \text{ lb}}{5}$$

$$= \frac{1127 \text{ lb}}{5} = 225.4 \text{ lb}$$

Try these problems for practice.

1. In a given week a student worked in the college library for the following hours each day: Monday—2 hours, Tuesday—3 hours, Wednesday—$2\frac{1}{2}$ hours, Thursday—$3\frac{1}{4}$ hours, Friday—$2\frac{1}{4}$ hours, and Saturday—4 hours. What average amount of time does the student work per day?
2. On four weekly quizzes in her history class, a student scores 84, 74, 92, and 88 points. What is her average score?
3. A salesman sells the following amounts in successive weeks: $647.20, $705.17, $1205.65, $349.34, and $409.89. What is his average weekly sales?

The answers are on page 578.

23　　$32.7.\overline{)\$6.8.4}$

$$
\begin{array}{r}
\$\ 0.209 \\
327.\overline{)\$68.400} \\
65\ 4 \downarrow\downarrow \\
\hline
3\ 000 \\
2\ 943 \\
\hline
\end{array}
\qquad
\begin{array}{l}
2 \times 327 = 654 \\
\\
9 \times 327 = 2943
\end{array}
$$

$0.209 rounded to two decimal places is $0.21.

$6.84 ÷ 32.7 = $0.21 rounded to two decimal places.

Check:　32.7 × $0.21 = $6.867, which is approximately equal to $6.84. (The check will not be exact because we have rounded.)

Work the following application problem.

Linda purchased 23.5 yards of carpet for $152.28. What was the cost per yard?

Check your answer in **24**.

24　　$152.28 ÷ 23.5

$$23.5.\overline{)152.2.8}$$

$$
\begin{array}{r}
6.48 \\
235.\overline{)1522.80} \\
1410 \\
\hline
112\ 8 \\
94\ 0 \\
\hline
18\ 80 \\
18\ 80 \\
\hline
\end{array}
$$

$152.28 ÷ 23.5 = $6.48

Check:　23.5 × 6.48 = $152.28

Go to **25** for a set of practice problems on multiplication and division of decimal numbers.

---

3 Decimal Numbers

PROBLEM SET 2

25

3.2 Multiplication and Division of Decimals

The answers are on page 578.

A. Multiply.

1. $0.01 \times 0.001 =$ _____
2. $10 \times 0.01 =$ _____
3. $10 \times 2.15 =$ _____
4. $3 \times 0.02 =$ _____
5. $0.04 \times 0.2 =$ _____
6. $0.07 \times 0.2 =$ _____
7. $0.3 \times 0.3 =$ _____
8. $0.9 \times 0.8 =$ _____
9. $1.2 \times 0.7 =$ _____
10. $4.5 \times 0.002 =$ _____
11. $0.005 \times 0.012 =$ _____
12. $3.5 \times 1.2 =$ _____
13. $6.41 \times 0.23 =$ _____
14. $7.25 \times 0.301 =$ _____
15. $16.2 \times 0.031 =$ _____
16. $0.2 \times 0.3 \times 0.5 =$ _____
17. $0.5 \times 1.2 \times 0.04 =$ _____
18. $0.6 \times 0.6 \times 6.0 =$ _____
19. $1.2 \times 1.23 \times 0.01 =$ _____
20. $2.3 \times 1.5 \times 1.05 =$ _____
21. $1.2 \times 10 \times 0.12 =$ _____
22. $321.4 \times 0.25 =$ _____
23. $0.234 \times 0.005 =$ _____
24. $125 \times 2.3 =$ _____
25. $5.224 \times 0.00625 =$ _____
26. $0.1234 \times 0.0075 =$ _____
27. $425.6 \times 2.875 =$ _____

B. Divide.

1. $6.5 \div 0.005 =$ _____
2. $3.78 \div 0.30 =$ _____
3. $0.0405 \div 0.9 =$ _____
4. $6.5 \div 0.5 =$ _____
5. $0.378 \div 0.003 =$ _____
6. $40.5 \div 0.09 =$ _____
7. $3 \div 0.05 =$ _____
8. $12 \div 0.006 =$ _____
9. $10 \div 0.001 =$ _____
10. $2.59 \div 70 =$ _____
11. $1.232 \div 0.11 =$ _____
12. $44.22 \div 6.7 =$ _____
13. $57.57 \div 0.0303 =$ _____
14. $104.2 \div 0.0320 =$ _____
15. $1.111 \div 10.1 =$ _____

Date _____

Name _____

Course/Section _____

C. Divide and round as indicated.

Round to two decimal digits:

1. $10 \div 3 =$ _____
2. $1 \div 0.7 =$ _____
3. $5 \div 6 =$ _____
4. $0.07 \div 0.80 =$ _____
5. $2.0 \div 0.19 =$ _____
6. $2 \div 3 =$ _____
7. $3 \div 0.08 =$ _____
8. $0.17 \div 0.19 =$ _____
9. $0.023 \div 0.19 =$ _____
10. $345 \div 4.7 =$ _____
11. $12.3 \div 4.7 =$ _____
12. $0.16 \div 1.35 =$ _____
13. $2.37 \div 0.07 =$ _____
14. $4.27 \div 0.009 =$ _____
15. $6.5 \div 1.3 =$ _____

Round to nearest tenth:

1. $100 \div 3 =$ _____
2. $16 \div 15 =$ _____
3. $21.23 \div 98.7 =$ _____
4. $20 \div 3 =$ _____
5. $1 \div 4 =$ _____
6. $1 \div 8 =$ _____
7. $100 \div 9 =$ _____
8. $20 \div 0.07 =$ _____
9. $0.006 \div 0.04 =$ _____
10. $0.8 \div 0.05 =$ _____

Round to three decimal digits:

1. $10 \div 0.70 =$ _____
2. $0.04 \div 1.71 =$ _____
3. $0.09 \div 0.40 =$ _____
4. $0.091 \div 0.0014 =$ _____
5. $22.4 \div 6.47 =$ _____
6. $3.41 \div 0.25 =$ _____
7. $3.51 \div 0.92 =$ _____
8. $6.001 \div 2.001 =$ _____
9. $4.0 \div 0.007 =$ _____
10. $123.21 \div 0.1111 =$ _____

D. Solve.

1. Andy worked 37.4 hours at $3.25 per hour. How much money did he earn?

   _____

2. A color television set is advertised for $420. It can also be bought "on time" for 24 payments of $22.75 each. How much extra do you pay by purchasing it on the installment plan?

   _____

3. What is the cost of 12.3 gallons of gasoline at $1.849 per gallon?

   _____

4. What is the total cost of 26 square yards of carpet at $7.49 per square yard?

_____

5. Beverly earns $215.74 per week. What is her annual salary?

_____

6. Jim drove 173.9 miles on 9.4 gallons of gas. What was his mileage per gallon?

_____

7. Pete earns $11,856 per year. What is his monthly salary? Weekly salary?

_____

8. If a jet plane averaged 624.38 mph for a 4.25 hour trip, what distance did it travel?

_____

9. If one chair costs $26.79, what would be the cost of 156 chairs?

_____

10. Find the cost of

    (a) 3 dozen eggs at $0.89 per dozen   _____

    (b) 24 cans of soup at $0.36 per can   _____

    (c) 16 lbs of potatoes at $0.19 per lb   _____

    (d) 4 yards of fabric at $8.95 per yard   _____

_____

11. The telephone rate between Santa Barbara and Zanzibar is $10.75 for the first three minutes and $1.95 for each additional minute. Calculate the cost of an 11-minute call.

_____

12. A small private plane uses 8.5 gallons of gasoline per hour. How much gas will it use on a flight lasting 4.6 hours?

_____

13. If one chair costs $34.75, what would be the cost of a set of six chairs?

_____

14. Calculate the cost of 28 hexacentrics at $2.76 each.

_____

15. Calculate the cost of 15 dozen biners at $3.98 each.

_____

When you have had the practice you need, either return to the Preview on page 85 or continue in frame 26 with the study of decimal fractions.

26

OBJECTIVE

Change a fraction to a decimal number.

Change a decimal number to a fraction.

Writing A Fraction
As A Decimal Number

To convert a number from fraction form to decimal form, simply divide the numerator (top number) by the denominator (bottom number).

For example, to convert $\frac{5}{8}$ to decimal form, divide 5 by 8.

$$
\begin{array}{r}
0.625 \\
8\overline{)5.000} \quad \longleftarrow \text{Attach as many zeroes as needed.} \\
\underline{4\,8} \\
20 \\
\underline{16} \\
40 \\
\underline{40}
\end{array}
$$

$\frac{5}{8} = 0.625$

If the division has no remainder, the decimal number is called a *terminating* decimal. The fraction $\frac{5}{8} = 0.625$ is a terminating decimal number. If a decimal does not terminate, it is a repeating decimal. You may round it to any desired number of decimal places.

For example, $\frac{2}{13} =$ _____

Divide it out and round to three decimal digits.

Check your work in **27**.

27

$$
\begin{array}{r}
0.1538 \\
13\overline{)2.0000} \\
\underline{1\,3} \\
70 \\
\underline{65} \\
50 \\
\underline{39} \\
110 \\
\underline{104} \\
6
\end{array}
$$

$\frac{2}{13} = 0.154$ rounded to three decimal digits.

Convert the following fractions to decimal form and round to two decimal digits.

(a) $\frac{2}{3} =$ _____

(b) $\frac{5}{6} =$ _____

(c) $\frac{17}{7} =$ _____

(d) $\frac{7}{16} =$ _____

Our work is in **29**.

---

28   (a) $29\dfrac{5}{9} = 29 + \dfrac{5}{9} = 29 + 0.56 = 29.56$ square feet

$$
\begin{array}{r}
0.555 \\
9)\overline{5.000} \\
\underline{4\,5} \\
50 \\
\underline{45} \\
50 \\
\underline{45} \\
5
\end{array}
\qquad \dfrac{5}{9} = 0.56 \qquad \text{rounded}
$$

(b) $186\dfrac{3}{8} = 186 + \dfrac{3}{8} = 186 + 0.375 = 186.375$ yards

$$
\begin{array}{r}
0.375 \\
8)\overline{3.000} \\
\underline{2\,4} \\
60 \\
\underline{56} \\
40 \\
\underline{40}
\end{array}
$$

Writing Decimal Numbers As Fractions

Converting decimal numbers to fractions is fairly easy.

$0.4 = \dfrac{4}{10} \qquad \text{or} \qquad \dfrac{2}{5}$

$0.13 = \dfrac{13}{100}$

$0.275 = \dfrac{275}{1000} = \dfrac{11}{40} \qquad$ reduced to lowest terms

$0.035 = \dfrac{35}{1000} = \dfrac{7}{200} \qquad$ reduced to lowest terms

Follow this procedure:

1. Write the digits to the right of the decimal point as the numerator in the fraction.

    Example: $0.00325 = ? \qquad \dfrac{325}{?}$

2. In the denominator write 1 followed by as many zeros as there are decimal digits in the decimal number.

    Example: $0.00325 = \dfrac{325}{100000}$

    5 digits ⟶ 5 zeros

3. Reduce to lowest terms $\dfrac{325}{100000} = \dfrac{13}{4000}$

Write 0.036 as a fraction in lowest terms.

Check your work in **30**.

The human brain is a fantastic machine, isn't it?

Sure. It starts working the moment you're born and never stops — until you pick up a math book.

29   (a)

$$\begin{array}{r} 0.666 \\ 3\overline{)2.000} \\ \underline{1\ 8} \\ 20 \\ \underline{18} \\ 20 \\ \underline{18} \\ 2 \end{array}$$

$\dfrac{2}{3} = 0.67$ rounded to two decimal digits

Notice that in order to round to two decimal digits, we must carry the division out to at least three decimal digits.

(b)

$$\begin{array}{r} 0.833 \\ 6\overline{)5.000} \\ \underline{4\ 8} \\ 20 \\ \underline{18} \\ 20 \\ \underline{18} \\ 2 \end{array}$$

$\dfrac{5}{6} = 0.83$ rounded to two decimal digits

(c)

$$\begin{array}{r} 2.428 \\ 7\overline{)17.000} \\ \underline{14} \\ 3\ 0 \\ \underline{2\ 8} \\ 20 \\ \underline{14} \\ 60 \\ \underline{56} \\ 4 \end{array}$$

$\dfrac{17}{7} = 2.43$ rounded to two decimal digits

(d)

$$\begin{array}{r} 0.437 \\ 16\overline{)7.000} \\ \underline{6\ 4} \\ 60 \\ \underline{48} \\ 120 \\ \underline{112} \\ 8 \end{array}$$

$\dfrac{7}{16} = 0.44$ rounded to two decimal digits

To convert a mixed number to a decimal, change the fraction part to a decimal number and then add the whole number part.

$$37\frac{1}{4} = 37 + \frac{1}{4} = 37 + 0.25 = 37.25$$

$$\begin{array}{r} 0.25 \\ 4\overline{)1.00} \\ \underline{8} \\ 20 \\ \underline{20} \end{array}$$

Change the following mixed numbers to decimals. If the decimal does not terminate, round to two decimal digits.

(a) $29\frac{5}{9}$ square feet   (b) $186\frac{3}{8}$ yards

Check your work in **28**.

30

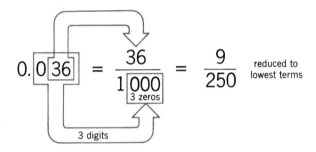

$$0.036 = \frac{36}{1000} = \frac{9}{250} \quad \text{reduced to lowest terms}$$

If the decimal number has a whole number portion, convert the decimal part to a fraction first and then add the whole number part. For example,

$$3.85 = 3 + 0.85$$

and

$$0.85 = \frac{85}{100} = \frac{17}{20} \quad \text{reduced to lowest terms.}$$

Therefore $3.85 = 3 + \frac{17}{20} = 3\frac{17}{20}$.

Write the following decimal numbers as fractions in lowest terms.

(a) 0.0075   (b) 2.08   (c) 3.11

Check your work in **31**.

---

31   (a) $0.0075 = \frac{75}{10000} = \frac{3}{400}$

(b) $2.08 = 2 + \frac{8}{100} = 2 + \frac{2}{25} = 2\frac{2}{25}$

(c) $3.11 = 3 + \frac{11}{100} = 3\frac{11}{100}$

Now turn to **32** for a set of practice problems on decimal fractions.

---

3 Decimal Numbers

32

3.3 Decimal Fractions

The answers are on page 579.

A. Write as decimal numbers (round to two decimal digits).

1. $\frac{1}{2} =$ _____
2. $\frac{1}{3} =$ _____
3. $\frac{2}{3} =$ _____
4. $\frac{1}{4} =$ _____

5. $\frac{2}{4} =$ _____
6. $\frac{3}{4} =$ _____
7. $\frac{1}{5} =$ _____
8. $\frac{2}{5} =$ _____

9. $\frac{3}{5} =$ _____
10. $\frac{4}{5} =$ _____
11. $\frac{1}{6} =$ _____
12. $\frac{5}{6} =$ _____

13. $\frac{1}{7} =$ _____
14. $\frac{2}{7} =$ _____
15. $\frac{3}{7} =$ _____
16. $\frac{1}{8} =$ _____

17. $\frac{3}{8} =$ _____
18. $\frac{5}{8} =$ _____
19. $\frac{7}{8} =$ _____
20. $\frac{1}{10} =$ _____

21. $\frac{2}{10} =$ _____
22. $\frac{3}{10} =$ _____
23. $\frac{1}{12} =$ _____
24. $\frac{2}{12} =$ _____

25. $\frac{3}{12} =$ _____
26. $\frac{5}{12} =$ _____
27. $\frac{7}{12} =$ _____
28. $\frac{11}{12} =$ _____

29. $\frac{1}{16} =$ _____
30. $\frac{3}{16} =$ _____
31. $\frac{5}{16} =$ _____
32. $\frac{7}{16} =$ _____

33. $\frac{9}{16} =$ _____
34. $\frac{11}{16} =$ _____
35. $\frac{13}{16} =$ _____
36. $\frac{15}{16} =$ _____

B. Write as a fraction in lowest terms.

1. $0.3 =$ _____
2. $0.75 =$ _____
3. $0.44 =$ _____

4. $0.8 =$ _____
5. $0.6 =$ _____
6. $0.025 =$ _____

7. $0.4 =$ _____
8. $1.3 =$ _____
9. $2.25 =$ _____

10. $2.05 =$ _____
11. $3.16 =$ _____
12. $1.125 =$ _____

13. $3.22 =$ _____
14. $2.04 =$ _____
15. $0.075 =$ _____

16. $10.875 =$ _____
17. $0.0007 =$ _____
18. $0.0012 =$ _____

19. $0.34 =$ _____
20. $11.0105 =$ _____
21. $6.0020 =$ _____

22. $4.115 =$ _____
23. $0.35 =$ _____
24. $0.955 =$ _____

C. Solve.

1. If Andy is paid $3.25 per hour, how much money will he receive for $3\frac{1}{4}$ hours of work?

_____

2. Find the cost of $1\frac{3}{4}$ lb of steak at $4.19 per lb.

_____

3. What is the total cost of $4\frac{1}{4}$ gallons of paint at $4.25 per gallon?

_____

Date

Name

Course/Section

4. What is the cost of $45\frac{1}{2}$ board feet of lumber that sells for $0.73 per board foot?

   _____

5. If seven avocados cost $3, what is the selling price of one?

   _____

6. A baseball player hit safely 46 times in 165 times at bat during a season. What was his batting average? (Convert the fraction $\frac{46}{165}$ to a decimal fraction.)

   _____

7. What is the cost of $3\frac{1}{2}$ liters of cleaning fluid that sells for $3.48 per liter?

   _____

8. What is the total cost of $19\frac{2}{3}$ square yards of carpet at $9.50 per square yard?

   _____

9. Larry's pay rate for overtime hours is $1\frac{1}{2}$ times his base pay rate of $3.46. What is his overtime rate?

   _____

10. What is the cost of $8\frac{3}{4}$ feet of walnut shelving at $2.19 per foot?

    _____

11. On four mathematics examinations, Keri scored 98, 82, 89, and 97. Calculate her average exam grade. (Add the scores and divide by the number of tests.)

    _____

When you have had the practice you need, turn to 33 for the Self-Test on decimals.

3 Decimal Numbers

The answers are on page 579.

1. Add: 3.6715 + 23.8 + 8.539.   _____

2. Subtract: 23.52 − 8.4592.   _____

3. Multiply: 5.97 × 23.8.   _____

4. Divide: 8.142 ÷ 3.45.   _____

5. Divide and round to two decimal places: $2856.13 ÷ 146.   _____

6. Write 0.56 as a fraction and reduce to lowest terms.   _____

7. Write $1\frac{5}{16}$ as a decimal number.   _____

8. Bruce had $467.18 in his checking account. He wrote checks for $246.98, $18.36, and $38.17. What was his new balance?   _____

9. What is the total cost of $24\frac{3}{4}$ square yards of carpet at $12.75 per square yard?   _____

10. A case of 15 crack tacks costs $23.85. What is the cost per crack tack?   _____

Date _____

Name _____

Course/Section _____

4 Percent

YOU CAN FOOL 100 % OF THE PEOPLE 57 % OF THE TIME; YOU CAN FOOL 26 % OF THE PEOPLE 100 % OF THE TIME; BUT YOU CAN'T FOOL 100% OF THE PEOPLE 100% OF THE TIME !

Perhaps that isn't exactly what the author of the above quotation said—but, in essence, that's what he meant.

Percents are a part of every business day. Employers and employees need to demonstrate skill in the use of percentages.

This unit is designed to show individuals how percentages are incorporated into everyday situations. Once you have completed this section, you will have very little difficulty with the rest of the text where percentages are used. Hopefully, you'll be 100% confident when using percents.

UNIT OBJECTIVE

When you complete this unit successfully, you will be able to write fractions and decimal numbers as percents, convert percents to decimal and fraction form, and solve problems involving percents.

PREVIEW 4   This is a sample test for Unit 4. All of the answers are placed immediately after the preview. Work as many of the problems as you can, then check your answers and look after the answers for further directions.

		Where to Go for Help	
		Page	Frame
1. Write $1\frac{7}{8}$ as a percent.	_____	123	5
2. Write 0.45 as a percent.	_____	122	4

3. Write $37\frac{1}{2}\%$ as a decimal.   _____   124   7

4. Write 44% as a fraction.   _____   125   9

5. Find 120% of 45.   _____   131   12

6. What percent of $6.00 is $2.50?   _____   136   19

7. 70% of what number is 56?   _____   138   22

8. A new calculator sells for $17.50 plus 6% sales tax. What is the total cost?   _____   131   12

★ If you cannot work any of these problems, start with frame 1 on page 121.
★ If you missed some of the problems, turn to page numbers indicated.
★ If all of your answers were correct, you are probably ready to proceed to Unit 5. (If you would like more practice on Unit 4 before turning to Unit 5, try the Self-Test on page 147.
★ Super-students—those who want to be certain they learn all of this material—will turn to frame 1 and begin work there.

Preview 4
Answers

1. 187.5%
2. 45%
3. 0.375
4. $\frac{11}{25}$
5. 54
6. $41\frac{2}{3}\%$ or   41.7% rounded
7. 80
8. $18.55

<table>
<tr><td>1</td><td>

OBJECTIVE

Write fractions and decimal numbers as percents.

Convert percents to decimal and fraction form.

</td></tr>
</table>

The word *percent* comes from the Latin words *per centum* meaning "by the hundred" or "for every hundred." A number expressed as a percent is being compared with a second number called the standard or *base* by dividing the base into 100 equal parts and writing the comparison number as so many hundredths of the base.

For example, what part of the base or standard length is length A?

We could answer the question with a fraction or a decimal or a percent. First, divide the base into 100 equal parts.

Then compare A with it. The length of A is 40 parts out of the 100 parts that make up the base.

A is $\dfrac{40}{100}$ or 0.40 or 40% of A.

$\dfrac{40}{100} = 40\%$ Thus 40% means 40 parts in 100 or $\dfrac{40}{100}$.

What part of this base is length B?

Answer with a percent.

Turn to **2** to check your answer.

2

B is $\dfrac{60}{100}$ or 60%.

Of course the compared number may be larger than the base. For example,

Base

C

In this case, divide the base into 100 parts and extend it in length.

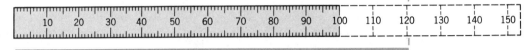

C

The length of C is 120 parts out of the 100 parts that make up the base.

C is $\dfrac{120}{100}$ or 120% of A.

Because our number system, and our money, is based on ten and multiples of ten, it is very handy to write comparisons in hundredths or percent.

What part of $1.00 is 50¢? Write your answer as a fraction, as a decimal, and as a percent.

Check in **3** for the answer.

---

3   50¢ is what part of 100¢?

$$\frac{50¢}{100¢} = \frac{50}{100} = 0.50 = 50\%$$

To find the answer as a percent, write it as a fraction with a denominator equal to 100.

We may also write 50¢ = $\frac{1}{2}$ of $1.00 or 50¢ = 0.50 of $1.00. Fractions, decimals, and percents are alternative ways to talk about a comparison of two numbers.

What percent of 10 is 2? Write 2 as a fraction of 10, rename it as a fraction with denominator equal to 100, then write as a percent.

When you have completed this, go to **4**.

---

4 $\dfrac{2}{10} = \dfrac{2 \times 10}{10 \times 10} = \dfrac{20}{100} = 20\%$

Writing A Decimal Number As A Percent

How do you rewrite a decimal number as a percent? The procedure is simply to multiply the decimal number by 100%. For example,

$0.60 = 0.60 \times 100\% = 60\%$

$$\begin{array}{r} 0.60 \\ \times 100\% \\ \hline 60.00\% = 60\% \end{array}$$

More examples

$0.375 = 0.375 \times 100\% = 37.5\%$
$3.4 = 3.4 \times 100\% = 340\%$
$0.02 = 0.02 \times 100\% = 2\%$

Any number larger than 1 is more than 100%.

Notice in each of these examples that **multiplication by 100% has the effect of moving the decimal point two digits to the right.**

SECTION 1: NUMBERS AND PERCENT

---

1 | **OBJECTIVE**

Write fractions and decimal numbers as percents.

Convert percents to decimal and fraction form.

The word *percent* comes from the Latin words *per centum* meaning "by the hundred" or "for every hundred." A number expressed as a percent is being compared with a second number called the standard or *base* by dividing the base into 100 equal parts and writing the comparison number as so many hundredths of the base.

For example, what part of the base or standard length is length *A*?

We could answer the question with a fraction or a decimal or a percent. First, divide the base into 100 equal parts.

Then compare *A* with it. The length of *A* is 40 parts out of the 100 parts that make up the base.

A is $\frac{40}{100}$ or 0.40 or 40% of *A*.

$\frac{40}{100} = 40\%$ Thus 40% means 40 parts in 100 or $\frac{40}{100}$.

What part of this base is length *B*?

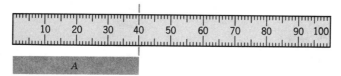

Answer with a percent.

Turn to **2** to check your answer.

---

2 |

B is $\frac{60}{100}$ or 60%.

Of course the compared number may be larger than the base. For example,

In this case, divide the base into 100 parts and extend it in length.

The length of C is 120 parts out of the 100 parts that make up the base.

C is $\frac{120}{100}$ or 120% of A.

Because our number system, and our money, is based on ten and multiples of ten, it is very handy to write comparisons in hundredths or percent.

What part of $1.00 is 50¢? Write your answer as a fraction, as a decimal, and as a percent.

Check in **3** for the answer.

---

3   50¢ is what part of 100¢?

$$\frac{50¢}{100¢} = \frac{50}{100} = 0.50 = 50\%$$

To find the answer as a percent, write it as a fraction with a denominator equal to 100.

We may also write $50¢ = \frac{1}{2}$ of $1.00 or $50¢ = 0.50$ of $1.00. Fractions, decimals, and percents are alternative ways to talk about a comparison of two numbers.

What percent of 10 is 2? Write 2 as a fraction of 10, rename it as a fraction with denominator equal to 100, then write as a percent.

When you have completed this, go to **4**.

---

4 $\dfrac{2}{10} = \dfrac{2 \times 10}{10 \times 10} = \dfrac{20}{100} = 20\%$

Writing A Decimal Number As A Percent

How do you rewrite a decimal number as a percent? The procedure is simply to multiply the decimal number by 100%. For example,

$0.60 = 0.60 \times 100\% = 60\%$
$$\begin{array}{r} 0.60 \\ \times 100\% \\ \hline 60.00\% = 60\% \end{array}$$

More examples

$0.375 = 0.375 \times 100\% = 37.5\%$
$3.4 = 3.4 \times 100\% = 340\%$
$0.02 = 0.02 \times 100\% = 2\%$

Any number larger than 1 is more than 100%.

Notice in each of these examples that **multiplication by 100% has the effect of moving the decimal point two digits to the right.**

$0.3\underset{\curvearrowright}{7}5 = 037.5\% = 37.5\%$

$3.4 \quad = 3.4\underset{\curvearrowright}{0} \quad = 340.\% = 340\%$

$0.0\underset{\curvearrowright}{2} = 002.\% \quad = 2\%$

Rewrite the following as percents.

(a) 0.70 = _____   (b) 1.25 = _____

(c) 0.001 = _____   (d) 3 = _____

Look in **5** for the answers.

---

5 (a) $0.7\underset{\curvearrowright}{0} = 070.\% = 70\%$ (b) $1.2\underset{\curvearrowright}{5} = 125.\% = 125\%$

(c) $0.00\underset{\curvearrowright}{1} = 000.1\% = 0.1\%$ (d) $3 = 3.0\underset{\curvearrowright}{0} = 300.\% = 300\%$

Writing A Fraction As A Percent

To rewrite a fraction as a percent, we can always rename it as a fraction with 100 as denominator.

$$\frac{3}{20} = \frac{3 \times 5}{20 \times 5} = \frac{15}{100} = 15\%$$

However, **the easiest way is to convert the fraction to decimal form by dividing and then moving the decimal point two digits to the right.**

$\dfrac{1}{2} = 0.5\underset{\curvearrowright}{0} = 050.\% = 50\%$

$$\begin{array}{r} 0.50 \\ 2\overline{)1.00} \end{array}$$

$\dfrac{3}{4} = 0.7\underset{\curvearrowright}{5} = 075.\% = 75\%$

$$\begin{array}{r} 0.75 \\ 4\overline{)3.00} \end{array}$$

$\dfrac{3}{20} = 0.1\underset{\curvearrowright}{5} = 015.\% = 15\%$

$$\begin{array}{r} 0.15 \\ 20\overline{)3.00} \\ 2\ 0 \\ \hline 1\ 00 \\ 1\ 00 \\ \hline \end{array}$$

$1\dfrac{7}{20} = 1.3\underset{\curvearrowright}{5} = 135.\% = 135\%$

$$\begin{array}{r} 1.35 \\ 20\overline{)27.00} \\ 20 \\ \hline 7\ 0 \\ 6\ 0 \\ \hline 1\ 00 \\ 1\ 00 \\ \hline \end{array}$$

Calculators are currently on sale and have been reduced $\frac{5}{16}$ of their regular price. What is $\frac{5}{16}$ as a percent?

Check your answer in **6**.

---

6 $\dfrac{5}{16} = 0.31\underset{\curvearrowright}{2}5 = 31.25\%$

This is often written as $31\dfrac{1}{4}\%$.

$$\begin{array}{r} 0.3125 \\ 16\overline{)5.0000} \\ 4\ 8 \\ \hline 20 \\ 16 \\ \hline 40 \\ 32 \\ \hline 80 \\ 80 \\ \hline \end{array}$$

Some fractions cannot be converted to exact decimals. For example,

$\frac{1}{3} = 0.333$ . . . where the 3s continue to repeat endlessly.

We can round to get an approximate percent,

$\frac{1}{3} = 0.333 = 33.3\%$ rounded

or convert it to a fraction with 100 as denominator.

$$\frac{1}{3} = \frac{?}{100} \qquad \text{gives} \qquad \frac{1}{3} = \frac{1 \times 33\frac{1}{3}}{3 \times 33\frac{1}{3}} = \frac{33\frac{1}{3}}{100} = 33\frac{1}{3}\%$$

Business expenses have increased by $\frac{1}{6}$. What is the percent increase?

Change $\frac{1}{6}$ to a percent and check your work in **8**.

---

7 (a) $\frac{7}{5} = 1.4 = 1.40 = 140\%$

(b) $\frac{2}{3} = \frac{?}{100} \qquad \frac{2}{3} = \frac{2 \times 33\frac{1}{3}}{3 \times 33\frac{1}{3}} = \frac{66\frac{2}{3}}{100} = 66\frac{2}{3}\%$

(c) $3\frac{1}{8} = 3.125 = 312.5\%$

(d) $\frac{5}{12} = \frac{?}{100} = \frac{5 \times 8\frac{1}{3}}{12 \times 8\frac{1}{3}} = \frac{40\frac{5}{3}}{100} = \frac{41\frac{2}{3}}{100} = 41\frac{2}{3}\%$

Changing A Percent To A Decimal

In order to use percent in solving business problems, it is often necessary to change a percent to a decimal number. The procedure is to divide by 100%. For example,

$$50\% = \frac{50\%}{100\%} = \frac{50}{100} = 0.50 \qquad 100\overline{)50.0}^{\,0.5}$$

$$5\% = \frac{5\%}{100\%} = \frac{5}{100} = 0.05 \qquad 100\overline{)5.00}^{\,0.05}$$

$$0.2\% = \frac{0.2\%}{100\%} = \frac{0.2}{100} = 0.002 \qquad 100\overline{)0.200}^{\,0.002}$$

Notice that in each of these examples **division by 100% has the effect of moving the decimal point two digits to the left and dropping the percent sign.**

$50\% = 50.\% = .50 = 0.50$

$5\% = 05.\% = .05 = 0.05$

$0.2\% = 00.2\% = .002 = 0.002$

Fractions may be part of the percent number. If so, it is easiest to change to a decimal number and round if necessary.

$6\frac{1}{2}\% = 6.5\% = 06.5\% = 0.065$

$33\frac{1}{3}\% = 33.3\% = 0.333$ rounded

Now you try a few: Write these as decimal numbers.

(a) $4\% = $ _____ (b) $0.5\% = $ _____

(c) $16\frac{2}{3}\% = $ _____ (d) $79\frac{1}{4}\% = $ _____

Our answers are in **9**.

---

8 $\dfrac{1}{6} = \dfrac{?}{100}$ $\dfrac{1}{6} = \dfrac{1 \times \boxed{16\frac{2}{3}}}{6 \times \boxed{16\frac{2}{3}}}$ $\dfrac{16\frac{2}{3}}{100} = 16\frac{2}{3}\%$ To get this result ask "What number times 6 equals 100?"

                                                                       Answer: $100 \div 6 = 16\frac{2}{3}$

Rewrite the following fractions as percents.

(a) $\dfrac{7}{5} =$ _____   (b) $\dfrac{2}{3} =$ _____

(c) $3\dfrac{1}{8} =$ _____   (d) $\dfrac{5}{12} =$ _____

Go to **7** to check your answers.

---

9   (a) $4\% = 04.\% = 0.04$

       (b) $0.5\% = 00.5\% = 0.005$

       (c) $16\frac{2}{3}\% = 16.7\% = 0.167$ rounded

       (d) $79\frac{1}{4}\% = 79.25\% = 0.7925$

Changing A Percent To A Fraction   To change a percent to a fraction, divide by 100% and reduce to lowest terms.

$$36\% = \frac{36\%}{100\%} = \frac{\overset{9}{\cancel{36}}}{\underset{25}{\cancel{100}}} = \frac{9}{25}$$ 36% means 36 hundredths or

$$\frac{36}{100} \quad \text{or} \quad \frac{9}{25}$$

$$12\frac{1}{2}\% = \frac{12\frac{1}{2}\%}{100\%} = \frac{12\frac{1}{2}}{100} = \frac{\frac{25}{2}}{100} = \frac{25}{200} = \frac{1}{8}$$

Note that $\dfrac{\frac{25}{2}}{100} = \dfrac{25}{2} \div 100 = \dfrac{25}{2} \times \dfrac{1}{100} = \dfrac{25}{200}$

$$125\% = \frac{125\%}{100\%} = \frac{125}{100} = \frac{5}{4} = 1\frac{1}{4}$$

Try these for practice:

(a) $72\% =$ _____   (b) $16\frac{1}{2}\% =$ _____

(c) $240\% =$ _____   (d) $7\frac{1}{2}\% =$ _____

You will find the answers in **10**.

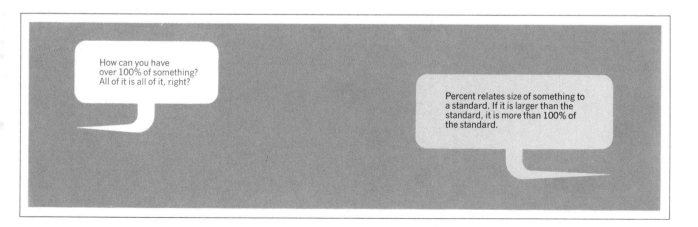

How can you have over 100% of something? All of it is all of it, right?

Percent relates size of something to a standard. If it is larger than the standard, it is more than 100% of the standard.

10   (a) $72\% = \dfrac{72\%}{100\%} = \dfrac{\overset{18}{\cancel{72}}}{\underset{25}{\cancel{100}}} = \dfrac{18}{25}$

(b) $16\tfrac{1}{2}\% = \dfrac{16\tfrac{1}{2}\%}{100\%} = \dfrac{16\tfrac{1}{2}}{100} = \dfrac{\tfrac{33}{2}}{100} = \dfrac{33}{200}$

(c) $240\% = \dfrac{240\%}{100\%} = \dfrac{\overset{12}{\cancel{240}}}{\underset{5}{\cancel{100}}} = \dfrac{12}{5} = 2\tfrac{2}{5}$

(d) $7\tfrac{1}{2}\% = \dfrac{7\tfrac{1}{2}\%}{100\%} = \dfrac{7\tfrac{1}{2}}{100} = \dfrac{\tfrac{15}{2}}{100} = \dfrac{15}{200} = \dfrac{3}{40}$

SUMMARY   **1. Change a decimal to a percent:**

Move the decimal point two digits to the right and add the % sign.

$0.28 = 028.\% = 28\%$

2. Change a fraction to a percent:

A. For fractions that convert exactly to a decimal: Change the fraction to a decimal, move the decimal point two digits to the right, and add the % sign.

$\dfrac{9}{20} = 0.45 = 045.\% = 45\%$

B. For fractions that do not convert exactly to a decimal: Convert to a fraction with 100 as the denominator and change to percent.

$\dfrac{5}{6} = \dfrac{?}{100} \qquad \dfrac{5}{6} = \dfrac{5 \times 16\tfrac{2}{3}}{6 \times 16\tfrac{2}{3}} = \dfrac{83\tfrac{1}{3}}{100} = 83\tfrac{1}{3}\%$

3. Change a percent to a decimal:

Move the decimal point two digits to the left and delete the % sign.

$17\% = 17.\% = 0.17$

$7\tfrac{1}{2}\% = 7.5\% = 07.5\% = 0.075$

$8\tfrac{1}{3}\% = 8.3\% = 08.3\% = 0.083 \text{ rounded}$

4. Change a percent to a fraction:

Divide by 100% and reduce to lowest terms.

$56\% = \dfrac{56\%}{100\%} = \dfrac{56}{100} = \dfrac{14}{25}$

Now turn to **11** for a set of practice problems on what you have learned in this unit so far.

ALIQUOT PARTS OF 1

An *aliquot part* is any number that can be divided evenly into another number. In business, we frequently use aliquot parts of 1 in percent, decimal, and fraction form. The following is a short list of aliquot parts that you will find very helpful, if memorized.

Fraction	Decimal	Percent
$\frac{1}{2}$	0.50	50%
$\frac{1}{3}$	0.333	$33\frac{1}{3}\%$
$\frac{2}{3}$	0.666	$66\frac{2}{3}\%$
$\frac{1}{4}$	0.25	25%
$\frac{3}{4}$	0.75	75%
$\frac{1}{5}$	0.20	20%
$\frac{2}{5}$	0.40	40%
$\frac{3}{5}$	0.60	60%
$\frac{4}{5}$	0.80	80%
$\frac{1}{6}$	0.1666	$16\frac{2}{3}\%$
$\frac{5}{6}$	0.833	$83\frac{1}{3}\%$
$\frac{1}{8}$	0.125	$12\frac{1}{2}\%$
$\frac{3}{8}$	0.375	$37\frac{1}{2}\%$
$\frac{5}{8}$	0.625	$62\frac{1}{2}\%$
$\frac{7}{8}$	0.875	$87\frac{1}{2}\%$
$\frac{1}{10}$	0.10	10%
$\frac{3}{10}$	0.30	30%
$\frac{7}{10}$	0.70	70%
$\frac{9}{10}$	0.90	90%
$\frac{10}{10}$	1.00	100%
$\frac{1}{12}$	0.0833	$8\frac{1}{3}\%$
$\frac{1}{16}$	0.0625	$6\frac{1}{4}\%$
$\frac{1}{20}$	0.05	5%

4 Percent

The answers are on page 579.

4.1 Numbers and Percent

A. Write each number as a percent.

1. $0.40 =$ _____
2. $0.10 =$ _____
3. $0.95 =$ _____

4. $0.03 =$ _____
5. $0.3 =$ _____
6. $0.015 =$ _____

7. $0.6 =$ _____
8. $0.01 =$ _____
9. $1.2 =$ _____

10. $4.56 =$ _____
11. $2.25 =$ _____
12. $7.75 =$ _____

13. $0.003 =$ _____
14. $3.0 =$ _____
15. $0.8 =$ _____

16. $5.5 =$ _____
17. $4 =$ _____
18. $6.04 =$ _____

19. $10 =$ _____
20. $0.335 =$ _____

B. Write each fraction as a percent.

1. $\frac{1}{5} =$ _____
2. $\frac{3}{4} =$ _____
3. $\frac{7}{10} =$ _____

4. $\frac{7}{20} =$ _____
5. $\frac{3}{2} =$ _____
6. $\frac{1}{4} =$ _____

7. $\frac{1}{10} =$ _____
8. $\frac{1}{2} =$ _____
9. $\frac{3}{8} =$ _____

10. $\frac{3}{5} =$ _____
11. $\frac{7}{4} =$ _____
12. $\frac{11}{5} =$ _____

13. $1\frac{4}{5} =$ _____
14. $\frac{9}{10} =$ _____
15. $\frac{1}{3} =$ _____

16. $2\frac{1}{6} =$ _____
17. $\frac{2}{3} =$ _____
18. $\frac{11}{16} =$ _____

19. $\frac{23}{12} =$ _____
20. $3\frac{3}{10} =$ _____

C. Write each percent as a decimal number.

1. $7\% =$ _____
2. $3\% =$ _____
3. $56\% =$ _____

4. $15\% =$ _____
5. $1\% =$ _____
6. $7\frac{1}{2}\% =$ _____

7. $90\% =$ _____
8. $200\% =$ _____
9. $0.3\% =$ _____

10. $0.07\% =$ _____
11. $0.25\% =$ _____
12. $150\% =$ _____

13. $1\frac{1}{2}\% =$ _____
14. $6\frac{1}{3}\% =$ _____
15. $\frac{1}{2}\% =$ _____

16. $12\frac{1}{4}\% =$ _____
17. $125\frac{1}{2}\% =$ _____
18. $66\frac{2}{3}\% =$ _____

19. $30\frac{1}{2}\% =$ _____
20. $8\frac{1}{2}\% =$ _____

Date _____

Name _____

Course/Section _____

D. **Write each percent as a fraction in lowest terms.**

1. $10\% = \underline{\hspace{1cm}}$ 2. $65\% = \underline{\hspace{1cm}}$ 3. $50\% = \underline{\hspace{1cm}}$

4. $20\% = \underline{\hspace{1cm}}$ 5. $25\% = \underline{\hspace{1cm}}$ 6. $8\% = \underline{\hspace{1cm}}$

7. $90\% = \underline{\hspace{1cm}}$ 8. $135\% = \underline{\hspace{1cm}}$ 9. $3\% = \underline{\hspace{1cm}}$

10. $12\% = \underline{\hspace{1cm}}$ 11. $\frac{1}{2}\% = \underline{\hspace{1cm}}$ 12. $0.03\% = \underline{\hspace{1cm}}$

13. $4.5\% = \underline{\hspace{1cm}}$ 14. $220\% = \underline{\hspace{1cm}}$ 15. $1\frac{1}{2}\% = \underline{\hspace{1cm}}$

16. $33\frac{1}{3}\% = \underline{\hspace{1cm}}$ 17. $7\frac{3}{4}\% = \underline{\hspace{1cm}}$ 18. $6\frac{1}{2}\% = \underline{\hspace{1cm}}$

19. $16\frac{2}{3}\% = \underline{\hspace{1cm}}$ 20. $3\frac{1}{8}\% = \underline{\hspace{1cm}}$

When you have had the practice you need, either return to the Preview for this unit on page 119 or continue in frame **12** with the study of problems involving percent.

SECTION 2: PERCENT PROBLEMS

12

> **OBJECTIVE**
>
> Solve the three types of percent problems.

In all of your work with percent you will find that there are three basic types of problems. These three form the basis for all percent problems that arise in business, industry, science, or other areas. All of these problems involve three quantities:

1. The *base* or *total* amount or standard used for a comparison.
2. The *percentage* or part being compared with the base or total.
3. The *percent* or *rate* indicating the relationship of the percentage to the base, the part to the total.

All three basic percent problems involve finding one of these three quantities when the other two are known.

In every problem follow these five steps:

Step 1 *Translate* the problem sentence into a math statement. Certain words and phrases appear in most percent problems. They are signals alerting you to the mathematical operations to be done. Here are the two most common signal words in percent problems.

Signal Words	Translate as
Is, is equal to, equals, will be	=
of	×

Use a □ , letter of the alphabet, or ? for the unknown quantity you are asked to find.

For example, the question

30% of what number is 16? should be translated

$$30\% \times \square = 16$$

or $\qquad$ $30\% \times \square = 16$

In this case, 30% is the percent or rate; □ , the unknown quantity, is the total or base; and 16 is the percentage or part of the total.

Step 2 It will be helpful if you *label* which numbers are the base or total (T), the percent $(\%)$, and the part (P).

Step 3 *Rearrange* the equation so that the unknown quantity is alone on the left of the equal sign and the other quantities are on its right.

The following *Equation Finder* may help you to do this arranging.

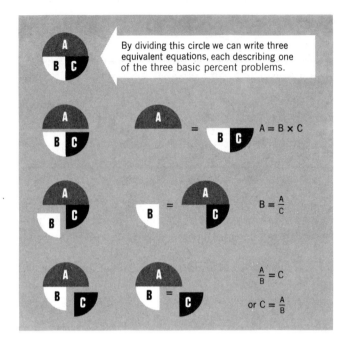

By dividing this circle we can write three equivalent equations, each describing one of the three basic percent problems.

$A = B \times C$

$B = \dfrac{A}{C}$

$\dfrac{A}{B} = C$

or $C = \dfrac{A}{B}$

The three equations $A = B \times C$, $B = \dfrac{A}{C}$, and $C = \dfrac{A}{B}$ are all equivalent.

The equation $\quad 30\% \times \square = 16$

becomes $\quad\quad\quad\quad \square = 16 \div 30\% = \dfrac{16}{30\%} = \dfrac{16}{0.30}$

Step 4 *Solve* the problem by doing arithmetic.

> **Be Careful**

Never do arithmetic, never multiply or divide, with percent numbers. All percents must be rewritten as fractions or decimals before you can use them in a multiplication or division.

Step 5 *Check* by putting the answer number you have found back into the original problem or equation to see if it makes sense. If possible, use the answer to calculate one of the other numbers in the equation as a check.

Now let's look very carefully at each type of problem. We'll explain each, give examples, show you have to solve them, and work through a few together.

Turn to **13**.

What does "%" mean? Where did that goofy-looking symbol come from?

It means "100". The word "percent" come from the Latin words meaning "for every hundred." It started as 100 then loo then $\frac{0}{0}$ in the 17th century and finally o/o or %.

P-Type
Problems

13

P-*Type problems* are usually stated in the form.

Find 30% of 50.

or   What is 30% of 50?

or   30% of 50 is what number?

Step 1
Translate: 30% × 50 = □

Step 2
Label: % × T = P

Step 3
Rearrange: □ = 30% × 50

Complete the calculation and find □.

□ = ___?___

(a) 150   Go to **14**.
(b) 15   Go to **15**.
(c) 1500   Go to **16**.

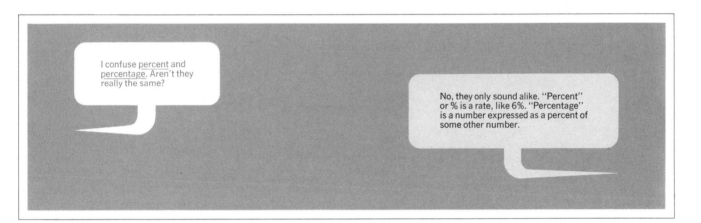

14   You answered that 30% of 50 = 150, and this is not correct.

Remember, *never* multiply by a percent number. Before you multiply 30% × 50, you must write 30% as a decimal number. (If you need help with this, turn to **7**.)

Use this hint to solve the problem, then return to **13** and choose a better answer.

15   Right you are!

Step 4   Solve: Never multiply by a percent number. Percent numbers are not arithmetic numbers. Before you multiply write the percent number as a decimal number.

30% = 30.% = 0.30

30% × 50 = 0.30 × 50 = 15

Don't be intimidated by numbers. If the problem involves very large or very complex numbers, reduce it to a simpler problem. The problem

Find $14\frac{7}{32}$% of 6.4

may look difficult until you realize that it is essentially the same problem as

Find 10% of 6

which is fairly easy.

Before you begin any actual arithmetic problem involving percent you should

(a) Have a plan for solving the problem based on a simpler problem.
(b) Always change percents to decimals or fractions before multiplying or dividing with them. Never use percent numbers in arithmetic operations.

Now, try this problem:

Find $8\frac{1}{2}\%$ of 160.

Check your answer in **17**.

---

16   Your answer is incorrect.

Beware! ⟩ *Never* multiply by a percent number. Before you multiply $30\% \times 50$, you must write 30% as a decimal number. If you need help in writing a percent as a decimal number, turn to frame **7**; otherwise return to **13** and try again.

---

17   **Step 1.**   Translate: $8\frac{1}{2}\% \times 160 = \square$

Step 2.   Label: $\% \times T = P$

Step 3.   Rearrange: $\square = 8\frac{1}{2}\% \times 160$

Step 4.   Solve: $8\frac{1}{2}\% = 8.5\% = 08.5\% = 0.085$

$$\square = 0.085 \times 160 = 13.6$$

Now try these for practice:

(a)   Find 2% of 140.   _____

(b)   35% of 20 =   _____

(c) $7\frac{1}{4}\%$ of \$1000 =   _____

(d)   What is $5\frac{1}{3}\%$ of 3.3?   _____

(e)   120% of 15 is what number?   _____

The step-by-step answers are in **18**.

---

18   (a)   2% of 140 = ?

$2\% \times 140 = \square$

$\square = 2\% \times 140$

$\square = 0.02 \times 140$ $2\% = 02.\% = 0.02$

$\square = 2.8$

(b)   35% of 20 = ?

$35\% \times 20 = \square$

$\square = 0.35 \times 20 = 7$ $35\% = 35.\% = 0.35$

(c) $7\frac{1}{4}\%$ of \$1000 = ?

$7\frac{1}{4}\% \times \$1000 = \square$

$\square = 7\frac{1}{4}\% \times \$1000$

Then

$\square = 0.0725 \times \$1000$ $\qquad$ $7\frac{1}{4}\% = 7.25\% = 07.25\% = 0.0725$

or $\qquad\qquad\qquad\qquad$ 0.0725
$\qquad\qquad\qquad\qquad\qquad$ $\times 1000$
$\qquad\qquad\qquad\qquad\qquad$ $\overline{}$
$\square = \$72.50$ $\qquad\qquad\qquad$ 72.5000

(d) $5\frac{1}{3}\%$ of $3.3 = ?$

$5\frac{1}{3}\% \times 3.3 = \square$

$\square = 5\frac{1}{3}\% \times 3.3$

$\square = \dfrac{16}{300} \times 3.3 = \dfrac{16}{300} \times \dfrac{33}{10} = \dfrac{528}{3000}$ $\qquad\qquad$ $5\frac{1}{3}\% = \dfrac{5\frac{1}{3}}{100} = \dfrac{\frac{16}{3}}{100} = \dfrac{16}{300}$

$\square = 0.176$

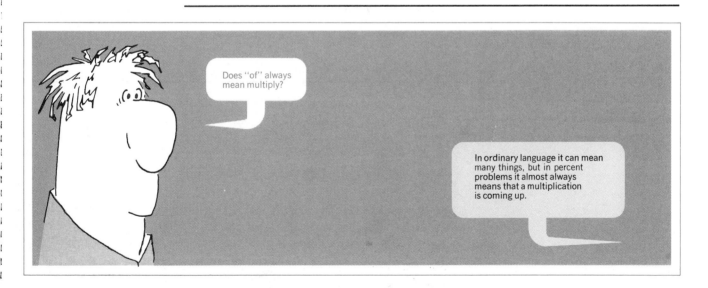

Note $\qquad$ Rounding $5\frac{1}{3}\%$ to a decimal number will result in an approximate answer. More decimal places in the rounded decimal number will result in greater accuracy.

$5\frac{1}{3}\% = 5.3\% = 05.3\% = 0.053$ rounded

$\square = 0.053 \times 3.3 = 0.1749 = 0.175$ rounded

(e) 120% of $15 = ?$

$120\% \times 15 = \square$

$\square = 120\% \times 15$

$\square = 1.2 \times 15$ $\qquad\qquad$ $120\% = 120.\% = 1.20$

$\square = 18$

Now, try the following P-type business problem.

A reasonable budget allows 25% of net income for housing. How much should be allowed for housing if the net monthly income is $789.51? Round your answer to the nearest cent.

Turn to 19 to check your answer.

19   What is 25% of $789.51?

$$\square = 25\% \times \$789.51$$

$$\square = 0.25 \times \$789.51 = \$197.3775 = \$197.38 \text{ rounded}$$

%-Type
Problems

A *%-type problem* requires that you find the rate or percent. Problems of this kind are usually stated:

7 is what percent of 16?

or   Find what percent 7 is of 16.

or   What percent of 16 is 7?

Step 1
Translate: $\square\% \times 16 = 7$

$$\square\% \times 16 = 7$$

Step 2
Label: $\% \times T = P$ All of the problem statements are equivalent to this equation.

Step 3
Rearrange: $\square\% = \dfrac{7}{16}$

To rearrange the equation and solve for $\square\%$ notice that it is in the form

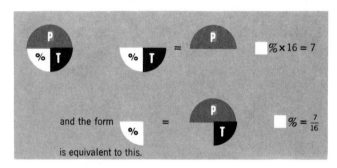

Therefore

$\square\% = \dfrac{7}{16}$ 16 is the total amount or base and 7 is the part of the base being described.

Solve this last equation.

Check your answer in **20**.

---

20   **Step 4**

$$\square\% = \frac{7}{16} = 0.4375 = 43.75\%$$

$$\square\% = 43.75\%$$

Step 5
Check: $43.75\% \times 16 = ?$

$$0.4375 \times 16 = 7$$

$$\begin{array}{r} 0.4375 \\ 16\overline{)7.0000} \\ \underline{6\ 4} \\ 60 \\ \underline{48} \\ 120 \\ \underline{112} \\ 80 \\ \underline{80} \end{array}$$

If you had trouble converting $\frac{7}{16}$ to a percent, you should review this process by turning back to frame **5**.

The solution to a %-type problem will be a fraction or decimal number that must be converted to a percent.

Try these problems for practice.

(a) What percent of 40 is 16? _____

(b) Find what percent of 25 is 65. _____

(c) $6.50 is what percent of $18.00? _____

(d) What percent of 2 is 3.5? _____

(e) $10\frac{2}{5}$ is what percent of 2.6? _____

Check your work in **21**.

21　(a)　$\square\% \times 40 = 16$　　　A %– Type problem $\% = \dfrac{P}{T}$

$$\square\% = \frac{16}{40}$$

$$\square\% = \frac{16}{40} = 0.40 = 40\%$$

$\square\% = 40\%$　　**Check:**　$40\% \times 40 = ?$

$0.40 \quad \times 40 = 16$

(b)　$\square\% \times 25 = 65$　　　$\% = \dfrac{P}{T}$

$$\square\% = \frac{65}{25}$$

$$\square\% = \frac{65}{25} = 2.60 = 260\%$$　　**Check:**　$260\% \times 25 = ?$

$2.60 \times 25 = 65$

The most difficult part of this problem is in deciding whether the percent needed is found from $\frac{65}{25}$ or $\frac{25}{65}$. There is no magic to it. If you read the problem very carefully you will see that it speaks of 65 as a part "of 25." The base or total is 25. The percentage or part is 65.

(c)　$\$6.50 = \square\% \times \$18.00$

or　　$\square\% \times \$18.00 = \$6.50$　　　$\% = \dfrac{P}{T}$

$$\square\% = \frac{6.50}{18.00}$$

Note　There are two ways to change $\dfrac{6.50}{18.00}$ to a percent: the exact method and the rounding method. (If you need help in writing a fraction as a percent, turn to frame **6**.)

Exact Method:　$\dfrac{6.50}{18.00} = \dfrac{6\frac{1}{2}}{18} = \dfrac{6\frac{1}{2} \times 5\frac{5}{9}}{18 \times 5\frac{5}{9}} = \dfrac{36\frac{1}{9}}{100} = 36\frac{1}{9}\%$　　$\square\% = 36\frac{1}{9}\%$

Rounding Method:　$\dfrac{6.50}{18.00} = 0.361 = 36.1\%$ rounded

$\square\% = 36.1\%$ rounded

Check: $36\frac{1}{9}\% \times \$18 = ?$　or　**Check:** $36.1\% \times \$18 = ?$

$\dfrac{325}{900} \times 18 = 6.50$　　　　　　　$0.361 \quad \times \$18 = 6.498$

$= 6.50$ rounded

(d) $\square\% \times 2 = 3.5$ $\% = \dfrac{P}{T}$

   $\square\% = \dfrac{3.5}{2}$

   $\square\% = \dfrac{3.5}{2} = 1.75 = 175\%$

   $\square\% = 175\%$ **Check:** $175\% \times 2 = ?$

                                          $1.75 \ \times 2 = 3.5$

(e) $10\frac{2}{5} = \square\% \times 2.6$

   $\square\% \times 2.6 = 10\frac{2}{5}$ $\% = \dfrac{P}{T}$

   $\square\% = \dfrac{10\frac{2}{5}}{2.6}$

   $\square\% = \dfrac{10\frac{2}{5}}{2.6} = \dfrac{10.4}{2.6} = 4.00 = 400\%$

   $\square\% = 400\%$ **Check:** $400\% \times 2.6 = ?$

                                          $4.00 \ \times 2.6 = 10.4 = 10\frac{2}{5}$

Let's try a %-type business problem.

If your original hourly pay rate was $5.75 per hour and you received an increase to $6.44, what would be your percent increase?

Hint: First calculate the amount of increase by subtraction.

Then calculate the percent increase using the amount of increase and the original pay rate.

When calculating the percent increase or decrease, always use the amount of increase or decrease and the original amount.

Check your answer in 22 .

---

22   Amount of increase $= \$6.44 - 5.75 = \$0.69$

Next, calculate the percent increase.

$0.69 is what percent of $5.75

$\$0.69 = \square\% \times \$5.75$

   $\square\% = \dfrac{\$0.69}{\$5.75}$

$\square\% = 0.12 = 12\%$

*T-*Type
Problem   A *T-type problem* requires that you find the total, given the percent and the percentage or part. Typically they are stated like this:

        8.7 is 30% of what number?

or   Find a number such that 30% of it is 8.7.

or   8.7 is 30% of a number. Find the number.

or   30% of what number is equal to 8.7?

Step 1
Translate: $30\% \times \square = 8.7$

or $30\% \times \square = 8.7$

Step 2
Label: $\% \times T = P$

Step 3
Rearrange: $\square = \dfrac{8.7}{30\%}$

138   4 PERCENT

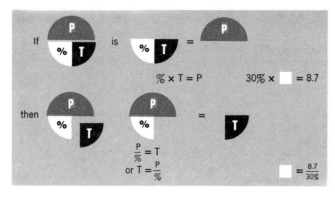

$$\% \times T = P \qquad 30\% \times \boxed{} = 8.7$$

$$\frac{P}{\%} = T$$
$$\text{or } T = \frac{P}{\%} \qquad\qquad \boxed{} = \frac{8.7}{30\%}$$

The rearranged problem is

$$\boxed{} = \frac{8.7}{30\%}$$

Solve this problem.

Check your answer in **23**.

---

23 $\boxed{} = \dfrac{8.7}{30\%} = \dfrac{8.7}{0.30}$

$$
\begin{array}{r}
29. \\
0.30\overline{)8.70} \\
\underline{6\ 0} \\
2\ 70 \\
\underline{2\ 70}
\end{array}
$$

(Remember, never multiply or divide by a percent number. Always convert the percent number to a decimal number.)

$\boxed{} = 29$ **Check:** $30\% \times 29 = ?$

$$0.30 \times 29 = 8.7$$

> **Important**

We cannot divide by 30%. We must change the percent to a decimal number before we do the division.

Here are a few practice problems to test your mental muscles.

(a) 16% of what number is equal to 5.76?
(b) 41 is 5% of what number?
(c) Find a number such that $12\frac{1}{2}\%$ of it is $26\frac{1}{4}$.
(d) 2 is 8% of a number. Find the number.
(e) 125% of what number is 35?

Check your answers against ours in **24**.

24 (a) $16\% \times \square = 5.76$ $T = \dfrac{P}{\%}$

$\square = \dfrac{5.76}{16\%}$ $16\% = 0.16$

$\square = \dfrac{5.76}{16\%} = \dfrac{5.76}{0.16} = 36$

$$\begin{array}{r} 36. \\ 0.16\overline{)5.76} \\ \underline{4\ 8} \\ 96 \\ \underline{96} \end{array}$$

$\square = 36$ **Check:** $16\% \times 36 = ?$

$0.16 \times 36 = 5.76$

(b) $41 = 5\% \times \square$ $T = \dfrac{P}{\%}$

$\square = \dfrac{41}{5\%}$ $5\% = 0.05$

$\square = \dfrac{41}{5\%} = \dfrac{41}{0.05} = 820$

$$\begin{array}{r} 820. \\ 0.05\overline{)41.00} \\ \underline{40} \\ 1\ 0 \\ \underline{1\ 0} \\ 0 \\ \underline{0} \end{array}$$

$\square = 820$ **Check:** $5\% \times 820 = ?$

$0.05 \times 820 = 41$

(c) $26\tfrac{1}{4} = 12\tfrac{1}{2}\% \times \square$ $T = \dfrac{P}{\%}$

or $\square = \dfrac{26\tfrac{1}{4}}{12\tfrac{1}{2}\%}$ $12\tfrac{1}{2}\% = 12.5\% = 0.125$

$\square = \dfrac{26\tfrac{1}{4}}{12\tfrac{1}{2}\%} = \dfrac{26.25}{0.125} = 210$

$$\begin{array}{r} 210. \\ 0.125\overline{)26.250} \\ \underline{25\ 0} \\ 1\ 25 \\ \underline{1\ 25} \\ 0 \\ \underline{0} \end{array}$$

$\square = 210$ **Check:** $12\tfrac{1}{2}\% \times 210 = ?$

$0.125 \times 210 = 26.25$

$= 26\tfrac{1}{4}$

(d) $2 = 8\% \times \square$ $T = \dfrac{P}{\%}$

or $\square = \dfrac{2}{8\%}$ $8\% = 0.08$

$\square = \dfrac{2}{8\%} = \dfrac{2}{0.08} = 25$

$$\begin{array}{r} 25. \\ 0.08\overline{)2.00} \\ \underline{1\ 6} \\ 40 \\ \underline{40} \end{array}$$

$$\square = 25 \qquad\qquad \textbf{Check:} \quad 8\% \times 25 = ?$$
$$0.08 \times 25 = 2$$

(e) $125\% \times \square = 35$ $\qquad\qquad T = \dfrac{P}{\%}$

$\qquad$ or $\qquad \square = \dfrac{35}{125\%}$ $\qquad\qquad 125\% = 1.25$

$$\square = \dfrac{35}{125\%} = \dfrac{35}{1.25} = 28 \qquad 1.25\overline{)35.00}$$

$$\begin{array}{r} 28. \\ \underline{25\ 0} \\ 10\ 00 \\ \underline{10\ 00} \end{array}$$

$$\square = 28 \qquad\qquad \textbf{Check:} \quad 125\% \times 28 = ?$$
$$1.25 \times 28 = 35$$

Next, try the following T-type business problem.

The sales tax on a coat was \$4.71. If the tax rate was 6%, what was the price of the coat?

Check your answer in **25**.

---

25 $\quad$ 6% of what is \$4.71?

$$6\% \times \square = \$4.71$$

$$\square = \dfrac{\$4.71}{6\%}$$

$$\square = \dfrac{\$4.71}{0.06} = \$78.50$$

A REVIEW $\qquad$ Let's review these five steps for solving percent problems:

Step 1. $\quad$ **Translate** the problem sentence into a math equation.

Step 2. $\quad$ **Label** the numbers as base or total (T), percentage or part (P), and percent (%).

Step 3. $\quad$ **Rearrange** the math equation so that the unknown quantity is alone on the left. Use the Equation Finder.

Step 4. $\quad$ **Solve** the problem by doing the arithmetic. Always change percent numbers to decimal numbers first.

Step 5. $\quad$ **Check** the answer if you can by putting it back into the original problem to see if it is correct.

Are you ready for a bit of practice on the three basic kinds of percent problems? Wind your mind and turn to **26** for a problem set.

HOW TO MISUSE PERCENT

1. In general you cannot add, subtract, multiply, or divide percent numbers. Percent helps you compare two numbers; it cannot be used in the normal arithmetic operations.

   For example, if 60% of class 1 earned A grades and 50% of class 2 earned A grades, what was the total percent of A grades for the two classes?

   The answer is that you cannot tell unless you know the number of students in each class.

2. In advertisements designed to trap the unwary, you might hear that "children had 23% fewer cavities when they used . . . ," or "50% more doctors smoke . . . ."

   Fewer than what? Fewer than the worst dental health group the advertiser could find? Fewer than the national average?

   More than what? More than a year ago? More than nurses? More than other adults? More than infants?

   There must be some reference or base given in order for the percent number to have any meaning at all.

   BEWARE of people who misuse percent!

4 Percent

4.2 Percent Problems

The answers are on page 579.

A. Solve.

1. 4 is ____% of 5.

2. 15 is ____% of 75.

3. 6% of $150 = ____

4. What percent of 25 is 16? ____

5. 20% of what number is 3? ____

6. 25% of 428 = ____

7. 8 is what percent of 8? ____

8. 120% of 45 is ____

9. 35% of 60 is ____

10. 9 is 15% of ____

11. 8 is ____% of 64.

12. 3% of 5000 = ____

13. 100% of what number is 59? ____

14. 2.5% of what number is 2? ____

15. What percent of 54 is 36? ____

16. 60 is ____% of 12.

17. 17 is 17% of ____

18. 13 is what percent of 25? ____

19. 74% of what number is 370? ____

20. $8\frac{1}{2}$% of $250 is ____

B. Solve.

1. 75 is $33\frac{1}{3}$% of ____

2. $137\frac{1}{2}$% of 5640 is ____

3. What percent of 10 is 2.5? ____

4. 21 = $116\frac{2}{3}$% of ____

5. 6% of $3.29 is ____

6. 63 is ____% of 35.

7. 12.5% of what number is 20? ____

8. $33\frac{1}{3}$% of $8.16 = ____

9. 9.6 is what percent of 6.4? ____

10. What % of 28 is 3.5? ____

11. What percent of 7.5 is 2? ____

12. $37\frac{1}{2}$% of 12 is ____

13. $.75 is ____% of $37.50

14. 0.516 is what percent of 7.74? ____

15. $6\frac{1}{4}$% of 280 = ____

16. $2\frac{1}{4}$ is what percent of 9? ____

17. 1.28 is ____% of 0.32

18. 42.7 is 10% of ____

19. 260% of 8.5 is ____

20. 4.75% of what number is 76? ____

C. Solve.

1. The population of Boomville increased from 48,200 to 63,850 in two years. What was the percent increase?

Date

Name

Course/Section

2. The Smiths bought a house for $74,500 and made a down payment of 20%. What was the actual amount of the down payment?

_____

3. If your salary is $760 per month and you get a 6% raise, what is your new salary?

_____

4. California state sales tax is 6%. What tax would you pay on a purchase of $7.60 in Los Angeles?

_____

5. A carpenter earns $6.95 per hour. If he receives a $7\frac{1}{2}$% pay raise, what is his new pay rate?

_____

6. If 6.7% of your salary is withheld for Social Security, what amount is withheld from your monthly earnings of $645.00?

_____

7. Linda purchased a classic car for $6840. A year later she sold it for $7866. What was the percent increase in value?

_____

8. The climbing nut workers received a pay raise of $0.27 per hour. If the raise was 6%, what was their original pay rate? What is their new rate?

_____

9. The FHA requires a minimum down payment of 3% on all houses appraised under $25,000. What is the minimum down payment on a house selling for $22,950?

_____

10. A reasonable estimate of annual home maintenance expenses is $1\frac{1}{2}$% of the cost of the house. What is the maintenance expense on a house valued at $79,750?

_____

11. Rosemary purchased a new car for $5895. The first year it depreciated 22%. What was the car's value after the first year?

_____

12. The Super Charge Company estimates that 1.8% of its total charges must be sent to the Go Get'em Collection Agency. With total monthly charges of $2,547,800, what amount must be sent to the agency?

_____

13. The Tan family deposits 5% of their income in savings each month. If they deposited $45.17 last month, what was their income?

_____

14. The Dow Jones Industrial Average closed at 971.90. A year earlier, it closed at 931.52. What was the percent increase rounded to the hundredth of a percent?

_____

15. If you answered 37 problems correctly on a 40 question test, what percent score would you have?

_____

16. A marathon runner weighing a normal 146 pounds weighed 138 pounds after his 26 mile race. What percentage of his body weight did he lose?

_____

17. How many problems did Trish answer correctly if she received a grade of 90% on a test of 30 questions?

_____

18. If you purchased a lot for $5400 and later sold it for 30% more than its purchase price, what was the selling price?

_____

19. A calculator originally sold for $16.50, but has been reduced 20%. What is the reduction and the reduced price?

_____

20. Calculate the sales tax on a purchase of $56.72. Use a 6% tax rate.

_____

When you have had the practice you need, turn to **27** for the Self-Test on percent problems.

4 Percent

The answers are on page 580.

1. Write $\frac{5}{16}$ as a percent.   _____

2. Write 1.27 as a percent.   _____

3. Write 2.3% as a decimal.   _____

4. Write 35% as a fraction and reduce to lowest terms.   _____

5. What percent of $7.50 is $1.35?   _____

6. What is 26% of $1856?   _____

7. 132% of what is 356.4?   _____

8. A building is valued at $76,000 but insured for 80% of its value. For how much is it insured?   _____

9. Lou earned $5.80 per hour before he received a raise to $6.09. What was his rate of pay increase?   _____

10. Howard Huge accrued 205,000 shares of Up-In-The-Airlines Co. stock. If this is 82% of the total stock, what is the total number of shares?   _____

Date   _____

Name   _____

Course/Section   _____

5 Mathematics of Buying and Selling

THIS REFRIGERATOR HAS BEEN MARKED UP ONLY 50%

Most retail businesses buy merchandise from suppliers, wholesalers, or manufacturers, reprice it, and sell it to consumers. This repricing involves adding a markup to the purchase cost. The markup is not usually related to the number of dents or flaws in the merchandise—as it appears in the above cartoon. It is, in fact, the amount added to the original cost to obtain the actual selling price.

This unit is designed to help you to understand how merchandise is bought, repriced, and sold to consumers.

Ready? Good! Sharpen your pencil, make sure you have plenty of scratch paper, and get ready to "mark up" quite a bit of it as you learn about the mathematics of buying and selling.

UNIT OBJECTIVE

After successfully completing this unit, you will be able to calculate cash and trade discounts, the single equivalent trade discount rate, markup based on cost or selling price, markdown, sales tax, and excise tax. You will also be able to calculate an inventory valuation by the specific identification, average cost, FIFO, and LIFO methods.

PREVIEW 5   This is a sample test for Unit 5. All of the answers are placed immediately after the test. Work as many of the problems as you can, then check your answers and look after the answers for further directions.

		Where to Go for Help	
		Page	Frame
1.	Calculate the net price on a desk listed at $150 and sold with trade discounts of 15/10/5.	151	1

149

2. An invoice for $239.76, dated March 17, offers cash discount terms 2/10prox. What amount is due if the bill is paid April 7?   _____   161   11

3. Oscar's Hardware made the following purchases of trash cans.

   February 17   125 @ $10.00
   May 5   150 @   10.50
   July 26   75 @   11.00
   September 9   150 @   11.20

   The year-end inventory revealed 162 trash cans. Calculate the inventory value by the average cost method.   _____   169   19

4. A shirt costs $12.50 and is marked up $3.25. Calculate the (a) selling price and (b) rate of markup based on cost.

   (a)   _____

   (b)   _____   179   26

5. A stove costs $343.96 and is marked up 20% based on selling price. What is the (a) selling price and (b) markup?

   (a)   _____

   (b)   _____   179   26

6. A chair originally sold for $249.98 but has been marked down 15%. What is the reduced price?   _____   193   38

7. If the sales tax rate is 6%, what is the total cost of a purchase of $357.63?   _____   194   40

★ If you cannot work any of these problems, start with frame 1 on page 151.
★ If you missed some of the problems, turn to the page numbers indicated above.
★ If all of your answers were correct, you are probably ready to proceed to Unit 6. (If you would like more practice on Unit Five before turning to Unit 6, try the Self-Test on page 199.)
★ Super-students—those who want to be certain they learn all of this material—will turn to frame 1 and begin work there.

3. $1726.92	6. $212.48	
2. $234.96	5. (a) $429.95 (b) $85.99	
1. $109.01	4. (a) $15.75 (b) 26%	7. $379.09

Preview 5
Answers

SECTION 1: TRADE DISCOUNTS

1 | **OBJECTIVE**
|
| Calculate trade discounts and the single equivalent discount rate.

In this unit we will be primarily dealing with retail businesses.

Retail firms purchase goods from manufacturers, wholesalers, or suppliers and, in turn, sell the merchandise to individual consumers.

The initial ordering of merchandise by the retailer is recorded on an invoice. The *invoice* is a record of the merchandise ordered along with shipping and billing information. The invoice is a bill.

A sample is shown below.

---

(2) #41-386

(1) POTPOURRI PAINT, INC.
654 Chartreuse Avenue
Mobile, Alabama

(3) Sold To: (4) Ship To:

Hobby Hut   Hobby Hut
514 Fox Avenue   514 Fox Avenue
Oklahoma City, Oklahoma   Oklahoma City, Oklahoma

(5) Date: 3/4/84   (7) Delivery Date: 3/7/84

(6) Via: UPS   (8) Terms: 2/10, n/30

(9) Quantity	(10) Description	(11) Unit Price	(12) Amount
10 doz.	Spray Enamel, 13 oz., White	$9.48	$ 94.80
12 doz.	Spray Enamel, 13 oz., Red	9.35	112.20
5 doz.	Clear Varnish Spray, 13 oz.	9.23	46.15
	(13) Total list price		253.15
	(14) Less trade discounts 15/10/5		69.17
	(15) Net		183.98
	(16) Freight		23.45
	(17) Amount due		207.43

Locate the following items on the invoice shown.

1. Supplier's name and address.

2. Invoice number.

3. The buyer.

4. To whom the merchandise is shipped.

5. Invoice date.

6. Method of delivery.

7. Delivery date.

8. Terms of payment. This may include cash discounts (discussed in Section 2).

9. Quantity ordered.

10. A description of the items ordered.

11. Unit price. Unit price may be per dozen, per case, or per item.

12. Amount or extension. The amount is the product of the quantity and the unit price. For example, the first line in our invoice has

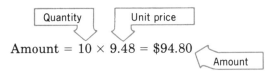

$$\text{Amount} = 10 \times 9.48 = \$94.80$$

13. Total list price is the sum total of the amount column.

14. A deduction for trade discounts. Trade discounts are discussed later in this section.

15. The net is the total list price minus the trade discounts. The *net* is the amount due after deductions.

16. Shipping or freight charge.

17. Total amount due, which is equal to the sum of the Net and the Freight.

Now continue in **2** for a discussion of discounts.

Discounts

2

Now that you're familiar with invoices, let's look at discounts. As you probably noticed, the invoice in **1** has several discounts. A *discount* is an amount deducted from the list price. Most discounts are stated in terms of a discount rate.

The discount is the product of the discount rate and the list price.

> Discount = rate × price

For example, if widgets list for $35.90 and are discounted 30%, the discount is

$$\text{Discount} = 30\% \times \$35.90$$
$$= 0.30 \times \$35.90 = \$10.77$$

The discount is $10.77.

Careful!

Remember to change the percent to a decimal number before multiplying. If you had trouble changing the percent to a decimal number, turn to page 126 for a review.

If you are purchasing one of these new widgets, you will want to calculate the actual cost. The actual or *net cost* is the difference between the list price and the discount. The net is the cost after any deductions.

Net cost = list price − discount

The net cost is sometimes called the "reduced price" or the "sale price." Complete the widget example by calculating the net cost.

Turn to **3** to check your work.

---

3   Net cost = list price − discount
= \$35.90 − 10.77 = \$25.13

The new cost of a widget is \$25.13—a bargain!

Now, you try a few problems. Calculate the discount and net cost.

1. A pair of jeans that originally cost \$35.95 is now on sale for 20% off. What is the reduced price?
2. A stereo system is reduced 32%. If the list price is \$749.50, what is the sale price?

Check your calculations in **4**.

---

4   1. Discount = 20% × \$35.95 = 0.20 × \$35.95 = \$7.19
Net cost = \$35.95 − 7.19 = \$28.76
2. Discount = 32% × \$749.50 = 0.32 × \$749.50 = \$239.84
Net cost = \$749.50 − 239.84 = \$509.66

Frequently the discount is only an intermediate result. Usually we are interested only in the amount we must pay, the net cost. In this case, there is a shortcut method for finding the net cost directly.

If the discount rate is 30% the amount you must pay will be 100% − 30%, or 70%. We are breaking the list price (100%) into two parts: the discount (30%) and the net cost (70%).

100%

Discount 30%	Net cost 70%

In the example of a widget that lists for \$35.90 but is on sale for 30% off, the net cost is

100% − 30%

Net cost = 70% × \$35.90 = 0.70 × \$35.90 = \$25.13

The formula for the shortcut method is

Net cost = (100% − discount rate) × list price

Work the following problems using the shortcut method.

1. A pair of jeans that originally cost $35.95 is now on sale for 20% off. What is the reduced price?
2. A stereo system is reduced 32%. If the list price is $749.50, what is the sale price?

Turn to **5** to check your answers.

---

5   1. Net cost = (100% − 20%) × $35.95.
              = 80% × $35.95 = 0.80 × $35.95 = $28.76
    2. Net cost = (100% − 32%) × $749.50
              = 68% × $749.50 = 0.68 × $749.50 = $509.66

These are the same answers you calculated using the other method in frame **4**.

TRADE DISCOUNTS

Manufacturers and wholesalers usually issue a catalogue of their products with the list prices. In addition to the catalogue, a separate sheet of discount rates, called trade discounts, is provided. *Trade discounts* are amounts deducted from the catalogue list price.

The use of the trade discount sheet has two primary advantages. The catalogue is expensive to print, and costly to revise. Alterations in price are easily made by changing the discount sheet. If the price is reduced, the discount is increased or another discount is added. If the price is increased, a discount is reduced or omitted.

A second advantage of discount sheets is it enables different discounts to be given to various types of businesses.

Trade discounts are often written as two or more successive discounts called *chain discounts*. A discount of 15% followed by a discount of 10% is written simply as 15/10. But successive discounts of 15% and 10% are *not* equal to a single discount of 25%. This is illustrated in the following example.

Example: An order of $500 offers trade discounts of 15/10. What is the net cost?

Step 1. Calculate the first discount.

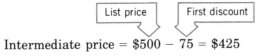

First discount = 15% × $500
                = 0.15 × $500 = $75

Next, subtract the discount from the list price. The result is called the *intermediate price.*

Intermediate price = $500 − 75 = $425

Step 2. Calculate the second discount. The second discount is calculated on the intermediate price, *not* the list price.

Second discount = 10% × $425
                 = 0.10 × $425 = $42.50

Finally, subtract the second discount from the intermediate price to obtain the *net* cost.

Net cost = $425 − 42.50 = $382.50

Important ⟩ Always use the intermediate price to calculate the second discount. Do not use the original list price when you are doing Step 2.

Successive trade discounts of 15% and 10% are not equal to a single discount of 25%. A discount rate of 25% yields a discount of

Discount = 25% × $500 = 0.25 × $500 = $125

Net = $500 − 125 = $375

Successive trade discounts of 15% and 10% yield a net cost of $382.50.

Your turn.

An item lists for $726.50 but offers trade discounts of 20/15. What is the net cost?

Check your work in **6**.

---

6   **Step 1.**   Calculate the first discount.

First discount = 20% × $726.50 = 0.20 × $726.50 = $145.30

Intermediate price = $726.50 − 145.30 = $581.20

Step 2.   Calculate the second discount.

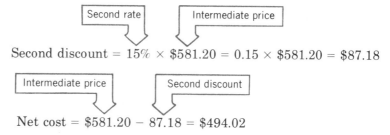

Second discount = 15% × $581.20 = 0.15 × $581.20 = $87.18

Net cost = $581.20 − 87.18 = $494.02

Now that you can calculate two discounts, try three. You guessed it; there are three steps.

Example: An item lists for $247.25 but carries trade discounts of 15/10/5. What is the net cost?

Step 1.   Calculate the first discount.

First discount = 15% × $247.25 = $37.09

First intermediate price = $247.25 − 37.09 = $210.16

Step 2. Calculate the second discount.

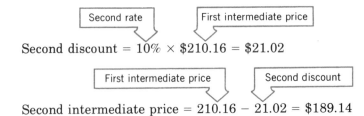

Second discount = 10% × $210.16 = $21.02

Second intermediate price = 210.16 − 21.02 = $189.14

Step 3. Calculate the third discount.

Third discount = 5% × $189.14 = $9.46

Net cost = $189.14 − 9.46 = $179.68

 **Remember** When calculating a trade discount, *always use the previous intermediate price.*

Your turn.

1. An order lists for $357 and carries trade discounts of 30/20/10. Calculate the net cost.
2. An item lists for $756.75 and carries trade discounts of 15/10/7½. What is the net cost?

Check your work in **7**.

How did trade discounts get their name? What is traded?

Nothing is traded. Trade discounts were originally discounts given to members of the same or a similar business or trade.

7 1. **Step 1.** First discount = 30% × $357 = 0.30 × $357 = $107.10
First intermediate price = $357 − 107.10 = $249.90

    Step 2. Second discount = 20% × $249.90 = 0.20 × $249.90 = $49.98
Second intermediate price = $249.90 − 49.98 = $199.92

    Step 3. Third discount = 10% × $199.92 = 0.10 × $199.92 = $19.99
Net cost = $199.92 − 19.99 = $179.93

2. **Step 1.** First discount = 15% × $756.75 = 0.15 × $756.75 = $113.51
First intermediate price = $756.75 − 113.51 = $643.24

    Step 2. Second discount = 10% × $643.24 = 0.10 × $643.24 = $64.32
Second intermediate price = $643.24 − 64.32 = $578.92

    Step 3. Third discount = 7½% × $578.92 = 7.5% × $578.92
= 0.075 × $578.92 = $43.42
Net cost = $578.92 − 43.42 = $535.50

Whew. That's a lot of work, even using a calculator. But there is an easier way using the shortcut method of calculating the net cost in frame **4**.

Example: An item lists for $500 but carries trade discounts of 15/10/5. Calculate the net cost.

The first discount rate is 15%. The intermediate price rate is 100% − 15% = 85%. The first intermediate price is

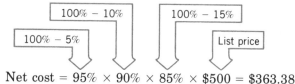

First intermediate price = 85% × $500 = 0.85 × $500 = $425

The second discount rate is 10%. The intermediate price rate is 100% − 10% = 90%. The second intermediate price is

Second intermediate price = 90% × $425 = 0.90 × $425 = $382.50

The third discount rate is 5%. The net cost rate will be 100% − 5% = 95%. The net cost is

Net cost = 95% × $382.50 = 0.95 × $382.50 = $363.375
= $363.38 rounded

All this may be reduced to just one step:

Net cost = 95% × 90% × 85% × $500 = $363.38

which can be punched into a calculator in one step.

Use the shorter method on the following problems.

1. An order lists for $357 and carries trade discounts of 30/20/10. Calculate the net cost.
2. An item lists for $756.75 and carries trade discounts of 15/10/7½. What is the net cost?

Go to **8** to check your work.

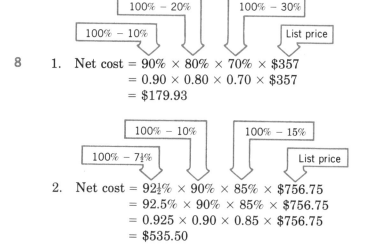

8   1.   Net cost = 90% × 80% × 70% × $357
= 0.90 × 0.80 × 0.70 × $357
= $179.93

2.   Net cost = 92½% × 90% × 85% × $756.75
= 92.5% × 90% × 85% × $756.75
= 0.925 × 0.90 × 0.85 × $756.75
= $535.50

I'm sure you will agree this method is much shorter than the first method.

Single Equivalent
Discount Rate

In Problem 1 above, the net cost is 90% × 80% × 70% × list price or 50.4% × list price. The company must pay 50.4% of the list price. The total discount rate is 100% − 50.4% = 49.6%. This is called the *single equivalent discount rate* and is equal to the three trade discounts of 30%, 20%, and 10%. To calculate the single equivalent discount rate, first calculate the rate at which the company must pay. Then subtract that amount from 100%.

Example: An item lists for $756.75 and carries trade discounts of 15/10/7½. Calculate the single equivalent discount rate and the net cost.

$$(100\% - 15\%) \times (100\% - 10\%) \times (100\% - 7\tfrac{1}{2}\%) = 85\% \times 90\% \times 92\tfrac{1}{2}\%$$
$$= 85\% \times 90\% \times 92.5\%$$
$$= 0.85 \times 0.90 \times 0.925 = 0.707625 = 70.7625\%$$

Single equivalent discount rate = 100% − 70.7625%
$$= 29.2375\%$$

Net cost = 70.7625% × $756.75 = $535.50

Your turn.

1. An item lists for $249 and carries trade discounts of 30/20/15. Calculate the net cost and single equivalent discount rate.

Check your work in 9.

---

9   $(100\% - 30\%) \times (100\% - 20\%) \times (100\% - 15\%)$
$$= 70\% \times 80\% \times 85\%$$
$$= 0.70 \times 0.80 \times 0.85 = 0.476 = 47.6\%$$

Net cost = 47.6% × $249 = $118.52

Single equivalent discount rate = 100% − 47.6% = 52.4%

For a set of practice problems on trade discounts, go to 10.

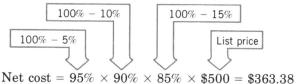

5 Mathematics of Buying and Selling

10   The answers are on page 580.

5.1 Trade Discounts   **A. Calculate the discount and net cost.**

	List Price	Discount Rate	Discount	Net Cost
1.	$200	4%	——————	——————
2.	$370	5%	——————	——————
3.	$120	12%	——————	——————
4.	$650	23%	——————	——————
5.	$275	$5\frac{1}{2}$%	——————	——————
6.	$463	$7\frac{3}{4}$%	——————	——————
7.	$137.95	6%	——————	——————
8.	$576.82	9%	——————	——————
9.	$198.27	$8\frac{1}{2}$%	——————	——————
10.	$849.58	$12\frac{1}{4}$%	——————	——————

B. Calculate the net cost.

	List Price	Trade Discounts	Net Cost
1.	$230	5/2	——————
2.	$150	10/5	——————
3.	$560	15/7	——————
4.	$750	12/8	——————
5.	$358	$15/8\frac{1}{2}$	——————
6.	$279	$12/7\frac{1}{2}$	——————
7.	$356.85	15/10/5	——————
8.	$423.98	20/15/10	——————
9.	$357.25	15/12/8	——————
10.	$406.32	15/10/10	——————

Date _____

Name _____

Course/Section _____

C. Calculate the single equivalent discount rate. Do not round.

	Trade Discounts	Single Equivalent Discount Rate
1.	15/10/5	_____
2.	20/15/10	_____
3.	15/10/10	_____
4.	15/12/8	_____
5.	25/15/10	_____
6.	$20/15/7\frac{1}{2}$	_____
7.	$15/10/8\frac{1}{2}$	_____
8.	10/5/5	_____
9.	20/18/10	_____
10.	15/13/10	_____

D. Calculate the single equivalent discount rate and net cost.

	List Price	Trade Discounts	Single Equivalent Discount Rate	Net Cost
1.	$240	20/10/5	_____	_____
2.	$350	15/10/8	_____	_____
3.	$500	10/5/2	_____	_____
4.	$750	20/15/10	_____	_____
5.	$849	10/8/5	_____	_____
6.	$495	15/12/10	_____	_____
7.	$256.75	10/7/3	_____	_____
8.	$382.49	20/10/10	_____	_____
9.	$423.51	30/20/10	_____	_____
10.	$157.14	30/25/15	_____	_____

When you have had the practice you need, either return to the Preview on page 149 or continue in 11 with cash discounts.

SECTION 2: CASH DISCOUNTS

11	**OBJECTIVE**
	Calculate cash discounts.

CASH DISCOUNTS

If your business receives a bill that is due in 30 days, should you pay it early? Probably not. The money could be put to work for your business. Firms handling thousands of dollars in purchases each month find that leaving the money in an interest-bearing account is more profitable than paying the bills early, unless there is some incentive to pay early.

Many companies offer *cash discounts* to encourage early payment. A discount written "2/10, n/30" means that a 2% discount is given if the bill is paid within 10 days, and the net (total) is due in 30 days.

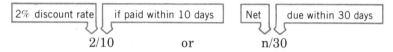

| 2% discount rate | | if paid within 10 days | | Net | | due within 30 days |

     2/10   or   n/30

The payment date is calculated from the date of the invoice. For example, a cash discount of 2/10 with an invoice date of March 2 offers a discount of 2% for 10 days *after* the invoice date. The last day of the discount period is March 12. Don't count the first day.

Another example: An invoice for $500, dated October 17 offers discount terms 2/10, n/30. If the bill is paid October 25, what amount should be paid?

Because the bill is paid within 10 days of the invoice date, a 2% discount is allowed.

Discount = 2% × $500 = 0.02 × $500 = $10
Amount paid = $500 − 10 = $490

Your turn. Work the following problem.

An invoice for $278, dated November 5 offers discount terms 2/10, n/30. If the bill is paid November 10, what is the amount paid?

Check your work in **12**.

How do you read 1/10, n/30?

Read it "One ten, net thirty." If you see that on a bill it means you get a 1% discount if the bill is paid within 10 days, and the net amount is due in 30 days.

12   Because November 10 is within 10 days of the invoice date, a 2% discount is given.

Discount = 2% × $278 = 0.02 × $278 = $5.56
Amount paid = $278 − 5.56 = $272.44

A business may offer several cash discount periods such as 4/10, 2/30, n/60. That is, a 4% discount if paid within 10 days, a 2% discount if paid within 30 days, and the net is due within 60 days. If the bill is paid within 10 days, only a 4% discount is taken, *not* both a 4% and 2% discount.

Try a few more problems:

1. An invoice for $155, dated February 5 offers discount terms 4/10, 2/30, n/60. What amount is due if the bill is paid on February 21?
2. A bill for $257.83, dated April 27, offers discount terms 3/10, 2/30, n/60. What amount is due if the bill is paid on June 5?

Check your answers in **13**.

Freight charges were added to my bill. Are freight charges discounted?

No.

13   1. February 21 is not within 10 days, but it is within 30 days. A 2% discount is given.

Discount = 2% × $155 = 0.02 × $155 = $3.10
Amount paid = $155 − 3.10 = $151.90

2. June 5 is more than 30 days past the billing date. No discount is given. The amount due is $257.83.

EOM AND PROX
DATING

EOM is short for *End of Month*. An example of EOM dating is 2/10EOM. This is a 2% discount if paid within 10 days after the end of the month, or by the tenth of the next month.

Example: An invoice for $275 dated May 17 offers discount terms 2/10EOM, n/60. What amount is due if the bill is paid on June 7?

A 2% discount is offered until June 10.

Discount = 2% × $275 = 0.02 × $275 = $5.50
Amount paid = $275 − 5.50 = $269.50

Another abbreviation for monthly dating is "prox." *Prox* is short for the Latin word *proximo,* which means the *next* month. A discount of 2/10prox is a 2% discount if paid by the tenth of the next month. The discount 2/10prox is the same as 2/10EOM.

Your turn.

1. A bill for $356, dated November 12, offers discount terms 3/10prox, n/60. What amount is due if the bill is paid on December 5?
2. An invoice for $759, dated December 18, offers discount terms 2/10EOM, n/60. What amount is due if the bill is paid on February 7?

Go to **14** to check your answers.

14

1. A 3% discount is offered until December 10.

    Discount = 3% × $356 = 0.03 × $356 = $10.68
    Amount paid = $356 − $10.68 = $345.32

2. A 2% discount is offered until January 10. February 7 is beyond the discount period. Since no discount is allowed, the amount due is $759.

With both EOM and prox dating, bills dated on or after the 26th of the month are granted a month's extension. For example, an invoice dated August 26 with terms 2/10EOM offers a discount of 2% until October 10.

Try a few more problems.

1. An invoice for $472, dated July 29, offers discount terms 3/10prox, n/90. When is the end of the discount period? What amount is due if the bill is paid on August 25?
2. A bill for $56.75, dated October 25, offers discount terms 2/10EOM, n/60. What amount is due if the bill is paid on December 5?

Check your work in 15.

What happens if I get a bill that is marked 5/10, n/30 and I don't pay the net amount by the due date?

Usually there is a delinquency charge.

15 1. Since the billing date is after the 26th, an extra month's extension is added to the discount period. The discount period ends September 10. If the bill is paid August 25, a 3% discount is taken.

    Discount = 3% × $472 = 0.03 × $472 = $14.16
    Amount paid = $472 − 14.16 = $457.84

2. The 2% discount period ends November 10. Remember, only bills dated on or after the 26th are granted a month's extension.

    If the bill is paid on December 5, no discount is allowed. The amount due is $56.75.

ROG DATING

ROG is the abbreviation for *receipt-of-goods*. With ROG dating, the discount period starts with the delivery date rather than the invoice date. This is designed to allow extra time when transportation is slow. With all other cash discount terms, the discount period starts with the invoice date.

Example: An invoice for $257.86, dated July 15, offers discount terms 3/10 ROG. If the goods were received August 7, when is the end of the discount period? What amount is due if the bill is paid August 15?

The discount period starts upon the receipt-of-goods on August 7. The discount period is 10 days, or until August 17. If the bill is paid on August 15, a 3% discount is allowed.

    Discount = 3% × $257.86 = 0.03 × $257.86 = $7.74
    Amount paid = $257.86 − 7.74 = $250.12

Work the following problems.

1. An invoice for $132.75, dated October 12, offers discount terms 4/15ROG. The goods were received on November 12. What amount is due if the bill is paid November 30?
2. An invoice for $576.89, dated September 5, offers discount terms 2/10ROG. The goods were received on October 2. What amount is due if the bill is paid October 10?

Check your answers in **16**.

---

16   1. The discount period starts November 12 and ends 15 days later on November 27. Since November 30 is beyond the discount period, the amount due is $132.75.
2. The discount period starts October 2 and ends October 12. Since October 10 is within the discount period, a 2% discount is allowed.

Discount = 2% × $576.89 = 0.02 × $576.89 = $11.54
Amount due = $576.89 − 11.54 = $565.35

EXTRA DATING

"Extra," "ex," or "X" are used to denote extra dating. *Extra dating* indicates that the discount period is extended for a certain number of days. Extra dating is sometimes used to encourage purchases. It is often used in seasonal businesses. For example, an air-conditioner supplier may use extra dating during the winter months to encourage off-seasonal purchases.

Example: An invoice for $846.23, dated October 3, offers discount terms 3/10-60X. What amount is due if the bill is paid December 5?

The discount of 3% is offered for 10 days plus 60 extra days for a total of 70 days. December 5 is within the discount period.

Discount = 3% × $846.23 = 0.03 × $846.23 = $25.39
Amount paid = $846.23 − 25.39 = $820.84

Try the following problems:

1. An invoice for $152.98, dated January 15, offers discount terms 2/15-30X. What amount is due if the bill is paid February 20?
2. An invoice for $576, dated March 12, offers discount terms 3/10-60X. What amount is due if the bill is paid May 15?

Check your work in **17**.

What happens when the last day of the discount period is a non-business day?

The discount is extended until the next business day.

17   1. The discount period is for 15 + 30 = 45 days.
           February 20 is within the discount period.

           Discount = 2% × $152.98 = 0.02 × $152.98 = $3.06
           Amount paid = $152.98 − 3.06 = $149.92

        2. The discount period is for 10 + 60 = 70 days.
           May 15 is within the discount period.

           Discount = 3% × $576 = 0.03 × $576 = $17.28
           Amount paid = $576 − 17.28 = $558.72

That wasn't too difficult, was it?

For more practice on cash discounts, go to 18.

Mathematics of Buying and Selling

PROBLEM SET 2

18   The answers are on page 580.

5.2 Cash Discounts   **A. Calculate the cash discount and net cost.**

	Invoice Date	Invoice Amount	Terms	Date Paid	Dis-count	Net Cost
1.	4/3	$200	2/10, n/30	4/10	————	————
2.	8/17	$350	3/10, n/30	8/23	————	————
3.	2/14	$500	3/10, 2/30, n/60	3/2	————	————
4.	6/3	$250	4/10, 2/30, n/60	7/1	————	————
5.	10/21	$700	3/10, n/60	11/10	————	————
6.	3/18	$625	1/10, n/30	4/9	————	————
7.	9/9	$475	3/10EOM, n/60	10/5	————	————
8.	11/7	$842	2/10prox, n/60	12/9	————	————
9.	5/2	$749	3/10-60X	6/15	————	————
10.	8/12	$128	2/10-30X	9/10	————	————
11.	12/5	$451	2/15-30X	1/10	————	————
12.	7/20	$742	2/15-60X	9/7	————	————
13.	3/8	$850	2/10, 1/30, n/60	4/2	————	————
14.	4/12	$560	2/10prox	5/4	————	————
15.	5/6	$125	2/10EOM	6/15	————	————
16.	8/2	$ 35	3/10-30X	9/5	————	————
17.	10/12	$235	2/10-45X	11/15	————	————
18.	12/2	$972	2/10, n/30	1/5	————	————
19.	6/14	$720	4/10, 2/30, n/60	7/11	————	————
20.	7/2	$635	3/10, 1/30, n/60	7/11	————	————

Date _____

Name _____

Course/Section _____

B. Calculate the cash discount and net cost.

	Invoice Date	Invoice Amount	Terms	Goods Received	Date Paid	Discount	Net Cost
1.	6/2	$300	3/15ROG	7/5	7/10	————	————
2.	2/15	$450	2/10ROG	3/20	3/25	————	————
3.	8/14	$400	2/10, n/30	8/17	8/20	————	————
4.	1/10	$500	3/10, n/30	1/15	1/17	————	————
5.	11/5	$275	3/10ROG	12/15	12/22	————	————
6.	5/28	$236	2/15ROG	7/5	8/10	————	————
7.	9/29	$842.95	3/10EOM	10/5	11/5	————	————
8.	10/30	$263.19	4/10prox	11/3	12/7	————	————
9.	3/12	$486.37	3/10, 2/30, n/60	3/15	4/10	————	————
10.	7/9	$126.18	4/10, 2/30, n/60	7/12	7/15	————	————
11.	9/6	$239.42	3/10-30X	9/11	10/10	————	————
12.	4/11	$418.36	2/15-30X	4/15	5/7	————	————
13.	2/7	$253.25	2/10ROG	2/15	2/20	————	————
14.	4/5	$455.29	3/10ROG	4/15	4/21	————	————
15.	11/21	$632.45	2/10prox	11/29	12/5	————	————
16.	6/12	$816.21	2/10EOM	6/15	7/2	————	————
17.	10/13	$126.38	2/10ROG	10/25	11/1	————	————
18.	7/2	$ 29.56	2/10, n/30	7/4	8/2	————	————
19.	9/12	$ 72.96	3/10ROG	9/20	9/27	————	————
20.	8/10	$645.26	2/10-30X	8/13	9/9	————	————

When you have had the practice you need, either return to the Preview on page 149 or continue in 19 with inventory evaluation.

19

> **OBJECTIVE**
>
> You will be able to calculate an inventory value using the specific identification, average cost, FIFO, and LIFO methods.

An *inventory* is a procedure for listing the individual items of merchandise in stock and the value of each item. This is an important process and is usually required at least once a year.

The first step is a physical inventory where the quantity of each item is counted. This total is then multiplied by the item's value to obtain the total value. Sounds simple, doesn't it? Well, there are a few details. A problem arises when determining the value for each item. If one item is purchased from several different suppliers at several times during the year, each time at a different price, which cost is used? The cost is determined by one of several *inventory valuation methods*.

First, let's look at an example of this problem.

The Calculator Store made the following purchases of RC calculators:

February 17	150	@	$12.50
April 25	120	@	12.00
July 9	130	@	11.50
September 7	100	@	11.75

At their year-end inventory, The Calculator Store had 163 RC calculators in stock. What value does the store place on the calculators? The highest price? The lowest price? The first price? The last price? The average price?

The value placed on the calculators depends on the inventory evaluation method used by the store. There are four basic methods used. For a discussion of the first method, turn to **20**.

SPECIFIC
IDENTIFICATION

20

The first method is *specific identification*. In this method, each item is valued at its actual purchase cost. This means each item must be tagged with its original cost. Because of the time required with this method, it is generally used only by businesses with few items to inventory. It is not used by businesses with large physical inventories.

In The Calculator Store example, each of the 163 RC calculators must be tagged with the original purchase cost to use the specific identification method.

In the inventory, the following information was compiled:

RC Calculators

Number	Cost
5	$12.50
13	12.00
50	11.50
95	11.75
163	

Next, the total cost is calculated.

RC Calculators

Number	Cost	Total Cost	
5	$12.50	$ 62.50	← 5 × $12.50
13	12.00	156.00	← 13 × 12.00
50	11.50	575.00	← 50 × 11.50
95	11.75	1116.25	← 95 × 11.75
163		$1909.75	

Total

The total inventory value of the RC calculators is $1909.75. This figure is accurate, but a lot of time is required to tag and check the purchase cost of each calculator. The three other methods are quicker.

Now, it's your turn.

Work the following problem using the specific identification method.

A television store made the following purchases of the XC 1000 model television sets.

January 7	35	@	$350
April 11	50	@	365
August 15	65	@	360
October 5	50	@	370

The year-end inventory revealed a total of 57 televisions in stock.

Number	Cost
10	$350
12	365
15	360
20	370
57	

Determine the inventory value of the TVs by the specific identification method.

Check your work in 21.

21

Number	Cost	Total Cost	
10	$350	$ 3,500	←10 × $350
12	365	4,380	←12 × 365
15	360	5,400	←15 × 360
20	370	7,400	←20 × 370
57		$20,680	

Total

AVERAGE COST

A second method of inventory evaluation uses the average cost. This method does not require finding the purchase cost of each item, but uses an averaging method.

Example: The Calculator Store made the following purchases of RC calculators:

February 17	150	@	$12.50
April 25	120	@	12.00
July 9	130	@	11.50
September 7	100	@	11.75

At the year-end inventory, 163 RC calculators were in stock. Calculate the inventory value by the average cost method.

First, find the total number of calculators purchased and their total cost.

Number	Cost	Total Cost
150	$12.50	$1875.00 ←150 × $12.50
120	12.00	1440.00 ←120 × 12.00
130	11.50	1495.00 ←130 × 11.50
100	11.75	1175.00 ←100 × 11.75
500		$5985.00

Total number of calculators

Total cost

Next, find the average cost of an RC calculator by dividing the total cost by the total number of calculators.

Total cost

$$\text{Average cost} = \frac{\$5985.00}{500} = \$11.97$$

Total number of calculators

Finally, multiply the number of calculators in stock times the average cost. This gives the inventory value.

In stock   Average cost

Inventory value = 163 × $11.97 = $1951.11

Try the following problem.

A television store had the following purchases of the XC 1000 model TV.

January 7	35 @	$350
April 11	50 @	365
August 15	65 @	360
October 5	50 @	370

The year-end inventory counted 57 televisions in stock. Determine the inventory value of the TVs by the average cost method.

Our work is in **22**.

22

Number	Cost	Total Cost
35	$350	$12,250 ←35 × $350
50	365	18,250 ←50 × 365
65	360	23,400 ←65 × 360
50	370	18,500 ←50 × 370
200		$72,400

Total TVs

Total cost

$$\text{Average cost} = \frac{\$72,400}{200} = \$362$$

Total TVs

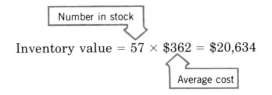

$$\text{Inventory value} = 57 \times \$362 = \$20,634$$

FIFO

Another inventory evaluation method is FIFO. FIFO is short for *First-In First-Out*. This method assumes the first items purchased are the first ones sold. **This means that the items in stock at inventory time are the last ones received.** The individual value of the last received items is used in the inventory evaluation procedure.

Example: The Calculator Store made the following purchases of RC calculators.

February 17	150	@	$12.50
April 25	120	@	12.00
July 9	130	@	11.50
September 7	100	@	11.75

At the year-end inventory, 163 RC calculators were in stock. Calculate the inventory value by the FIFO method.

Since there are 163 RC calculators in stock, the **last** 100 were purchased September 7 at $11.75 apiece. The remaining 63 must be from the previous order on July 9 at $11.50 apiece.

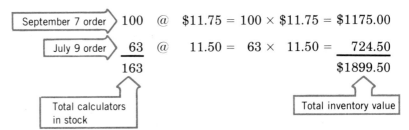

The total inventory value using the FIFO method is $1899.50.

Note that the inventory value is based on the cost of the *last* 163 calculators purchased. As in the example above, the total inventory may exceed the last order. In this case, the purchase price of prior order(s) must be used. Notice that the inventory value found by this method is different from the value obtained in either of the other two methods.

Work the following problem.

A television store had the following purchases of the XC 1000 model TVs.

January 7	35	@	$350
April 11	50	@	365
August 15	65	@	360
October 5	50	@	370

The year-end inventory counted 57 televisions in stock. Determine the inventory value by the FIFO method.

Check your calculations in **23**.

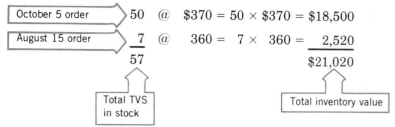

23

The inventory value using the FIFO method is $21,020.

LIFO

The last inventory evaluation method is LIFO. LIFO is short for *Last-In First-Out*. This is the opposite of the FIFO method. LIFO assumes the last items purchased are the first ones sold. **The items in stock at inventory time are the first items received.** The inventory value is based on the cost of the first items received.

Example: The Calculator Store made the following purchases of RC calculators.

February 17	150 @	$12.50
April 25	120 @	12.00
July 9	130 @	11.50
September 7	100 @	11.75

At the year-end inventory, 163 RC calculators were in stock. Calculate the inventory value by the LIFO method.

Since there are 163 RC calculators in stock, the first 150 were purchased February 17 at $12.50 a piece. The remaining 13 were purchased in the second order on April 25 at $12.00 per item.

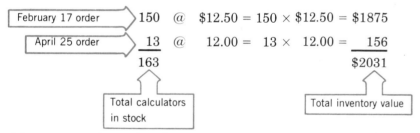

The total inventory value using the LIFO method is $2031.

Work our television problem using the LIFO method.

A television store had the following purchases of the XC 1000 model TVs.

January 7	35 @	$350
April 11	50 @	365
August 15	65 @	360
October 5	50 @	370

The year-end inventory counted 57 TVs in stock. Determine the inventory value by the LIFO method.

Our work is in 24.

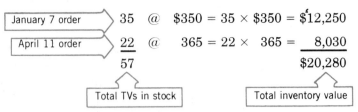

January 7 order	35	@	$350	= 35 × $350 =	$12,250
April 11 order	22	@	365	= 22 × 365 =	8,030
	57				$20,280

Total TVs in stock   Total inventory value

The inventory value using the LIFO method is $20,280.

Although we worked the same problems, the total inventory value was different with each method. The inventory value depends on the evaluation method used.

For a set of practice problems on inventory evaluation, go to 25.

PROBLEM SET 3

25   The answers are on page 580.

5.3 Inventory
     Evaluation

A. Calculate the inventory value by the specific identification method.

1. A furniture store had the following purchases of tables.

January 16	50	@	$125
March 5	60	@	150
August 25	70	@	145
October 7	20	@	155

    The year-end inventory counted a total of 62 tables in stock with the following costs.

Number	Cost
12	$125
21	150
17	145
12	155

    Determine the inventory value by the specific identification method.

2. A furniture store made the following purchases of lamps.

February 3	200	@	$29.50
April 7	300	@	29.75
May 25	250	@	30.05
July 17	350	@	30.20
August 5	300	@	30.10
September 23	250	@	30.04
November 5	350	@	30.15

    The year-end inventory counted a total of 432 lamps in stock with the following costs.

Number	Cost
23	$29.75
13	30.05
52	30.20
118	30.10
153	30.04
73	30.15

    Determine the inventory value by the specific identification method.

Date

Name

Course/Section

3. A store made the following purchases of dishes.

March 11	90	@	$10
May 17	140	@	11
August 24	160	@	13
October 5	110	@	12

The year-end inventory counted a total of 122 dishes in stock with the following costs.

Number	Cost
15	$10
25	11
42	13
40	12

Determine the inventory value by the specific identification method.

B. Calculate the inventory value.

1. A furniture store had the following purchases of tables.

January 16	50	@	$125
March 5	60	@	150
August 25	70	@	145
October 7	20	@	155

The year-end inventory counted a total of 62 tables in stock. Calculate the inventory value by the following methods.

(a) Average cost method

(b) FIFO method

(c) LIFO method

2. A furniture store made the following purchases of lamps.

February 3	200	@	$29.50
April 7	300	@	29.75
May 25	250	@	30.05
July 17	350	@	30.20
August 5	300	@	30.10
September 23	250	@	30.04
November 5	350	@	30.15

The year-end inventory counted a total of 432 lamps in stock. Calculate the inventory value by the following methods.

(a) Average cost method

(b) FIFO method

(c) LIFO method

3. A store made the following purchases of dishes.

March 11	90 @	$10
May 17	140 @	11
August 24	160 @	13
October 5	110 @	12

The year-end inventory counted a total of 122 dishes in stock. Determine the inventory value by the following methods.

(a) Average cost method

(b) FIFO method

(c) LIFO method

4. The Computer Store made the following purchases of PQ model 10 computers:

February 12	172 @	$246.50
May 5	118 @	235.20
July 23	120 @	196.75
November 17	90 @	162.50

The year-end inventory counted a total of 128 computers in stock. Determine the inventory by the following methods:

(a) Average cost method

(b) FIFO method

(c) LIFO method

When you have had the practice you need, either return to the Preview on page 149 or continue in **26** with markup.

SECTION 4: MARKUP

26

OBJECTIVE

Calculate cost, selling price, markup, and rate of markup based on cost and selling price.

An important part of any retail business is the proper pricing of its merchandise. The *cost* is the original price of the merchandise paid by the business. To the cost is added an additional amount called the *markup* to cover the business' expenses and profit. The *selling price* is the business' price to individual consumers. The selling price is the sum of cost plus the markup.

Selling price = cost + markup

For example, if a radio's cost is $27.50 and the markup is $8.25, the selling price is easily calculated by the above formula.

Cost   Markup

Selling price = $27.50 + $8.25 = $35.75

It's easy; you try one.

The cost of a suit is $75.26 and is marked up $41.39. Calculate the selling price.

Check your answer in **27**.

---

27   Selling price = cost + markup
            = $75.26 + 41.39 = $116.65

In markup problems where the cost and selling price are known, the markup can be calculated using another form of the basic markup equation.

Markup = selling price − cost

For example, if the selling price is $129.95 and the cost is $95.50, the markup is

Selling price   Cost

Markup = $129.95 − 95.50 = $34.45

If the selling price and the markup are known, the cost can be calculated using the following markup equation.

Cost = selling price − markup

All three of these basic markup equations are equivalent.

Complete the following problems.

1. The cost of a calculator is $32.50. If the selling price is $41.95, calculate the markup.
2. The cost of a radial arm saw is $195.25 and the markup is $54.72. What is the selling price?
3. The selling price of vinyl flooring is $13.75 per square yard. If the markup is $5.90, what is the cost?

Check your work in **28**.

28

1. Markup = selling price − cost
   = $41.95 − 32.50 = $9.45
2. Selling price = cost + markup
   = $195.25 + 54.72 = $249.97
3. Cost = selling price − markup
   = $13.75 − 5.90 = $7.85

MARKUP BASED ON COST

There are two basic methods to calculate markup and rate of markup. The first is based on cost and the second is based on selling price. Markup *based on cost* is generally used by manufacturers, wholesalers, and some retailers who take inventory at cost.

Finding the Markup When Cost and Rate Are Known

There are four basic markup problems using markup based on cost. The first involves calculating the markup when the rate of markup and the cost are known. The markup may be calculated using the following equation.

Markup = rate × cost   (based on cost)

This is the same type of percent equation as in Unit 4. The cost is the base, the rate is the percent, and the markup is the percentage. If you need a review of the basic percent equation, return to page 131. Otherwise continue here.

Example: A new bicycle costs $75.30 and is marked up 32% based on cost. Calculate the markup and selling price.

First, use the above formula to find the markup.

Rate   Cost

Markup = 32% × $75.30 = 0.32 × $75.30 = $24.096 = $24.10 rounded

Once the markup is known, the selling price may be found using the following formula.

Selling price = cost + markup
= $75.30 + 24.10 = $99.40

Your turn. Work the following problems.

1. A tool set costs $129.00 and is marked up 45% based on cost. Calculate the markup and selling price.
2. A radio's cost is $59.26 and is marked up 37.5% based on cost. What is the markup and selling price?

Go to 29 to check your work.

29   1.   Markup = rate × cost
                     = 45% × $129 = 0.45 × $129 = $58.05
              Selling price = cost + markup
                     = $129 + 58.05 = $187.05
      2.   Markup = rate × cost
                     = 37.5% × $59.26 = 0.375 × $59.26
                     = $22.2225 = $22.22 rounded
              Selling price = cost + markup
                     = $59.26 + 22.22 = $81.48

Finding the Cost When Markup And Rate Are Known

The second type of problem is calculating the cost when the markup and rate of markup are known. In the formula markup = rate × cost, the cost is the base or total. Solving for the cost is the same as solving for the total in a percent problem. This is a *T*-type percent problem. If you need a review of *T*-type percent problems, turn to page 138. Otherwise continue here.

Use the circle diagram to find the cost.

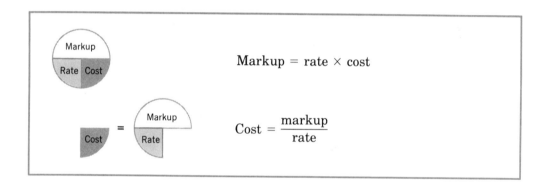

$$\text{Markup} = \text{rate} \times \text{cost}$$

$$\text{Cost} = \frac{\text{markup}}{\text{rate}}$$

Example: A place setting of china is marked up $35.40. If the rate of markup is 60% based on cost, find the cost and selling price.

First, use the formula for cost.

$$\text{Cost} = \frac{\text{markup}}{\text{rate}}$$

$$= \frac{\$35.40}{60\%} = \frac{\$35.40}{0.60} = \$59$$

Once the cost is known, the selling price is easily found by the following formula.

Selling price = cost + markup
            = $59 + 35.40 = $94.40

Work the following problems:

1.   A new chair is marked up $40.60. If the rate of markup is 28% based on cost, calculate the cost and selling price.
2.   A desk is marked up $36.40. If the rate of markup is 65% based on cost, what are the cost and selling price?

Check your answers in 30.

30

1. $\text{Cost} = \dfrac{\text{markup}}{\text{rate}} = \dfrac{\$40.60}{28\%} = \dfrac{\$40.60}{0.28} = \$145.00$

   Selling price = cost + markup
   $\qquad\qquad = \$145.00 + 40.60 = \$185.60$

2. $\text{Cost} = \dfrac{\text{markup}}{\text{rate}} = \dfrac{\$36.40}{65\%} = \dfrac{\$36.40}{0.65} = \$56.00$

   Selling price = cost + markup
   $\qquad\qquad = \$56.00 + 36.40 = \$92.40$

Finding the Rate When Markup and Cost Are Known

A third type of problem is finding the rate when the cost and markup are known. This is equivalent to the %-type percent problem.

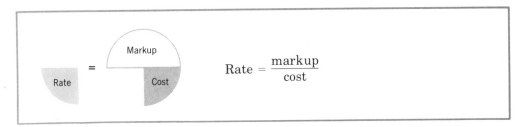

$$\text{Rate} = \frac{\text{markup}}{\text{cost}}$$

Example: A stereo system costs $365 and is marked up $131.40. What is the rate of markup based on cost?

$$\text{Rate} = \frac{\text{markup}}{\text{cost}} = \frac{\$131.40}{365} = 0.36 = 36\%$$

Work the following problems.

1. A lamp costs $28.30 and is marked up $8.49. What is the rate of markup based on cost?
2. A freezer costs $320 and sells for $400. Calculate the rate of markup based on cost. (**Hint:** First calculate the markup.)

Our work is in 31.

31

1. Rate $= \dfrac{\text{markup}}{\text{cost}} = \dfrac{\$8.49}{\$28.30} = 0.30 = 30\%$

2. Markup = selling price − cost
$= \$400 - 320 = \$80$

Rate $= \dfrac{\text{markup}}{\text{cost}} = \dfrac{\$80}{\$320} = 0.25 = 25\%$

In Problem 2, the markup and cost must be known to calculate the rate. If the selling price and the cost are given, the markup must first be calculated. If the selling price and the markup are given, then the cost must be calculated first.

Finding the Cost When the Selling Price and Rate Are Known

The last type of problem using markup based on cost is calculating the cost when the selling price and rate are known. A new formula is used for this problem.

$$\text{Cost} = \dfrac{\text{selling price}}{100\% + \text{rate}}$$

The algebraic derivation of this formula is included in the box on page 187.

Example: A weight lifting set sells for $24.30 and is marked up 35% based on cost. Calculate the cost and markup.

$\text{Cost} = \dfrac{\text{selling price}}{100\% + \text{rate}} = \dfrac{\$24.30}{100\% + 35\%} = \dfrac{\$24.30}{135\%} = \dfrac{\$24.30}{1.35} = \$18.00$

Once the cost is known, the markup is easily found.

Markup = selling price − cost
$= \$24.30 - 18.00 = \$6.30$

Work the following problems.

1. A chair sells for $278.46 and is marked up 56% based on cost. Calculate the cost and markup.
2. A desk sells for $349.83 and is marked up 38% based on cost. Find the cost and markup.

Check your work in **32**.

32

1. $\text{Cost} = \dfrac{\text{selling price}}{100\% + \text{rate}} = \dfrac{\$278.46}{100\% + 56\%} = \dfrac{\$278.46}{156\%}$

$= \dfrac{\$278.46}{1.56} = \$178.50$

Markup = selling price − cost
$= \$278.46 - 178.50 = \$99.96$

2. $\text{Cost} = \dfrac{\text{selling price}}{100\% + \text{rate}} = \dfrac{\$349.83}{100\% + 38\%} = \dfrac{\$349.83}{138\%}$

$= \dfrac{\$349.83}{1.38} = \$253.50$

Markup = selling price − cost
$= \$349.83 - 253.50 = \$96.33$

This may seem like a lot to remember, but there are only two basic formulas. The first formula

$$\text{Markup} = \text{rate} \times \text{cost}$$

or one of its equivalent forms is used in three of the four different problems. In the case where the first formula will not work, use the second formula based on cost

$$\text{Cost} = \frac{\text{selling price}}{100\% + \text{rate}}$$

It's really easy.

MARKUP BASED ON SELLING PRICE

The second basic method to calculate markup is *based on selling price*. This method is used by many retailers.

There are four types of markup problems using markup based on selling price. Three of the four problems use the fundamental markup formula based on selling price

$$\text{Markup} = \text{rate} \times \text{selling price}$$

This is a percent equation where the base is the selling price, the rate is the percent, and the markup is the percentage.

Finding the Markup When the Selling Price and Rate Are Known

The easiest problem is calculating the markup when the selling price and rate are known.

Example: A synthetic sleeping bag sells for $125.90. If the rate of markup is 30% based on selling price, what is the markup and cost?

Rate   Selling price

$$\text{Markup} = 30\% \times \$125.90 = 0.30 \times \$125.90 = \$37.77$$

Once the markup is known, the cost is easily found.

$$\text{Cost} = \text{selling price} - \text{markup}$$
$$= \$125.90 - 37.77 = \$88.13$$

Your turn. Work the following problems.

1. A wastepaper basket sells for $7.95 and is marked up 45% based on selling price. What is the markup and cost?
2. A carabiner has a selling price of $4.00 and is marked up 34½% based on selling price. Calculate the markup and cost.

Check your work in 33.

33 1. $\text{Markup} = \text{rate} \times \text{selling price}$
$$= 45\% \times \$7.95 = 0.45 \times \$7.95$$
$$= \$3.5775 = \$3.58 \text{ rounded}$$

$\text{Cost} = \text{selling price} - \text{markup}$
$$= \$7.95 - 3.58 = \$4.37$$

2. Markup = rate × selling price
   $$= 34\tfrac{1}{2}\% \times \$4.00 = 34.5\% \times \$4.00$$
   $$= 0.345 \times \$4.00 = \$1.38$$

   Cost = selling price − markup
   $$= \$4.00 - 1.38 = \$2.62$$

Finding the Selling
Price When the Markup
and Rate Are Known

A second type of problem is calculating the selling price when the markup and rate of markup based on selling price are known. In the formula,

$$\text{markup} = \text{rate} \times \text{selling price}$$

we must solve for the selling price or total. This is a *T*-type percent problem with selling price as the base.

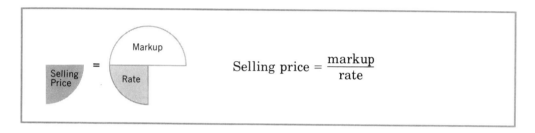

$$\text{Selling price} = \frac{\text{markup}}{\text{rate}}$$

Example: A record is marked up $2.20, which is 40% based on selling price. What is the selling price and cost?

$$\text{Selling price} = \frac{\text{markup}}{\text{rate}} = \frac{\$2.20}{40\%} = \frac{\$2.20}{0.40} = \$5.50$$

Cost = selling price − markup
$$= \$5.50 - 2.20 = \$3.30$$

Work the following problems.

1. A sewing machine is marked up $157.50, which is 35% based on selling price. Calculate the selling price and cost.
2. An automobile starter is marked up $8.28. If the rate of markup is 18% based on selling price, find the selling price and cost.

Our solutions are in 34.

---

34   1. $\text{Selling price} = \dfrac{\text{markup}}{\text{rate}} = \dfrac{\$157.50}{35\%} = \dfrac{\$157.50}{0.35} = \$450$

   Cost = selling price − markup
   $$= \$450 - 157.50 = \$292.50$$

   2. $\text{Selling price} = \dfrac{\text{markup}}{\text{rate}} = \dfrac{\$8.28}{18\%} = \dfrac{\$8.28}{0.18} = \$46.00$

   Cost = selling price − markup
   $$= \$46.00 - 8.28 = \$37.72$$

Finding the Rate When
the Markup and Selling
Price Are Known

The third type of problem is calculating the rate of markup based on selling price when the markup and selling price are known. This is a %-type percent problem.

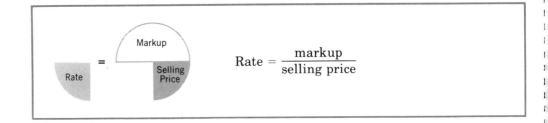

$$\text{Rate} = \frac{\text{markup}}{\text{selling price}}$$

Example: A calculator is marked up $3.85 and sells for $17.50. Calculate the rate of markup based on selling price.

$$\text{Rate} = \frac{\text{markup}}{\text{selling price}} = \frac{\$3.85}{\$17.50} = 0.22 = 22\%$$

Work the following problems.

1. A desk costs $53.97 and sells for $89.95. What is the rate of markup based on selling price? (**Hint:** First find the markup.)
2. A dress is marked up $25.68 and sells for $42.80. Calculate the rate of markup based on selling price.

Check your answers in **35**.

---

35 1. Markup = selling price − cost
$$= \$89.95 - 53.97 = \$35.98$$

$$\text{Rate} = \frac{\text{markup}}{\text{selling price}} = \frac{\$35.98}{\$89.95} = 0.40 = 40\%$$

2. $$\text{Rate} = \frac{\text{markup}}{\text{selling price}} = \frac{\$25.68}{\$42.80} = 0.60 = 60\%$$

In Problem 1, the markup must be known in order to calculate the rate. When the selling price and the cost are given, the markup must be calculated first. Also, if the cost and markup are known, the selling price must be found.

Finding the Selling Price When the Cost and Rate Are Known

The last type of markup problem involves calculating the selling price when the cost and rate of markup based on selling price are known. A new formula is necessary for this problem.

$$\text{Selling price} = \frac{\text{cost}}{100\% - \text{rate}}$$

This formula is derived algebraically in the box on page 187.

Example: A new business math book costs the bookstore $9.24 and is marked up 23% based on selling price. Calculate the selling price and markup.

$$\text{Selling price} = \frac{\text{cost}}{100\% - \text{rate}} = \frac{\$9.24}{100\% - 23\%} = \frac{\$9.24}{77\%}$$

$$= \frac{\$9.24}{0.77} = \$12.00$$

Markup = selling price − cost
= $12.00 − 9.24 = $2.76

Your turn. Work the following problems.

1. A microwave oven costs $260.64 and is marked up 28% based on selling price. Find the selling price and markup.
2. A microcomputer costs $5677.52 and is marked up 16% based on selling price. Calculate the selling price and markup.

Our work is in 36.

For all you algebraically-minded people, here is the derivation of the formulas in frames 31 and 35.

**FORMULA FOR FINDING THE COST WHEN
THE SELLING PRICE AND RATE BASED ON COST ARE KNOWN**

Selling price = cost + markup

Markup = rate × cost

Substitute the bottom equation into the top equation.

Selling price = cost + (rate × cost) = cost × 1 + cost × rate

Selling price = cost × (1 + rate)

Selling price = cost × (100% + rate)   (since rate is a percent use 100% for 1)

$$\frac{\text{selling price}}{100\% + \text{rate}} = \text{cost}$$

**FORMULA FOR FINDING THE SELLING PRICE
WHEN THE COST AND RATE BASED ON SELLING PRICE ARE KNOWN**

Cost = selling price − markup

Markup = rate × selling price

Substitute the bottom equation into the top equation.

Cost = selling price − (rate × selling price)

Cost = selling price × 1 − (selling price × rate)

Cost = selling price × (1 − rate)

Cost = selling price × (100% − rate)

$$\frac{\text{cost}}{100\% - \text{rate}} = \text{selling price}$$

36

1. Selling price $= \dfrac{\text{cost}}{100\% - \text{rate}} = \dfrac{\$260.64}{100\% - 28\%} = \dfrac{\$260.64}{72\%}$

$= \dfrac{\$260.64}{0.72} = \$362.00$

Markup = selling price − cost
$= \$362.00 - 260.64 = \$101.36$

2. Selling price $= \dfrac{\text{cost}}{100\% - \text{rate}} = \dfrac{\$5677.52}{100\% - 16\%} = \dfrac{\$5677.52}{84\%}$

$= \dfrac{\$5677.52}{0.84} = \$6758.95$ rounded

Markup = selling price − cost
$= \$6758.95 - 5677.52 = \$1081.43$

> **Important**

When working markup problems first, always check whether the markup is based on cost or based on selling price. Only the two formulas based on cost can be used on the markup based on cost problems. Also, only the two formulas based on selling price can be used on markup based on selling price problems. Here are the formulas:

Markup Based on Cost	**Markup Based on Selling Price**
Markup = rate × cost	Markup = rate × selling price
Cost $= \dfrac{\text{selling price}}{100\% + \text{rate}}$	Selling price $= \dfrac{\text{cost}}{100\% - \text{rate}}$

For a set of practice problems on markup, go to frame 37.

5 Mathematics of Buying and Selling

PROBLEM SET 4

37

5.4 Markup

The answers are on page 581.

A. Find the unknown.

	Cost	Markup	Selling Price
1.	$142.76	$29.35	———
2.	$18.95	$2.50	———
3.	$284.72	———	$328.48
4.	———	$10.25	$86.43
5.	$532.85	———	$632.11
6.	$8.50	$1.29	———
7.	———	$2.49	$38.91
8.	———	$109.27	$923.52
9.	$126.98	———	$150.12
10.	$241.15	$52.88	———
11.	$89.50	———	$105.29
12.	———	$114.63	$457.20
13.	$24.89	$12.55	———
14.	———	$23.89	$98.72
15.	———	$14.89	$56.32
16.	$516.27	———	$829.37
17.	$32.19	$25.50	———
18.	$562.50	———	$927.75
19.	———	$12.98	$27.25
20.	$355.82	———	$562.79

Date

Name

Course/Section

B. Find the unknowns. Use markup based on cost.

	Cost	Markup	Selling Price	Rate
1.	$25.00	——	——	15%
2.	$250.00	——	——	17%
3.	——	$44.20	——	26%
4.	——	$17.50	——	35%
5.	$300.00	$141.00	——	——
6.	$725.00	——	$833.75	——
7.	——	——	$476.00	19%
8.	——	——	$248.50	42%
9.	——	$54.64	——	23%
10.	$237.50	——	——	20%
11.	$487.50	$146.25	——	——
12.	——	——	$204.93	15%
13.	——	$29.56	——	18%
14.	$384.50	——	$522.92	——
15.	——	——	$260.85	25%
16.	$156.26	——	——	32%
17.	$152.75	——	——	24%
18	$182.95	$102.45	——	——
19.	$62.50	——	$90.00	——
20.	$29.50	——	——	44%
21.	——	——	$346.89	86%
22.	——	$608.83	——	72%
23.	——	——	$346.68	35%
24.	——	——	$20.65	18%
25.	——	——	$973.56	40%

C. Find the unknowns. Use markup based on selling price.

	Cost	Markup	Selling Price	Rate
1.	——	——	$289.96	20%
2.	——	——	$56.50	18%
3.	——	$40.14	——	16%
4.	——	$39.76	——	26%
5.	——	$36.11	$157.00	——
6.	——	$73.71	$245.70	——
7.	$29.61	——	——	30%
8.	$76.88	——	——	37%
9.	——	$20.05	——	38%
10.	——	——	$945.58	40%
11.	——	$19.80	$82.50	——
12.	$26.16	——	——	20%
13.	$94.17	$34.83	——	——
14.	$99.60	——	——	60%
15.	——	——	$18.50	24%
16.	——	$42.89	——	30%
17.	——	——	$289.70	36%
18.	——	$438.29	$952.80	——
19.	——	——	$92.50	22%
20.	$805.65	——	——	18%
21.	$511.68	$341.12	——	——
22.	——	$162.36	——	45%
23.	$557.09	——	——	15%
24.	$171.47	——	——	35%
25.	——	$129.02	——	40%

When you have had the practice you need, either return to the preview test on page 149 or continue in 38 with markdown, sales tax, and excise tax.

Date

Name

Course/Section

38

> **OBJECTIVE**
>
>
> Calculate markdown, markdown rate, sales tax, and excise tax.

Markdown is simply a discount on the selling price.

Oftentimes, it is necessary for a business to reduce the price or *mark down* its merchandise. This may be necessary to keep prices competitive, to move old merchandise, or to close out a line of merchandise due to changes in style or models.

If the markdown rate is given, the markdown is calculated the same as a discount.

> Markdown = rate × selling price
>
> Reduced selling price = selling price − markdown

Example: A new printing calculator originally sold for $129.90 but has been marked down 30%. Calculate the markdown and the reduced selling price.

Markdown = 30% × $129.90 = 0.30 × $129.90 = $38.97

Reduced selling price = $129.90 − 38.97 = $90.93

Try a few.

1. A new TV lists for $575.68 but is marked down 18%. What is the markdown and reduced selling price?
2. A coffee maker originally sold for $29.95 but has been marked down 28%. Calculate the markdown and sale price.

Check your work in 39.

39

1. Markdown = 18% × $575.68 = 0.18 × $575.68
   = $103.6224 = $103.62 rounded

   Reduced selling price = $575.68 − 103.62 = $472.06

2. Markdown = 28% × $29.95 = 0.28 × $29.95
   = $8.386 = $8.39 rounded

   Reduced selling price = $29.95 − 8.39 = $21.56

A second type of markdown problem involves calculating the markdown rate when the markdown is known. This is simply another percent problem. It is a %-type percent problem.

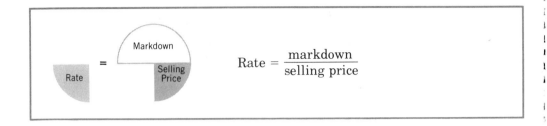

$$\text{Rate} = \frac{\text{markdown}}{\text{selling price}}$$

Example: A chair originally sold for $65 but has been marked down $13. Calculate the markdown rate.

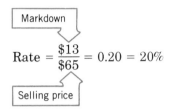

$$\text{Rate} = \frac{\$13}{\$65} = 0.20 = 20\%$$

Work the following problems.

1. A lamp originally sold for $25 but has been marked down $3. Find the rate of markdown.
2. An automobile's original price was $6545 but has been marked down to $6021.40. Calculate the markdown rate. (**Hint:** First find the markdown.)

Our work is in **40**.

40
1. $\text{Rate} = \dfrac{\text{markdown}}{\text{selling price}} = \dfrac{\$3}{\$25} = 0.12 = 12\%$

2. Markdown = selling price − reduced price
$\qquad\quad = \$6545 - 6021.40 = \$523.60$

$\text{Rate} = \dfrac{\text{markdown}}{\text{selling price}} = \dfrac{\$523.60}{\$6545} = 0.08 = 8\%$

SALES TAX

Most states and many cities and counties levy a sales tax. The sales tax is collected by the business from the buyer at the time of purchase. Manufacturers, wholesalers, and suppliers selling merchandise to other businesses for resale do not collect sales tax. The sales tax is only collected at the last sale—the sale to the consumer. This is usually done at the retail store.

Sales tax calculations may be made using a table or simply by multiplying by the tax rate.

Sales tax = rate × purchase price

The answer is rounded to the nearest cent.

Example: A file cabinet was purchased for $59.95. If the sales tax rate is 6%, calculate the sales tax and the total price.

Sales tax = 6% × $59.95 = 0.06 × $59.95
= $3.597 = $3.60 rounded

Total price = $59.95 + 3.60 = $63.55

Try the following problems.

1. What is the sales tax and total price on a purchase of $357.84 with a tax rate of $5\frac{1}{2}\%$?
2. A textbook was purchased for $11.95. If the sales tax rate is 6%, calculate the sales tax and the total price.

Check your work in **41**.

---

41 1. Sales tax = $5\frac{1}{2}\%$ × $357.84 = 5.5% × $357.84
= 0.055 × $357.84 = $19.6812 = $19.68 rounded

Total price = $357.84 + 19.68 = $377.52

2. Sales tax = 6% × $11.95 = 0.06 × $11.95
= $0.717 = $0.72 rounded

Total price = $11.95 + 0.72 = $12.67

EXCISE TAX

An *excise tax* is a sales tax levied by the federal government on the manufacture, sale, or consumption of a commodity. Gasoline, tires, and jewelry are a few of the items subject to an excise tax.

Excise tax is calculated by the same method as sales tax.

Excise tax = rate × price

Try the following problem.

A new automobile sells for $5765.76. If the excise tax rate is 7%, calculate the excise tax.

Check your calculation in **42**.

What is excise tax?
...a tax on jogging?

No, it's not an exercise tax. An excise tax is a federal sales tax. How would you tax joggers, by the mile or by the blister?

42 Excise tax = 7% × $5765.76 = 0.07 × $5765.76
= $403.6032 = $403.60 rounded

Markdown, sales tax, and excise taxes are all applications of percent. If you are still having trouble with these calculations, turn to page 141 for a review of percent problems.

Otherwise, go to 43 for a set of practice problems on markdown, sales tax, and excise tax.

Mathematics of Buying and Selling

43

5.5 Markdown, Sales
Tax, and Excise Tax

The answers are on page 581.

A. Calculate the markdown and reduced price.

	Selling Price	Rate	Markdown	Reduced Price
1.	$249.75	20%	——	——
2.	$115.50	17%	——	——
3.	$72.90	12%	——	——
4.	$356.00	25%	——	——
5.	$19.95	18%	——	——
6.	$129.64	35%	——	——
7.	$89.50	40%	——	——
8.	$22.70	5%	——	——
9.	$856.50	23%	——	——
10.	$2965.50	37%	——	——
11.	$297.82	$27\frac{1}{2}\%$	——	——
12.	$452.95	32%	——	——
13.	$18.50	$22\frac{1}{2}\%$	——	——
14.	$29.92	17%	——	——
15.	$237.80	15%	——	——
16.	$755.89	30%	——	——
17.	$72.21	35%	——	——
18.	$89.30	38%	——	——
19.	$255.85	55%	——	——
20.	$195.32	28%	——	——

Date

Name

Course/Section

B. Calculate the unknowns.

	Original Price	Reduced Price	Markdown	Rate
1.	$250.00	————	$67.50	———
2.	167.00	————	21.71	———
3.	————	$197.44	49.36	———
4.	150.00	————	22.50	———
5.	267.50	160.50	————	———
6.	214.75	————	51.54	———
7.	————	281.73	93.91	———
8.	————	176.00	99.00	———
9.	856.50	582.42	————	———
10.	19.70	————	3.94	———

C. Calculate the tax and total price.

	Selling Price	Tax Rate	Tax	Total Price
1.	$14.50	6%	————	————
2.	75.00	2%	————	————
3.	187.56	$2\frac{1}{2}\%$	————	————
4.	249.65	5%	————	————
5.	836.76	6%	————	————
6.	125.75	3%	————	————
7.	34.98	$9\frac{3}{4}\%$	————	————
8.	453.16	3%	————	————
9.	122.48	5%	————	————
10.	32.72	6%	————	————
11.	9.56	$5\frac{1}{4}\%$	————	————
12.	735.42	4%	————	————

When you have had the practice you need, turn to **44** for the Self-Test on this unit.

5 Mathematics of Buying and Selling

44 Self-Test

The answers are on page 581.

1. A table lists for $275 and is sold with trade discounts of 20/10/7½. Calculate: (a) the single equivalent discount rate and (b) net cost.

   (a)

   (b)

2. Calculate the net cost.

	Invoice Date	Invoice Amount	Terms	Goods Received	Date Paid	Net Cost
(a)	2/5	$237.80	2/10, n/30	2/8	3/2	———
(b)	5/14	57.65	3/10, 2/30, n/60	5/19	6/5	———
(c)	8/4	842.97	2/10ROG	9/17	9/25	———
(d)	10/27	135.80	3/10prox	10/30	12/7	———
(e)	4/23	23.17	2/10EOM	4/25	5/6	———
(f)	7/17	562.41	2/10-30X	7/21	8/20	———

3. A furniture company made the following purchases of sofas.

January 19	75	@	$260
March 27	50	@	300
July 5	125	@	310
October 14	50	@	320

   The year-end inventory revealed 79 sofas in stock. Calculate the inventory value by the following methods:

Date _____

Name _____

Course/Section _____

(a) Average cost method

(b) FIFO method

(c) LIFO method

4. The furniture store in Problem 3 found the following sofas in their year-end inventory.

Number	Cost
12	$260
15	300
17	310
35	320
79	

Calculate the inventory value by the specific identification method.

5. Calculate the unknowns. Use markup based on cost.

	Cost	Markup	Selling Price	Rate
(a)	$247.85	$49.57	——	——
(b)	——	56.16	——	32%
(c)	——	——	$15.93	8%
(d)	150.00	——	——	27%

6. Calculate the unknowns. Use markup based on selling price.

	Cost	Markup	Selling Price	Rate
(a)	———	$99.98	$249.95	———
(b)	———	———	18.50	18%
(c)	$647.87	———	———	26%
(d)	———	49.99	———	20%

7. An automobile originally listed for $6745.26 but has been marked down 8%. Find the new price.

   _____

8. If the tax rate is 6%, what is the total price on a purchase of $237.47?

   _____

6 Payroll

I'M GOING TO PAY YOU
WITH CASH, SO I DON'T
HAVE TO HASSLE WITH
ALL THAT PAPERWORK!

Unfortunately, most businesses don't pay employees in cash. Those which do pay in cash are not totally removed from some paperwork. The federal government requires quarterly and annual reports from businesses concerning employees' records of wages and taxes withheld.

This chapter is designed to help individuals understand how the varying amounts of federal income tax, FICA, and other deductions are computed based on the gross pay. Once these withholdings have been made, the net pay can be determined.

So, roll up your sleeves, grab a pencil (and a calculator, if you have one), and get ready for some "taxing" but fun work!

UNIT OBJECTIVE

After you successfully complete this unit, you will be able to calculate wages using salary, hourly, piecework, differential piecework, straight commission, sliding scale commission, and salary plus commission. You will also be able to complete a payroll sheet, employees' earning record and quarterly payroll summary. In addition, you will also be able to calculate federal income tax and FICA.

PREVIEW 6   This is a sample test for Unit 6. All the answers are placed immediately after the test. Work as many of the problems as you can, then check your answers and look after the answers for further directions. Use the tax tables in the unit.

		Where to Go for Help	
		Page	Frame
1.	Minyon Baker earns an annual salary of $11,895. What is her biweekly salary?	207	1

203

2. The Levy Clothing Company pays its employees on a differential piecework basis for sewing zippers using the following schedule:

$$1-400 \quad @ \quad \$0.05$$
$$401-500 \quad @ \quad 0.06$$
$$\text{Over } 500 \quad @ \quad 0.075$$

What is Gene's wage for a day in which he had sewn 547 zippers? _____ 215 10

3. A salesman is paid a weekly salary of $150 plus a 3% commission on his total sales. What is his gross pay for a week in which he sold $3457.92? _____ 215 10

4. Complete the following hourly payroll sheet. All employees are paid "time and a half" for hours over 40 per week. 207 1

Name	M	T	W	T	F	Total Hours	Regular Hours	Regular Rate	O.T. Hours	O.T. Rate	Regular Pay	O.T. Pay	Gross Pay
Creger, Richard	8	7	8	7	7			4.16					
Deleza, Maryhelen	8	9	9	8	9			4.25					
McKinney, Michael	8	10	10	9	9			3.98					
Robinson, Dora	8	8	8	8	8			3.88					
TOTALS													

5. Complete the following employee earning record. Complete the first quarter totals. Use the percentage method to compute the federal income tax. 223 19

Name   Martinez, Michael   Social Security No.   468-68-1351

Address   3205 Elm Street   Marital Status   M   No. of Exemptions   3

Pay Period	Gross Pay	Federal Income Tax	FICA	Other Deductions	Total Deductions	Net Pay
Jan.	857.95			42.55		
Feb.	895.40			43.78		
March	885.56			43.27		
First quarter						

Preview 6
Answers

1. $457.50

2. $29.53

3. $253.74

4.

Name	M	T	W	T	F	Total Hours	Regular Hours	Regular Rate	O.T. Hrs.	O.T. Rate	Regular Pay	O.T. Pay	Gross Pay
Creger, Richard	8	7	8	7	7	37	37	4.16	0	—	153.92	0	153.92
Deleza, Maryhelen	8	9	9	8	9	43	40	4.25	3	6.375	170.00	19.13	189.13
McKinney, Michael	8	10	10	9	9	46	40	3.98	6	5.97	159.20	35.82	195.02
Robinson, Dora	8	8	8	8	8	40	40	3.88	0	—	155.20	0	155.20
TOTALS											638.32	54.95	693.27

5.

Pay Period	Gross Pay	Federal Income Tax	FICA	Other Deductions	Total Deductions	Net Pay
Jan.	857.93	48.96	57.48	42.55	148.99	708.94
Feb.	895.40	53.45	59.99	43.78	157.22	738.18
March	885.56	52.27	59.33	43.27	154.87	730.69
First Quarter	2638.89	154.68	176.80	129.60	461.08	2177.81

★ If you cannot work any of these problems, start with frame 1 on page 207.
★ If you missed some of the problems, turn to the page numbers indicated.
★ If all of your answers were correct, you are probably ready to proceed to Unit 7. (If you would like more practice on Unit 6 before turning to Unit 7, try the Self-Test on page 247.
★ Super-students—those who want to be certain they learn all of this material—will turn to frame 1 and begin work there.

SECTION 1: SALARY AND HOURLY

1

> **OBJECTIVES**
>
> Given an annual salary, compute the salary per pay period.
>
> Compute hourly wage, including overtime.

Salary   The most common method of paying white collar and management personnel is by salary. *Straight salary* is a fixed amount of money paid to an employee for certain assigned duties.

The number of hours worked or the productivity of the employee do not affect the salary, although they probably do affect the employee's continuation and possible promotion or increase in pay.

Another method of paying an employee, covered in Section 2, involves salary plus commission.

Salaries are often stated as an amount per year; however, few people are paid only once a year. Most annual salaries are converted to pay periods that are more frequent.

Try the following problem, calculating monthly salary from annual salary.

Keith is paid $10,365 per year. What is his monthly salary?

Check your answer in 2.

---

2   To find Keith's monthly salary, simply divide the annual salary by the number of months in a year—12.

$$\frac{\$10,365}{12} = \$863.75$$

Salary is usually paid according to a regular schedule.

Common pay periods are:

Weekly	52 times per year
Biweekly	Every other week or 26 times per year
Semimonthly	Twice a month or 24 times per year
Monthly	12 times per year

Work the following problem.

Judy's annual salary is $10,062. Compute her salary payments if she is paid (a) weekly, (b) biweekly, (c) semimonthly, or (d) monthly.

Go to 3 to check your work.

---

3   (a) Weekly: $\dfrac{\$10,062}{52} = \$193.50$

(b) Biweekly: $\dfrac{\$10,062}{26} = \$387.00$

(c) Semimonthly: $\dfrac{\$10,062}{24} = \$419.25$

(d) Monthly: $\dfrac{\$10,062}{12} = \$838.50$

Hourly Pay   The most common method of paying wages is based on the number of hours worked. Straight *hourly wages* are very easy to calculate. The gross pay is the number of hours worked times the pay rate per hour.

> Gross pay = hours × rate per hour

Patti worked 38 hours last week. If her pay rate is $6.78 per hour, what is her gross pay?

Hours | Hourly rate

Gross pay = 38 × $6.78 = $257.64

Your turn. Work the following problem.

Tim earns $6.57 per hour. What is his gross pay for a week in which he worked $38\frac{1}{4}$ hours?

Turn to **4** for the answer.

I never seem to remember the difference between biweekly and semimonthly. Can you help me?

I'll try. Bi means two, as in bicycle, so biweekly means every two weeks.

Semi means half as in semicircle or semipro, so semimonthly means every half a month, or twice a month. Remember bicycle and semicircle.

Better?

4 $38\frac{1}{4}$ × $6.57 = 38.25 × $6.57 = $251.3025 = $251.30 rounded

The easiest way to work this problem is to first change $38\frac{1}{4}$ to a decimal form, $38\frac{1}{4}$ = 38.25, then multiply. If you need a review of changing fractions to decimal numbers, return to page 111.

Most businesses use a payroll sheet to compute gross pay. The payroll sheet for straight hourly wages will include the employee's name, Social Security number, number of hours worked per day, total hours for the week, rate per hour, gross pay, and total gross pay.

The following is a sample payroll sheet. Compute the payroll sheet by calculating the total hours per week and the gross pay for each employee. Total the gross pay column vertically to complete the payroll sheet.

Name	Hours M T W T F	Total Hours	Rate per Hour	Gross Pay
Ewing, Vernon	7 8 6 7 8		3.00	
Holbert, Barbara	8 8 8 8 7		3.29	
Kolakowski, Sally	8 8 8 8 8		2.58	
Martinez, Angel	8 6 8 8 7		3.18	
Richmond, Ralph	6 8 7 7 8		2.94	
TOTAL				

Check your work in 5.

5

Name	Hours M T W T F	Total Hours	Rate per Hour	Gross Pay
Ewing, Vernon	7 8 6 7 8	36	3.00	108.00
Holbert, Barbara	8 8 8 8 7	39	3.29	128.31
Kolakowski, Sally	8 8 8 8 8	40	2.58	103.20
Martinez, Angel	8 6 8 8 7	37	3.18	117.66
Richmond, Ralph	6 8 7 7 8	36	2.94	105.84
TOTAL				563.01

Overtime   Employees are paid at a higher rate per hour for overtime. *Overtime* is the time worked over 40 hours per week. The usual overtime rate is $1\frac{1}{2}$ times the regular rate.

> Overtime rate = $1\frac{1}{2}$ × regular rate

For example, if the regular rate is $3.58, the overtime rate would be:

$1\frac{1}{2}$ × $3.58 = $5.37   Again, write $1\frac{1}{2}$ as 1.5 and multiply.
       = 1.5 × $3.58 = $5.37

For employees working over 40 hours per week, four steps are necessary to compute their gross pay.

Step 1. **Compute the regular pay.**

Regular pay = regular hours × regular rate per hour
(Note that the regular hours cannot exceed 40 for a week.)

Step 2. **Compute the overtime rate.**

Overtime rate = $1\frac{1}{2}$ × regular rate

Step 3. **Compute the overtime pay.**

Overtime pay = overtime hours × overtime rate

Step 4. **Compute the total wages.**

Total wages = regular pay + overtime pay

It's really quite easy. Let's try an example:

Pauline earns $3.86 per hour. Last week she worked 46 hours. Compute her regular pay, overtime rate, overtime pay, and total wages.

Step 1. Regular pay = 40 × $3.86 = $154.40

Step 2. Overtime rate = $1\frac{1}{2}$ × $3.86 = 1.5 × $3.86 = $5.79

Step 3. Overtime pay = 6 × $5.79 = $34.74

Step 4. Total wages = $154.40 + 34.74 = $189.14

Now you try one.

Dave worked 45 hours last week. His regular pay rate is $3.88 per hour. What is his total wage?

Work the problem carefully and check your work in **6**.

---

6 Regular pay = 40 × $3.88 = $155.20

Overtime rate = $1\frac{1}{2}$ × $3.88 = $5.82

Overtime pay = 5 × $5.82 = $29.10

Total wage = $155.20 + 29.10 = $184.30

Be Careful ⟩ When computing the overtime rate, you can end up with more than two decimal places. Carry all the decimals places and round only after you compute the overtime pay.

Work the following problem. Be careful with the decimal places in the overtime rate.

Anna earns $3.19 per hour. Compute her gross pay for a week in which she worked 47 hours.

Turn to **7**.

Regular pay = 40 × $3.19 = $127.60

Be Careful ⟩ Overtime rate = $1\frac{1}{2}$ × $3.19 = $4.785 ◀——— Don't round here.

Overtime pay = 7 × $4.785 = $33.495 = $33.50   Rounded.

Gross pay = $127.60 + 33.50 = $161.10

We can easily incorporate the overtime calculations in the payroll sheet by adding columns for overtime hours, overtime rate, overtime pay, and gross pay.

Complete the following payroll sheet. Each employee's calculations should be made as in the previous problem. Add columns to find the totals for regular pay, overtime pay, and gross wages.

Name	M T W T F	Total Hours	Reg-ular Hours	Reg-ular Rate	O.T. Hours	O.T. Rate	Reg-ular Pay	O.T. Pay	Gross Pay
Briggs, Courtney	8 9 7 9 9			3.56					
Henderson, Chester	10 8 8 9 9			2.98					
Kellerman, Mandy	10 9 10 9 8			3.79					
Tan, Richard	8 8 7 8 8			3.42					
Underwood, Carrie	8 9 8 10 8			3.16					
TOTALS									

Check your work in **8**.

8

Name	M T W T F	Total Hours	Reg-ular Hours	Reg-ular Rate	O.T. Hours	O.T. Rate	Reg-ular Pay	O.T. Pay	Gross Pay
Briggs, Courtney	8 9 7 9 9	42	40	3.56	2	5.34	142.40	10.68	153.08
Henderson, Chester	10 8 8 9 9	44	40	2.98	4	4.47	119.20	17.88	137.08
Kellerman, Mandy	10 9 10 9 8	46	40	3.79	6	5.685	151.60	34.11	185.71
Tan, Richard	8 8 7 8 8	39	39	3.42	0	—	133.38	0.00	133.38
Underwood, Carrie	8 9 8 10 8	43	40	3.16	3	4.74	126.40	14.22	140.62
TOTALS							672.98	76.89	749.87

Careful with the third employee above. The overtime rate is $1\frac{1}{2} \times \$3.79 = \$5.685$. Do not round the overtime rate.

Check your addition by adding the total regular pay and total overtime pay. This should equal the total gross pay. $672.98 + 76.89 = 749.87$

 Accuracy is essential! Whether you compute the payroll by hand or with a calculator, you should always check your answers. An incorrect answer will cost either the business or an employee money; it may possibly cost a job. Be careful.

Turn to **9** for a set of practice problems.

6 Payroll

9   The answers are on page 582.

6.1 Salary and Hourly

A. Convert the following annual salaries to the required pay period.

1. $15,000 = ——————— monthly   2. $15,000 = ——————— semimonthly

3. $16,575 = ——————— monthly   4. $ 6,695 = ——————— biweekly

5. $10,153 = ——————— weekly   6. $12,390 = ——————— semimonthly

7. $ 9,165 = ——————— biweekly   8. $ 8,736 = ——————— weekly

9. $ 9,180 = ——————— monthly   10. $11,010 = ——————— semimonthly

11. $19,565 = ——————— biweekly   12. $14,326 = ——————— weekly

B. Convert the following annual salaries to the various pay periods.

	Annual	Monthly	Semimonthly	Biweekly	Weekly
1.	$17,550	———	———	———	———
2.	14,586	———	———	———	———
3.	15,366	———	———	———	———
4.	9,204	———	———	———	———
5.	11,388	———	———	———	———
6.	8,034	———	———	———	———
7.	12,090	———	———	———	———
8.	6,162	———	———	———	———
9.	18,330	———	———	———	———
10.	9,438	———	———	———	———
11.	18,720	———	———	———	———
12.	28,080	———	———	———	———
13.	18,500	———	———	———	———
14.	22,500	———	———	———	———
15.	28,350	———	———	———	———

Date

Name

Course/Section

C. Complete the weekly payroll sheet.

Name	M	T	W	T	F	Total Hours	Rate per Hour	Gross Pay
1. Dick, Raymond	8	7	8	7	6		$3.14	
2. Eccard, Art	7	8	5	6	4		2.95	
3. Hunter, Anna	8	8	8	8	8		3.45	
4. Mohr, Debbie	8	7	8	8	8		3.15	
5. Nichols, Don	8	7	6	5	5		3.72	
6. Spybuck, Garland	7	8	8	8	7		3.58	
7. Taylor, Vinita	6	8	8	8	0		3.25	
8. Watson, G. D.	8	7	7	7	6		3.89	
TOTAL								

D. Complete the weekly payroll sheet.

Name	M T W T F	Total Hours	Regular Hours	Regular Rate	O.T. Hours	O.T. Rate	Regular Pay	O.T. Pay	Gross Pay
1. Allgood, Harrell	9 8 10 8 9			$3.24					
2. Bertone, Marcus	9 9 9 9 8			2.96					
3. Edwards, Lawrence	8 8 8 8 8			3.46					
4. Gilmore, Marylina	9 8 8 9 10			3.50					
5. Hawkins, Rita	9 8 8 9 9			3.88					
6. Kimball, M. A.	8 7 8 8 6			3.76					
7. Parke, Robert	9 8 9 10 7			3.24					
8. Whitley, Debra	9 8 10 10 10			3.96					
TOTALS									

E. Complete the weekly payroll sheet.

Name	M T W T F	Total Hours	Regular Hours	Regular Rate	O.T. Hours	O.T. Rate	Regular Pay	O.T. Pay	Gross Pay
1. Canaan, Robbye	8 7 6 5 4			$5.92					
2. Culpepper, Lori	8 7 9 9 9			6.35					
3. Moore, Charlene	9 8 9 8 9			5.80					
4. Schein, Don	9 10 9 7 9			6.15					
5. Schein, Pat	9 10 10 9 10			6.45					
TOTALS									

When you have had the practice you need, either return to the Preview on page 203 or continue in 10 with the study of piecework and commission.

SECTION 2: PIECEWORK AND COMMISSION

10

> **OBJECTIVE**
>
> Compute gross pay using piecework, differential piecework, straight commission, sliding scale commission, and salary plus commission.

Piecework

Some businesses pay their employees based on actual production. One such method is by *piecework,* where each employee's pay is based on the number of pieces completed during his shift. The gross pay is the product of the number of pieces and the rate per piece.

> Gross pay = number of pieces × rate per piece

Use this technique in the following problem.

Galyn works for the Clean Climbing Company, manufacturing climbing nuts. If he earns $0.12 per climbing nut and completed 237 in a day, what is his pay?

Turn to 11.

11

Gross pay = 237 × $0.12 = $28.44

Piecework calculations can be easily incorporated in a payroll sheet. This payroll sheet will include the employee's name, number of pieces completed per day, total pieces completed per week, rate per piece, and gross pay.

Complete the following payroll sheet by finding the total pieces and gross pay. Don't forget to total the gross pay column.

Name	Pieces Completed M T W T F	Total Pieces	Rate	Gross Pay
Benton, Naomi	40 38 35 42 41		0.94	
Curtis, Roddy	35 33 38 40 36		1.15	
Lacefield, Larry	26 27 22 24 25		1.21	
Shultz, Caroline	56 56 57 58 55		0.54	
Whitfield, Alan	96 87 85 89 88		0.35	
TOTAL				

Check your work in 12.

Name	Pieces Completed M T W T F					Total Pieces	Rate	Gross Pay
Benton, Naomi	40	38	35	42	41	196	0.94	184.24
Curtis, Roddy	35	33	38	40	36	182	1.15	209.30
Lacefield, Larry	26	27	22	24	25	124	1.21	150.04
Shultz, Caroline	56	56	57	58	55	282	0.54	152.28
Whitfield, Alan	96	87	85	89	88	445	0.35	155.75
TOTAL								851.61

Differential Piecework

Differential piecework is one method used by some businesses as an incentive to increase production. This method uses a scale in which the rate per piece increases as the number of pieces completed increases. For example, the Gibson Box Works pays its employees for each box constructed according to the following schedule:

$$1-100 \text{ boxes} \quad @ \quad \$0.29 \text{ each}$$
$$101-150 \text{ boxes} \quad @ \quad \$0.31 \text{ each}$$
$$151 \text{ and up} \quad @ \quad \$0.34 \text{ each}$$

If Ray constructed 162 boxes in a day, compute his gross pay.

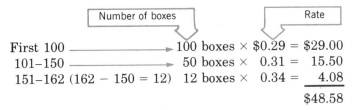

First 100 ──────────→ 100 boxes × $0.29 = $29.00
101–150 ──────────→ 50 boxes × 0.31 = 15.50
151–162 (162 − 150 = 12) 12 boxes × 0.34 = 4.08
                                            $48.58

Note. Ray is not paid at the $0.34 rate for all 162 boxes, only for those he constructs after he has completed 150 boxes.

Try the following problem.

The Ferchau Sign Painting Company pays its employees for each medium size sign by the following daily schedule:

$$1-12 \text{ signs} \quad @ \quad \$2.55 \text{ each}$$
$$13-16 \text{ signs} \quad @ \quad 2.65 \text{ each}$$
$$17 \text{ and up signs} \quad @ \quad 2.78 \text{ each}$$

Calculate Frank's gross wage if he painted 18 signs in one day.

Carefully work it out, then turn to 13.

---

13   First 12 signs ──────→ 12 × $2.55 = $30.60
Signs 13–16 ──────→ 4 × 2.65 = 10.60
Signs 17–18 ──────→ 2 × 2.78 = 5.56
Gross wages ──────────→ $46.76

In a differential piecework payroll sheet, the rate column is deleted, since the rate varies.

Steele's Cabinet Shop pays its employees for each cabinet made per week according to the following schedule:

$$1-25 \text{ cabinets} \quad @ \quad \$6.45 \text{ each}$$
$$26-30 \text{ cabinets} \quad @ \quad 6.67 \text{ each}$$
$$31 \text{ and up cabinets} \quad @ \quad 6.98 \text{ each}$$

Complete the following payroll sheet for Steele's Cabinet Shop.

Name	Cabinets per Day M T W T F	Total Cabinets	Gross Wages
Bullock, Janey	6 6 5 6 5		
Killian, Robert	6 5 5 5 4		
O'Leary, Jerry	6 7 7 6 7		
Sykora, Daniel	4 4 5 5 4		
TOTAL			

Turn to 14 to check your work.

14

Name	Cabinets per Day M T W T F	Total Cabinets	Gross Wages
Bullock, Janey	6 6 5 6 5	28	181.26
Killian, Robert	6 5 5 5 4	25	161.25
O'Leary, Jerry	6 7 7 6 7	33	215.54
Sykora, Daniel	4 4 5 5 4	22	141.90
TOTAL			699.95

Bullock, Janey

$25 \times \$6.45 = \$161.25$
$3 \times 6.67 = \underline{20.01}$
$\$181.26$

Killian, Robert

$25 \times \$6.45 = \$161.25$

O'Leary, Jerry

$25 \times \$6.45 = \$161.25$
$5 \times 6.67 = 33.35$
$3 \times 6.98 = \underline{20.94}$
$\$215.54$

Sykora, Daniel

$22 \times \$6.45 = \$141.90$

If you are having trouble with these arithmetic calculations, return to page 89 to review the addition and multiplication of decimal numbers.

Commission

Another procedure used for calculating wages is *commission*. A commission is usually paid to people selling, buying, or leasing merchandise. There are several different commission plans. The easiest to calculate is straight commission. In *straight commission* the commission is a percent of the total sales.

Commission = rate of commission × sales

Try the following problem:

Richard earns a 6% commission on his weekly sales. What was his commission for a week in which he sold $2847.65 in merchandise?

Check your answer in 15.

15

Change the percent

$6\% \times \$2847.65 = 0.06 \times \$2847.65 = \$170.859 = \$170.86$ rounded

Remember

Always change a percent to a decimal number before multiplying. If you are having difficulty changing 6% to a decimal, look at frame 7 on page 124.

Sliding Scale Commission

A second method of calculating commissions uses a *sliding scale*. Much like differential piecework, sliding scale commission rewards employees for increased production. The commission rate increases as sales increase. For example, the Wilton Sporting Good Supply Company pays its employees on the following weekly schedule:

4% on sales up to $3000
5% on sales from $3000 to $5000
6% on sales over $5000

What was Keith's gross wages for a week in which he had a sales total of $5473.86?

First $3000	$4\% \times \$3000 = 0.04 \times \$3000 = \$120.00$
$3000 to $5000	$5\% \times 2000 = 0.05 \times 2000 = 100.00$
Over $5000	$6\% \times 473.86 = 0.06 \times 473.86 = \underline{\quad 28.43}$
	$\$248.43$

Your turn. Try this problem:

The Greenwell Novelty Company pays its salespeople on the following monthly schedule:

7%   on sales up to $8000
7½% on sales from $8000 to $10,000
8¼% on sales from $10,000

Complete the following payroll sheet.

Name	Sales	Gross Wages
Brown, Leland	9,567.42	
Greenwell, Donald	7,526.98	
Morton, Laura	10,846.29	
TOTAL		

Check your work in 16.

16

Name	Sales	Gross Wages
Brown, Leland	9,567.42	677.56
Greenwell, Donald	7,526.98	526.89
Morton, Laura	10,846.29	779.82
TOTAL		1984.27

Brown, Leland

$7\% \times \$8000.00 = 0.07 \times \$8000.00 = \$560.00$
$7\frac{1}{2}\% \times 1567.42 = 0.075 \times 1567.42 = \underline{117.56}$ rounded

Total $= \$677.56$

Greenwell, Donald

$7\% \times \$7526.98 = 0.07 \times \$7526.98 = \$526.89$ rounded

Morton, Laura

$7\% \times \$8000.00 = 0.07 \times \$8000.00 = \$560.00$
$7\frac{1}{2}\% \times 2000.00 = 0.075 \times 2000.00 = 150.00$
$8\frac{1}{4}\% \times 846.29 = 0.0825 \times 846.29 = \underline{69.82}$ rounded

$\$779.82$

If you had trouble changing fractional percents such as $7\frac{1}{2}\%$ and $8\frac{1}{4}\%$ to decimal numbers, review this in frame 10 on page 126.

Salary Plus Commission

A third method of calculating commissions is *salary plus commission*. The salary assures employees of a regular income and the commission acts as an incentive plan. With this method, the commission is calculated and added to the salary to compute the total pay.

Try the following problem.

Rita's Boutique pays each salesperson a salary plus a commission of 2% on her total sales for the week. Complete the following payroll sheet.

Name	Weekly Salary	Sales	Commission	Gross Pay
Decker, Pauline	$125.50	$847.56		
Everett, Linda	117.75	502.45		
Garvin, Beverly	132.25	675.49		
Jones, Anita	121.40	517.49		
TOTAL				

To find the gross pay, calculate the commission and add it to the salary. Don't forget to total the gross pay column.

Check your work in 17.

17

Name	Weekly Salary	Sales	Commission	Gross Pay
Decker, Pauline	$125.50	$847.56	$16.95	$142.45
Everett, Linda	117.75	502.45	10.05	127.80
Garvin, Beverly	132.25	675.49	13.51	145.76
Jones, Anita	121.40	517.49	10.35	131.75
TOTAL				547.76

Be Careful Never multiply by a percent. *Always* change the percent to a decimal number first.

Accuracy is very important! *Always* check your answer whether you work it by hand or with a calculator. Calculators rarely malfunction, but fingers do. It's easy to hit the wrong keys–be sure to check your work.

Turn to 18 for a set of practice problems on piece rate and commission.

Be Careful

Payroll

18   The answers are on page 583.

6.2   Piecework and Commission

A.   Complete the following payroll sheet.

Name	Pieces Completed M T W T F	Total Pieces	Rate	Gross Pay
1. Anderson, Robert	65 72 71 68 66		$0.52	
2. Burns, Thomas	62 71 65 62 64		0.49	
3. Dutton, Charles	75 79 75 76 81		0.47	
4. Jones, Roberta	76 74 75 75 76		0.51	
5. Perry, Jacqulyn	42 45 45 46 43		0.75	
6. Soliz, Elizabeth	82 83 84 84 81		0.48	
7. Tuttle, Wesley	95 96 92 93 91		0.46	
8. Williams, Velma	25 26 26 25 24		1.25	
TOTAL				

B.   The Copeland Bearing Company pays its employees on a piecework basis using the following schedule:

1–300 bearings   @ $0.40 each
301–400 bearings   @   0.44 each
401 and up   @   0.49 each

Complete the following payroll sheet.

Name	Bearings Per Day M T W T F	Total	Gross Wages
1. Adeleke, Joel	65 62 68 69 61		
2. Hutte, Patsy	68 70 72 71 74		
3. Jones, Steve	62 64 63 62 64		
4. King, Richard	55 52 51 59 60		
5. Minor, Mark	69 70 75 76 74		
6. Myatt, Patti	82 81 85 87 82		
7. Reimler, Edward	75 76 77 77 76		
8. Tarrant, Betty	85 79 76 84 88		
TOTAL			

Date _____

Name _____

Course/Section _____

C. The Colson Realty Company pays its salespeople a straight 3% commission. Complete the company's monthly payroll sheet.

	Name	Total Sales	Gross Pay
1.	Biggs, Bobby	$25,700	
2.	Cao, Hung	42,500	
3.	Cummings, Sandra	35,500	
4.	Diglovanni, John	28,900	
5.	Jenkins, Dorothy	42,300	
6.	Morgan, Geraldine	39,950	
7.	Pack, Kathy	24,200	
8.	Walsh, Thomas	41,500	
	TOTAL		

D. The Martinez Office Products Company pays its salespeople on the following weekly schedule:

6% on sales up to $3000
7% on sales from $3000 to $4000
$8\frac{1}{2}$% on sales over $4000

Complete the weekly payroll sheet.

	Name	Total Sales	Gross Pay
1.	Brooks, Mendel	$4241.76	
2.	Hampton, Charles	2975.46	
3.	Johnson, Leonard	3152.34	
4.	Kauble, Cecelia	3851.85	
5.	Martin, Jim	3247.98	
6.	Norton, Thomas	4289.55	
7.	Sitter, David	2457.26	
8.	Whibbey, Betty	4557.86	
	TOTAL		

E. The Platt Speed Shop pays its salespeople a weekly salary plus a $2\frac{1}{2}$% commission on their sales. Complete the weekly payroll sheet.

	Name	Salary	Sales	Commission	Gross Pay
1.	Caldwell, Rita	$150	$547.98		
2.	Crownover, Vernon	145	425.56		
3.	Devilbiss, Virgil	150	398.26		
4.	Gaines, Russell	165	502.23		
5.	Houston, Larry	160	655.93		
6.	Morrell, Pamela	155	602.57		
7.	Platt, Bill	250	125.57		
8.	Weiner, Michael	170	509.98		
	TOTAL				

When you have had the practice you need, either return to the Preview on page 203 or continue in 20 with the study of federal income tax and FICA.

SECTION 3: FEDERAL INCOME TAX AND FICA

19

> **OBJECTIVE**
>
> Compute federal income tax by the wage bracket and percentage methods.
> Compute FICA.

Federal Income Tax

After an employee's gross wages are calculated, the employer must make various deductions in order to determine his or her "take-home" or net pay. One of the largest deductions, required by U.S. tax laws, is federal income tax.

To help employers compute the amount of federal income tax withholding, the Internal Revenue Service (IRS) supplies businesses with tax tables in Circular E.

Each employee is required to file a W4 form that declares the number of exemptions he claims. This form must be filed before income tax can be computed.

Below is a sample W4 form.

| Form **W-4** (Rev. December 1978) Department of the Treasury Internal Revenue Service | **Employee's Withholding Allowance Certificate** (Use for Wages Paid After December 31, 1978) This certificate is for income tax withholding purposes only. It will remain in effect until you change it. If you claim exemption from withholding, you will have to file a new certificate on or before April 30 of next year. |

Type or print your full name | **Your social security number**

Home address (number and street or rural route) | Marital Status | ☐ Single ☐ Married
City or town, State, and ZIP code | | ☐ Married, but withhold at higher Single rate
 | | Note: *If married, but legally separated, or spouse is a nonresident alien, check the single block.*

1 Total number of allowances you are claiming . . . .

2 Additional amount, if any, you want deducted from each pay (if your employer agrees) . . . . . . $

3 I claim exemption from withholding (see instructions). Enter "Exempt" . . . .

Under the penalties of perjury, I certify that the number of withholding allowances claimed on this certificate does not exceed the number to which I am entitled. If claiming exemption from withholding, I certify that I incurred no liability for Federal income tax for last year and I anticipate that I will incur no liability for Federal income tax for this year.

Signature ▶ _____ Date ▶ _____, 19_____

Go to 20 for the wage bracket method of computing income tax.

Wage Bracket Method

There are two principal methods of calculating federal income tax: the wage bracket and percentage methods. The percentage method is generally used by computers, because it requires less storage for tables.

20

The wage bracket is the easiest method to compute by hand for it only requires the use of a table. This method uses tables for weekly, biweekly, semimonthly, monthly, and daily payroll periods. For each payroll period, there is a separate table for single and married employees. This makes a total of 10 tables, weekly single, weekly married, biweekly single, biweekly married, and so on. In this section, on pages 225–228 we have included the tables for weekly single and weekly married.

To compute income tax by wage bracket method, use the following steps:

Step 1. Locate the correct payroll period and marital status table.

Step 2. Find the correct wage bracket in the two leftmost columns. This gives the correct row. When reading the column headings, be very careful. "But less than 280" means smaller than 280. That includes 279.99, but not 280.

Step 3. Move right in the row to the column with the correct number of exemptions. This figure is the income tax.

Madge Hall is single and declares one exemption. Calculate her federal income tax by the wage bracket method on her weekly wage of $279.36.

Step 1. Use the "single persons—weekly payroll period" table on pages 225–226.

Step 2. The current wage bracket is "at least 270, but less than 280," found on page 226.

Step 3. The correct tax amount in the 1 exemption column is $35.70.

It's easy. Now, you try one.

Fernando Hernandez earned $285.46 last week. He is married and declares four exemptions. Assume you are his employer, and compute his federal income tax by the wage bracket method.

Check your answer in 21.

SINGLE Persons—WEEKLY Payroll Period

(For Wages Paid After June 1983 and Before January 1985)

And the wages are—		And the number of withholding allowances claimed is—										
At least	But less than	0	1	2	3	4	5	6	7	8	9	10
		The amount of income tax to be withheld shall be—										
$0	$27	$0	$0	$0	$0	$0	$0	$0	$0	$0	$0	$0
27	28	.10	0	0	0	0	0	0	0	0	0	0
28	29	.20	0	0	0	0	0	0	0	0	0	0
29	30	.30	0	0	0	0	0	0	0	0	0	0
30	31	.40	0	0	0	0	0	0	0	0	0	0
31	32	.50	0	0	0	0	0	0	0	0	0	0
32	33	.70	0	0	0	0	0	0	0	0	0	0
33	34	.80	0	0	0	0	0	0	0	0	0	0
34	35	.90	0	0	0	0	0	0	0	0	0	0
35	36	1.00	0	0	0	0	0	0	0	0	0	0
36	37	1.10	0	0	0	0	0	0	0	0	0	0
37	38	1.30	0	0	0	0	0	0	0	0	0	0
38	39	1.40	0	0	0	0	0	0	0	0	0	0
39	40	1.50	0	0	0	0	0	0	0	0	0	0
40	41	1.60	0	0	0	0	0	0	0	0	0	0
41	42	1.70	0	0	0	0	0	0	0	0	0	0
42	43	1.90	0	0	0	0	0	0	0	0	0	0
43	44	2.00	0	0	0	0	0	0	0	0	0	0
44	45	2.10	0	0	0	0	0	0	0	0	0	0
45	46	2.20	0	0	0	0	0	0	0	0	0	0
46	47	2.30	0	0	0	0	0	0	0	0	0	0
47	48	2.50	.20	0	0	0	0	0	0	0	0	0
48	49	2.60	.30	0	0	0	0	0	0	0	0	0
49	50	2.70	.40	0	0	0	0	0	0	0	0	0
50	51	2.80	.50	0	0	0	0	0	0	0	0	0
51	52	2.90	.60	0	0	0	0	0	0	0	0	0
52	53	3.10	.80	0	0	0	0	0	0	0	0	0
53	54	3.20	.90	0	0	0	0	0	0	0	0	0
54	55	3.30	1.00	0	0	0	0	0	0	0	0	0
55	56	3.40	1.10	0	0	0	0	0	0	0	0	0
56	57	3.50	1.20	0	0	0	0	0	0	0	0	0
57	58	3.70	1.40	0	0	0	0	0	0	0	0	0
58	59	3.80	1.50	0	0	0	0	0	0	0	0	0
59	60	3.90	1.60	0	0	0	0	0	0	0	0	0
60	62	4.10	1.80	0	0	0	0	0	0	0	0	0
62	64	4.30	2.00	0	0	0	0	0	0	0	0	0
64	66	4.60	2.30	0	0	0	0	0	0	0	0	0
66	68	4.80	2.50	.20	0	0	0	0	0	0	0	0
68	70	5.00	2.70	.40	0	0	0	0	0	0	0	0
70	72	5.30	3.00	.70	0	0	0	0	0	0	0	0
72	74	5.50	3.20	.90	0	0	0	0	0	0	0	0
74	76	5.80	3.50	1.20	0	0	0	0	0	0	0	0
76	78	6.00	3.70	1.40	0	0	0	0	0	0	0	0
78	80	6.30	3.90	1.60	0	0	0	0	0	0	0	0
80	82	6.60	4.20	1.90	0	0	0	0	0	0	0	0
82	84	6.90	4.40	2.10	0	0	0	0	0	0	0	0
84	86	7.20	4.70	2.40	0	0	0	0	0	0	0	0
86	88	7.50	4.90	2.60	.30	0	0	0	0	0	0	0
88	90	7.80	5.10	2.80	.50	0	0	0	0	0	0	0
90	92	8.10	5.40	3.10	.80	0	0	0	0	0	0	0
92	94	8.40	5.60	3.30	1.00	0	0	0	0	0	0	0
94	96	8.70	5.90	3.60	1.20	0	0	0	0	0	0	0
96	98	9.00	6.10	3.80	1.50	0	0	0	0	0	0	0
98	100	9.30	6.40	4.00	1.70	0	0	0	0	0	0	0
100	105	9.80	6.90	4.50	2.10	0	0	0	0	0	0	0
105	110	10.50	7.60	5.10	2.70	.40	0	0	0	0	0	0
110	115	11.30	8.40	5.70	3.30	1.00	0	0	0	0	0	0
115	120	12.00	9.10	6.30	3.90	1.60	0	0	0	0	0	0
120	125	12.80	9.90	7.00	4.50	2.20	0	0	0	0	0	0
125	130	13.50	10.60	7.80	5.10	2.80	.50	0	0	0	0	0
130	135	14.30	11.40	8.50	5.70	3.40	1.10	0	0	0	0	0
135	140	15.00	12.10	9.30	6.40	4.00	1.70	0	0	0	0	0
140	145	15.80	12.90	10.00	7.10	4.60	2.30	0	0	0	0	0
145	150	16.50	13.60	10.80	7.90	5.20	2.90	.60	0	0	0	0
150	160	17.70	14.80	11.90	9.00	6.10	3.80	1.50	0	0	0	0
160	170	19.20	16.30	13.40	10.50	7.60	5.00	2.70	.40	0	0	0
170	180	20.70	17.80	14.90	12.00	9.10	6.20	3.90	1.60	0	0	0
180	190	22.20	19.30	16.40	13.50	10.60	7.70	5.10	2.80	.50	0	0
190	200	24.10	20.80	17.90	15.00	12.10	9.20	6.30	4.00	1.70	0	0
200	210	26.00	22.40	19.40	16.50	13.60	10.70	7.80	5.20	2.90	.60	0

And the wages are—		And the number of withholding allowances claimed is—										
At least	But less than	0	1	2	3	4	5	6	7	8	9	10
		The amount of income tax to be withheld shall be—										
$210	$220	$27.90	$24.30	$20.90	$18.00	$15.10	$12.20	$9.30	$6.50	$4.10	$1.80	$0
220	230	29.80	26.20	22.50	19.50	16.60	13.70	10.80	8.00	5.30	3.00	.70
230	240	31.70	28.10	24.40	21.00	18.10	15.20	12.30	9.50	6.60	4.20	1.90
240	250	33.60	30.00	26.30	22.70	19.60	16.70	13.80	11.00	8.10	5.40	3.10
250	260	35.50	31.90	28.20	24.60	21.10	18.20	15.30	12.50	9.60	6.70	4.30
260	270	37.40	33.80	30.10	26.50	22.80	19.70	16.80	14.00	11.10	8.20	5.50
270	280	39.30	35.70	32.00	28.40	24.70	21.20	18.30	15.50	12.60	9.70	6.80
280	290	41.70	37.60	33.90	30.30	26.60	23.00	19.80	17.00	14.10	11.20	8.30
290	300	44.20	39.50	35.80	32.20	28.50	24.90	21.30	18.50	15.60	12.70	9.80
300	310	46.70	41.90	37.70	34.10	30.40	26.80	23.10	20.00	17.10	14.20	11.30
310	320	49.20	44.40	39.60	36.00	32.30	28.70	25.00	21.50	18.60	15.70	12.80
320	330	51.70	46.90	42.10	37.90	34.20	30.60	26.90	23.30	20.10	17.20	14.30
330	340	54.20	49.40	44.60	39.80	36.10	32.50	28.80	25.20	21.60	18.70	15.80
340	350	56.70	51.90	47.10	42.30	38.00	34.40	30.70	27.10	23.40	20.20	17.30
350	360	59.20	54.40	49.60	44.80	40.00	36.30	32.60	29.00	25.30	21.70	18.30
360	370	61.70	56.90	52.10	47.30	42.50	38.20	34.50	30.90	27.20	23.60	20.30
370	380	64.20	59.40	54.60	49.80	45.00	40.20	36.40	32.80	29.10	25.50	21.80
380	390	66.70	61.90	57.10	52.30	47.50	42.70	38.30	34.70	31.00	27.40	23.70
390	400	69.20	64.40	59.60	54.80	50.00	45.20	40.40	36.60	32.90	29.30	25.60
400	410	71.70	66.90	62.10	57.30	52.50	47.70	42.90	38.50	34.80	31.20	27.50
410	420	74.20	69.40	64.60	59.80	55.00	50.20	45.40	40.60	36.70	33.10	29.40
420	430	76.80	71.90	67.10	62.30	57.50	52.70	47.90	43.10	38.60	35.00	31.30
430	440	79.80	74.40	69.60	64.80	60.00	55.20	50.40	45.60	40.80	36.90	33.20
440	450	82.80	77.10	72.10	67.30	62.50	57.70	52.90	48.10	43.30	38.80	35.10
450	460	85.80	80.10	74.60	69.80	65.00	60.20	55.40	50.60	45.80	41.00	37.00
460	470	88.80	83.10	77.30	72.30	67.50	62.70	57.90	53.10	48.30	43.50	38.90
470	480	91.80	86.10	80.30	74.80	70.00	65.20	60.40	55.60	50.80	46.00	41.20
480	490	94.80	89.10	83.30	77.50	72.50	67.70	62.90	58.10	53.30	48.50	43.70
490	500	97.80	92.10	86.30	80.50	75.00	70.20	65.40	60.60	55.80	51.00	46.20
500	510	100.80	95.10	89.30	83.50	77.80	72.70	67.90	63.10	58.30	53.50	48.70
510	520	103.80	98.10	92.30	86.50	80.80	75.20	70.40	65.60	60.80	56.00	51.20
520	530	106.80	101.10	95.30	89.50	83.80	78.00	72.90	68.10	63.30	58.50	53.70
530	540	109.80	104.10	98.30	92.50	86.80	81.00	75.40	70.60	65.80	61.00	56.20
540	550	113.20	107.10	101.30	95.50	89.80	84.00	78.20	73.10	68.30	63.50	58.70
550	560	116.60	110.10	104.30	98.50	92.80	87.00	81.20	75.60	70.80	66.00	61.20
560	570	120.00	113.50	107.30	101.50	95.80	90.00	84.20	78.40	73.30	68.50	63.70
570	580	123.40	116.90	110.40	104.50	98.80	93.00	87.20	81.40	75.80	71.00	66.20
580	590	126.80	120.30	113.80	107.50	101.80	96.00	90.20	84.40	78.70	73.50	68.70
590	600	130.20	123.70	117.20	110.60	104.80	99.00	93.20	87.40	81.70	76.00	71.20
600	610	133.60	127.10	120.60	114.00	107.80	102.00	96.20	90.40	84.70	78.90	73.70
610	620	137.00	130.50	124.00	117.40	110.90	105.00	99.20	93.40	87.70	81.90	76.20
620	630	140.40	133.90	127.40	120.80	114.30	108.00	102.20	96.40	90.70	84.90	79.10
630	640	143.80	137.30	130.80	124.20	117.70	111.20	105.20	99.40	93.70	87.90	82.10
640	650	147.50	140.70	134.20	127.60	121.10	114.60	108.20	102.40	96.70	90.90	85.10
650	660	151.20	144.10	137.60	131.00	124.50	118.00	111.40	105.40	99.70	93.90	88.10
660	670	154.90	147.80	141.00	134.40	127.90	121.40	114.80	108.40	102.70	96.90	91.10
670	680	158.60	151.50	144.40	137.80	131.30	124.80	118.20	111.70	105.70	99.90	94.10
680	690	162.30	155.20	148.10	141.20	134.70	128.20	121.60	115.10	108.70	102.90	97.10
690	700	166.00	158.90	151.80	144.70	138.10	131.60	125.00	118.50	111.90	105.90	100.10
700	710	169.70	162.60	155.50	148.40	141.50	135.00	128.40	121.90	115.30	108.90	103.10
710	720	173.40	166.30	159.20	152.10	144.90	138.40	131.80	125.30	118.70	112.20	106.10
720	730	177.10	170.00	162.90	155.80	148.60	141.80	135.20	128.70	122.10	115.60	109.10
730	740	180.80	173.70	166.60	159.50	152.30	145.20	138.60	132.10	125.50	119.00	112.50
740	750	184.50	177.40	170.30	163.20	156.00	148.90	142.00	135.50	128.90	122.40	115.90
750	760	188.20	181.10	174.00	166.90	159.70	152.60	145.50	138.90	132.30	125.80	119.30
760	770	191.90	184.80	177.70	170.60	163.40	156.30	149.20	142.30	135.70	129.20	122.70
770	780	195.60	188.50	181.40	174.30	167.10	160.00	152.90	145.80	139.10	132.60	126.10
780	790	199.30	192.20	185.10	178.00	170.80	163.70	156.60	149.50	142.50	136.00	129.50
790	800	203.00	195.90	188.80	181.70	174.50	167.40	160.30	153.20	146.10	139.40	132.90
800	810	206.70	199.60	192.50	185.40	178.20	171.10	164.00	156.90	149.80	142.80	136.30
810	820	210.40	203.30	196.20	189.10	181.90	174.80	167.70	160.60	153.50	146.40	139.70
820	830	214.10	207.00	199.90	192.80	185.60	178.50	171.40	164.30	157.20	150.10	143.10
830	840	217.80	210.70	203.60	196.50	189.30	182.20	175.10	168.00	160.90	153.80	146.60
		37 percent of the excess over $840 plus—										
$840 and over		219.60	212.50	205.40	198.30	191.20	184.10	177.00	169.80	162.70	155.60	148.50

MARRIED Persons—WEEKLY Payroll Period

(For Wages Paid After June 1983 and Before January 1985)

And the wages are—		And the number of withholding allowances claimed is—										
At least	But less than	0	1	2	3	4	5	6	7	8	9	10
		The amount of income tax to be withheld shall be—										
$ 0	$47	$0	$0	$0	$0	$0	$0	$0	$0	$0	$0	$0
47	48	.20	0	0	0	0	0	0	0	0	0	0
48	49	.30	0	0	0	0	0	0	0	0	0	0
49	50	.40	0	0	0	0	0	0	0	0	0	0
50	51	.50	0	0	0	0	0	0	0	0	0	0
51	52	.60	0	0	0	0	0	0	0	0	0	0
52	53	.80	0	0	0	0	0	0	0	0	0	0
53	54	.90	0	0	0	0	0	0	0	0	0	0
54	55	1.00	0	0	0	0	0	0	0	0	0	0
55	56	1.10	0	0	0	0	0	0	0	0	0	0
56	57	1.20	0	0	0	0	0	0	0	0	0	0
57	58	1.40	0	0	0	0	0	0	0	0	0	0
58	59	1.50	0	0	0	0	0	0	0	0	0	0
59	60	1.60	0	0	0	0	0	0	0	0	0	0
60	62	1.80	0	0	0	0	0	0	0	0	0	0
62	64	2.00	0	0	0	0	0	0	0	0	0	0
64	66	2.30	0	0	0	0	0	0	0	0	0	0
66	68	2.50	.20	0	0	0	0	0	0	0	0	0
68	70	2.70	.40	0	0	0	0	0	0	0	0	0
70	72	3.00	.70	0	0	0	0	0	0	0	0	0
72	74	3.20	.90	0	0	0	0	0	0	0	0	0
74	76	3.50	1.20	0	0	0	0	0	0	0	0	0
76	78	3.70	1.40	0	0	0	0	0	0	0	0	0
78	80	3.90	1.60	0	0	0	0	0	0	0	0	0
80	82	4.20	1.90	0	0	0	0	0	0	0	0	0
82	84	4.40	2.10	0	0	0	0	0	0	0	0	0
84	86	4.70	2.40	0	0	0	0	0	0	0	0	0
86	88	4.90	2.60	.30	0	0	0	0	0	0	0	0
88	90	5.10	2.80	.50	0	0	0	0	0	0	0	0
90	92	5.40	3.10	.80	0	0	0	0	0	0	0	0
92	94	5.60	3.30	1.00	0	0	0	0	0	0	0	0
94	96	5.90	3.60	1.20	0	0	0	0	0	0	0	0
96	98	6.10	3.80	1.50	0	0	0	0	0	0	0	0
98	100	6.30	4.00	1.70	0	0	0	0	0	0	0	0
100	105	6.80	4.50	2.10	0	0	0	0	0	0	0	0
105	110	7.40	5.10	2.70	.40	0	0	0	0	0	0	0
110	115	8.00	5.70	3.30	1.00	0	0	0	0	0	0	0
115	120	8.60	6.30	3.90	1.60	0	0	0	0	0	0	0
120	125	9.20	6.90	4.50	2.20	0	0	0	0	0	0	0
125	130	9.80	7.50	5.10	2.80	.50	0	0	0	0	0	0
130	135	10.40	8.10	5.70	3.40	1.10	0	0	0	0	0	0
135	140	11.00	8.70	6.30	4.00	1.70	0	0	0	0	0	0
140	145	11.60	9.30	6.90	4.60	2.30	0	0	0	0	0	0
145	150	12.20	9.90	7.50	5.20	2.90	.60	0	0	0	0	0
150	160	13.10	10.80	8.40	6.10	3.80	1.50	0	0	0	0	0
160	170	14.30	12.00	9.60	7.30	5.00	2.70	.40	0	0	0	0
170	180	15.50	13.20	10.80	8.50	6.20	3.90	1.60	0	0	0	0
180	190	16.70	14.40	12.00	9.70	7.40	5.10	2.80	.50	0	0	0
190	200	18.40	15.60	13.20	10.90	8.60	6.30	4.00	1.70	0	0	0
200	210	20.10	16.80	14.40	12.10	9.80	7.50	5.20	2.90	.60	0	0
210	220	21.80	18.50	15.60	13.30	11.00	8.70	6.40	4.10	1.80	0	0
220	230	23.50	20.20	16.90	14.50	12.20	9.90	7.60	5.30	3.00	.70	0
230	240	25.20	21.90	18.60	15.70	13.40	11.10	8.80	6.50	4.20	1.90	0
240	250	26.90	23.60	20.30	17.10	14.60	12.30	10.00	7.70	5.40	3.10	.80
250	260	28.60	25.30	22.00	18.80	15.80	13.50	11.20	8.90	6.60	4.30	2.00
260	270	30.30	27.00	23.70	20.50	17.20	14.70	12.40	10.10	7.80	5.50	3.20
270	280	32.00	28.70	25.40	22.20	18.90	15.90	13.60	11.30	9.00	6.70	4.40
280	290	33.70	30.40	27.10	23.90	20.60	17.30	14.80	12.50	10.20	7.90	5.60
290	300	35.40	32.10	28.80	25.60	22.30	19.00	16.00	13.70	11.40	9.10	6.80
300	310	37.10	33.80	30.50	27.30	24.00	20.70	17.50	14.90	12.60	10.30	8.00
310	320	38.80	35.50	32.20	29.00	25.70	22.40	19.20	16.10	13.80	11.50	9.20
320	330	40.50	37.20	33.90	30.70	27.40	24.10	20.90	17.60	15.00	12.70	10.40
330	340	42.20	38.90	35.60	32.40	29.10	25.80	22.60	19.30	16.20	13.90	11.60
340	350	43.90	40.60	37.30	34.10	30.80	27.50	24.30	21.00	17.70	15.10	12.80
350	360	45.60	42.30	39.00	35.80	32.50	29.20	26.00	22.70	19.40	16.30	14.00
360	370	47.30	44.00	40.70	37.50	34.20	30.90	27.70	24.40	21.10	17.90	15.20
370	380	49.30	45.70	42.40	39.20	35.90	32.60	29.40	26.10	22.80	19.60	16.40
380	390	51.50	47.40	44.10	40.90	37.60	34.30	31.10	27.80	24.50	21.30	18.00
390	400	53.70	49.50	45.80	42.60	39.30	36.00	32.80	29.50	26.20	23.00	19.70
400	410	55.90	51.70	47.50	44.30	41.00	37.70	34.50	31.20	27.90	24.70	21.40

And the wages are—		And the number of withholding allowances claimed is—										
At least	But less than	0	1	2	3	4	5	6	7	8	9	10
		The amount of income tax to be withheld shall be—										
$410	$420	$58.10	$53.90	$49.60	$46.00	$42.70	$39.40	$36.20	$32.90	$29.60	$26.40	$23.10
420	430	60.30	56.10	51.80	47.70	44.40	41.10	37.90	34.60	31.30	28.10	24.80
430	440	62.50	58.30	54.00	49.80	46.10	42.80	39.60	36.30	33.00	29.80	26.50
440	450	64.70	60.50	56.20	52.00	47.80	44.50	41.30	38.00	34.70	31.50	28.20
450	460	66.90	62.70	58.40	54.20	50.00	46.20	43.00	39.70	36.40	33.20	29.90
460	470	69.40	64.90	60.60	56.40	52.20	47.90	44.70	41.40	38.10	34.90	31.60
470	480	71.90	67.10	62.80	58.60	54.40	50.10	46.40	43.10	39.80	36.60	33.30
480	490	74.40	69.60	65.00	60.80	56.60	52.30	48.10	44.80	41.50	38.30	35.00
490	500	76.90	72.10	67.30	63.00	58.80	54.50	50.30	46.50	43.20	40.00	36.70
500	510	79.40	74.60	69.80	65.20	61.00	56.70	52.50	48.30	44.90	41.70	38.40
510	520	81.90	77.10	72.30	67.50	63.20	58.90	54.70	50.50	46.60	43.40	40.10
520	530	84.40	79.60	74.80	70.00	65.40	61.10	56.90	52.70	48.40	45.10	41.80
530	540	86.90	82.10	77.30	72.50	67.70	63.30	59.10	54.90	50.60	46.80	43.50
540	550	89.40	84.60	79.80	75.00	70.20	65.50	61.30	57.10	52.80	48.60	45.20
550	560	91.90	87.10	82.30	77.50	72.70	67.90	63.50	59.30	55.00	50.80	46.90
560	570	94.70	89.60	84.80	80.00	75.20	70.40	65.70	61.50	57.20	53.00	48.80
570	580	97.50	92.10	87.30	82.50	77.70	72.90	68.10	63.70	59.40	55.20	51.00
580	590	100.30	94.90	89.80	85.00	80.20	75.40	70.60	65.90	61.60	57.40	53.20
590	600	103.10	97.70	92.30	87.50	82.70	77.90	73.10	68.30	63.80	59.60	55.40
600	610	105.90	100.50	95.10	90.00	85.20	80.40	75.60	70.80	66.00	61.80	57.60
610	620	108.70	103.30	97.90	92.50	87.70	82.90	78.10	73.30	68.50	64.00	59.80
620	630	111.50	106.10	100.70	95.30	90.20	85.40	80.60	75.80	71.00	66.20	62.00
630	640	114.30	108.90	103.50	98.10	92.80	87.90	83.10	78.30	73.50	68.70	64.20
640	650	117.10	111.70	106.30	100.90	95.60	90.40	85.60	80.80	76.00	71.20	66.40
650	660	119.90	114.50	109.10	103.70	98.40	93.00	88.10	83.30	78.50	73.70	68.80
660	670	123.10	117.30	111.90	106.50	101.20	95.80	90.60	85.80	81.00	76.20	71.30
670	680	126.40	120.10	114.70	109.30	104.00	98.60	93.20	88.30	83.50	78.70	73.80
680	690	129.70	123.30	117.50	112.10	106.80	101.40	96.00	90.80	86.00	81.20	76.30
690	700	133.00	126.60	120.30	114.90	109.60	104.20	98.80	93.40	88.50	83.70	78.80
700	710	136.30	129.90	123.60	117.70	112.40	107.00	101.60	96.20	91.00	86.20	81.30
710	720	139.60	133.20	126.90	120.50	115.20	109.80	104.40	99.00	93.60	88.70	83.80
720	730	142.90	136.50	130.20	123.80	118.00	112.60	107.20	101.80	96.40	91.20	86.30
730	740	146.20	139.80	133.50	127.10	120.80	115.40	110.00	104.60	99.20	93.80	88.80
740	750	149.50	143.10	136.80	130.40	124.10	118.20	112.80	107.40	102.00	96.60	91.30
750	760	152.80	146.40	140.10	133.70	127.40	121.00	115.60	110.20	104.80	99.40	94.10
760	770	156.10	149.70	143.40	137.00	130.70	124.30	118.40	113.00	107.60	102.20	96.90
770	780	159.40	153.00	146.70	140.30	134.00	127.60	121.30	115.80	110.40	105.00	99.70
780	790	162.70	156.30	150.00	143.60	137.30	130.90	124.60	118.60	113.20	107.80	102.50
790	800	166.00	159.60	153.30	146.90	140.60	134.20	127.90	121.50	116.00	110.60	105.30
800	810	169.30	162.90	156.60	150.20	143.90	137.50	131.20	124.80	118.80	113.40	108.10
810	820	172.60	166.20	159.90	153.50	147.20	140.80	134.50	128.10	121.80	116.20	110.90
820	830	175.90	169.50	163.20	156.80	150.50	144.10	137.80	131.40	125.10	119.00	113.70
830	840	179.20	172.80	166.50	160.10	153.80	147.40	141.10	134.70	128.40	122.10	116.50
840	850	182.50	176.10	169.80	163.40	157.10	150.70	144.40	138.00	131.70	125.40	119.30
850	860	185.80	179.40	173.10	166.70	160.40	154.00	147.70	141.30	135.00	128.70	122.30
860	870	189.20	182.70	176.40	170.00	163.70	157.30	151.00	144.60	138.30	132.00	125.60
870	880	192.90	186.00	179.70	173.30	167.00	160.60	154.30	147.90	141.60	135.30	128.90
880	890	196.60	189.50	183.00	176.60	170.30	163.90	157.60	151.20	144.90	138.60	132.20
890	900	200.30	193.20	186.30	179.90	173.60	167.20	160.90	154.50	148.20	141.90	135.50
900	910	204.00	196.90	189.80	183.20	176.90	170.50	164.20	157.80	151.50	145.20	138.80
910	920	207.70	200.60	193.50	186.50	180.20	173.80	167.50	161.10	154.80	148.50	142.10
920	930	211.40	204.30	197.20	190.10	183.50	177.10	170.80	164.40	158.10	151.80	145.40
930	940	215.10	208.00	200.90	193.80	186.80	180.40	174.10	167.70	161.40	155.10	148.70
940	950	218.80	211.70	204.60	197.50	190.30	183.70	177.40	171.00	164.70	158.40	152.00
950	960	222.50	215.40	208.30	201.20	194.00	187.00	180.70	174.30	168.00	161.70	155.30
960	970	226.20	219.10	212.00	204.90	197.70	190.60	184.00	177.60	171.30	165.00	158.60
970	980	229.90	222.80	215.70	208.60	201.40	194.30	187.30	180.90	174.60	168.30	161.90
980	990	233.60	226.50	219.40	212.30	205.10	198.00	190.90	184.20	177.90	171.60	165.20
990	1,000	237.30	230.20	223.10	216.00	208.80	201.70	194.60	187.50	181.20	174.90	168.50
1,000	1,010	241.00	233.90	226.80	219.70	212.50	205.40	198.30	191.20	184.50	178.20	171.80
1,010	1,020	244.70	237.60	230.50	223.40	216.20	209.10	202.00	194.90	187.80	181.50	175.10
1,020	1,030	248.40	241.30	234.20	227.10	219.90	212.80	205.70	198.60	191.50	184.80	178.40
1,030	1,040	252.10	245.00	237.90	230.80	223.60	216.50	209.40	202.30	195.20	188.10	181.70
1,040	1,050	255.80	248.70	241.60	234.50	227.30	220.20	213.10	206.00	198.90	191.80	185.00
1,050	1,060	259.50	252.40	245.30	238.20	231.00	223.90	216.80	209.70	202.60	195.50	188.40
		37 percent of the excess over $1,060 plus—										
$1,060 and over		261.40	254.20	247.10	240.00	232.90	225.80	218.70	211.50	204.40	197.30	190.20

21   **Step 1.**   Use the married weekly table.

        Step 2.   The weekly wage is $285.46. It's in the row "at least 280, but less than 290."

        Step 3.   Move right in the row to 4 exemptions. The tax is $20.60.

Isn't it easy? But remember, check each answer. Accuracy is very important in any business problem, especially payroll.

Try the following problem.

Calculate the income tax for the following employees using the wage bracket tables.

Name	Marital Status	Exemptions	Weekly Wage	Income Tax
Childless, Mollie	Single	1	$187.56	
Ercanbrack, Gerald	Married	3	188.47	
Stephens, Brenda	Married	4	203.28	
Tinsley, Maynard	Single	0	195.15	

Turn to 22 when you have completed the payroll sheet above.

What is the difference between net and gross?

The gross is before deductions and the net is after deductions. The net is what is caught in the net after the deductions slip through.

22

Name	Marital Status	Exemptions	Weekly Wage	Income Tax
Childless, Mollie	Single	1	$187.56	$19.30
Ercanbrack, Gerald	Married	3	188.47	9.70
Stephens, Brenda	Married	4	203.28	9.80
Tinsley, Maynard	Single	0	195.15	24.10

Percentage Method   Many employers prefer to use the percentage method, especially those using computer payrolls. This method uses two tables.

    Step 1.   Use the "Percentage Method Income Tax Withholding Table" at the top of the next page. Find the amount of withholding allowance for the payroll period. Multiply this amount by the number of exemptions and subtract from gross wages.

Percentage Method Income Tax Withholding Table

Payroll period	One withholding allowance
Weekly	$19.23
Biweekly	38.46
Semimonthly	41.66
Monthly	83.33
Quarterly	250.00
Semiannually	500.00
Annually	1,000.00
Daily or miscellaneous (each day of the payroll period)	2.74

Step 2. In the second set of tables on page 232 labeled "Tables for Percentage Method of Withholding," find the correct table for the pay period and marital status.

Step 3. Locate the correct wage bracket for the amount found in **Step 1** (not the gross wage) in the two leftmost columns. Perform the calculation on the right.

Let's try an example:

Marjorie Geist is married and declares two exemptions. Compute her income tax by the percentage method for a month in which she earned $1,107.49.

Step 1. Since she is paid monthly, the withholding allowance from the table above is $83.33. For two exemptions the allowance is 2 × $83.33 = $166.66. Subtract $166.66 from her gross wages.

$$\begin{array}{r} \$1,107.49 \\ -166.66 \longleftarrow 2 \times \$83.33 \\ \hline \$940.83 \end{array}$$

Step 2. Locate the second table for monthly payroll, married, on page 232.

TABLE 4. If the Payroll Period With Respect to an Employee is Monthly

(b) MARRIED person—

If the amount of wages is:		The amount of income tax to be withheld shall be:			
Not over $200		0			
Over—	But not over—				of excess over—
$200	—$800	12%			—$200
$800	—$1,598	$72.00	plus	17%	—$800
$1,598	—$1,967	$207.66	plus	22%	—$1,598
$1,967	—$2,408	$288.84	plus	25%	—$1,967
$2,408	—$2,850	$399.09	plus	28%	—$2,408
$2,850	—$3,733	$522.85	plus	33%	—$2,850
$3,733		$814.24	plus	37%	—$3,733

$940.83 falls in this wage bracket → ($800 row)

Step. 3.   Her wage is "over $800, but not over $1,598." Her income tax is

$72.00 + 17% of excess over $800
  72.00 + 17% × ($940.83 − 800)
  72.00 + 17% × 140.83
  72.00 + 0.17 × 140.83
  72.00 + 23.94 rounded
$95.94

"Of the excess over $800" means her wage (from Step 1, not her gross wage) minus $800. The statement 17% of means "17% ×," or "17% times." (This is just like the percent problems in Unit 4. For a review, take a quick look at page 131.)

This method is a little longer than the wage bracket method. But with a little practice, it's as easy.

Try the following problem:

Fernando Hernandez earned $285.46 last week. He is married and declares four exemptions. Compute his income tax by the percentage method.

Check your work in **23**.

Tables for Percentage Method of Withholding

(For Wages Paid After June 1983 and Before January 1985)

TABLE 1. If the Payroll Period with Respect to an Employee is Weekly

(a) SINGLE person—including head of household:

If the amount of wages is:

Not over $27 0

Over—	But not over—		of excess over—
$27	—$79	. . . . . . .12%	—$27
$79	—$183	. . . . .$6.24 plus 15%	—$79
$183	—$277	. . . . .$21.84 plus 19%	—$183
$277	—$423	. . . . .$39.70 plus 25%	—$277
$423	—$535	. . . . .$76.20 plus 30%	—$423
$535	—$637	. . . . .$109.80 plus 34%	—$535
$637		. . . .$144.48 plus 37%	—$637

(b) MARRIED person—

If the amount of wages is:

Not over $46 0

Over—	But not over—		of excess over—
$46	—$185	. . . . . .12%	—$46
$185	—$369	. . . . . .$16.68 plus 17%	—$185
$369	—$454	. . . . . .$47.96 plus 22%	—$369
$454	—$556	. . . . . .$66.66 plus 25%	—$454
$556	—658	. . . . . .$92.16 plus 28%	—$556
$658	—$862	. . . . . .$120.72 plus 33%	—$658
$862		. . . .$188.04 plus 37%	—$862

TABLE 2. If the Payroll Period With Respect to an Employee is Biweekly

(a) SINGLE person—including head of household:

If the amount of wages is:

Not over $54 0

Over—	But not over—		of excess over—
$54	—$158	. . . . . .12%	—$54
$158	—$365	. . . . . .$12.48 plus 15%	—$158
$365	—$554	. . . . . .$43.53 plus 19%	—$365
$554	—$846	. . . . . .$79.44 plus 25%	—$554
$846	—$1,069	. . . .$152.44 plus 30%	—$846
$1,069	—$1,273	. . . .$219.34 plus 34%	—$1,069
$1,273		. . . .$288.70 plus 37%	—$1,273

(b) MARRIED person—

If the amount of wages is:

Not over $92 0

Over—	But not over—		of excess over—
$92	—$369	. . . . . .12%	—$92
$369	—$738	. . . . .$33.24 plus 17%	—$369
$738	—$908	. . . . .$95.97 plus 22%	—$738
$908	—$1,112	. . . .$133.37 plus 25%	—$908
$1,112	—$1,315	. . . .$184.37 plus 28%	—$1,112
$1,315	—$1,723	. . . .$241.21 plus 33%	—$1,315
$1,723		. . . .$375.85 plus 37%	—$1,723

TABLE 3. If the Payroll Period With Respect to an Employee is Semimonthly

(a) SINGLE person—including head of household:

If the amount of wages is:

Not over $58 0

Over—	But not over—		of excess over—
$58	—$171	. . . . . .12%	—$58
$171	—$396	. . . . .$13.56 plus 15%	—$171
$396	—$600	. . . . .$47.31 plus 19%	—$396
$600	—$917	. . . . .$86.07 plus 25%	—$600
$917	—$1,158	. . . .$165.32 plus 30%	—$917
$1,158	—$1,379	. . . .$237.62 plus 34%	—$1,158
$1,379		. . . .$312.76 plus 37%	—$1,379

(b) MARRIED person—

If the amount of wages is:

Not over $100 0

Over—	But not over—		of excess over—
$100	—$400	. . . . . .12%	—$100
$400	—$799	. . . . .$36.00 plus 17%	—$400
$799	—$983	. . . . .$103.83 plus 22%	—$799
$983	—$1,204	. . . .$144.31 plus 25%	—$983
$1,204	—$1,425	. . . .$199.56 plus 28%	—$1,204
$1,425	—$1,867	. . . .$261.44 plus 33%	—$1,425
$1,867		. . . .$407.30 plus 37%	—$1,867

TABLE 4. If the Payroll Period With Respect to an Employee is Monthly

(a) SINGLE person—including head of household:

If the amount of wages is:

Not over $117 0

Over—	But not over—		of excess over—
$117	—$342	. . . . . .12%	—$117
$342	—$792	. . . . .$27.00 plus 15%	—$342
$792	—$1,200	. . . .$94.50 plus 19%	—$792
$1,200	—$1,833	. . . .$172.02 plus 25%	—$1,200
$1,833	—$2,317	. . . .$330.27 plus 30%	—$1,833
$2,317	—$2,758	. . . .$475.47 plus 34%	—$2,317
$2,758		. . . .$625.41 plus 37%	—$2,758

(b) MARRIED person—

If the amount of wages is:

Not over $200 0

Over—	But not over—		of excess over—
$200	—$800	. . . . . .12%	—$200
$800	—$1,598	. . . .$72.00 plus 17%	—$800
$1,598	—$1,967	. . . .$207.66 plus 22%	—$1,598
$1,967	—$2,408	. . . .$288.84 plus 25%	—$1,967
$2,408	—$2,850	. . . .$399.09 plus 28%	—$2,408
$2,850	—$3,733	. . . .$522.85 plus 33%	—$2,850
$3,733		. . . .$814.24 plus 37%	—$3,733

23 **Step 1.** Since he is paid weekly, the withholding allowance is $19.23. For four exemptions, the allowance is 4 × $19.23 = $76.92.

$$\begin{array}{r} \$285.46 \\ -76.92 \quad \longleftarrow 4 \times \$19.23 \\ \hline \$208.54 \end{array}$$

Step 2. Use the weekly pay period, married table.

TABLE 1. If the Payroll Period with Respect to an Employee is Weekly

(b) MARRIED person—

If the amount of wages is:		The amount of income tax to be withheld shall be:		
Not over $46		0		
Over—	But not over—			of excess over—
$46	—$185	12%		—$46
$185	—$369	$16.68	plus 17%	—$185
$369	—$454	$47.96	plus 22%	—$369
$454	—$556	$66.66	plus 25%	—$454
$556	—$658	$92.16	plus 28%	—$556
$658	—$862	$120.72	plus 33%	—$658
$862		$188.04	plus 37%	—$862

Use this wage bracket ➤ $185

Step 3. His wage is "over $185, but not over $369." The tax will be

$16.68 + 17% of excess over $185
16.68 + 17% × ($208.54 − 185)
16.68 + 17% × 23.54
16.68 + 4.00 rounded
$20.68

"Well, that's a little strange. In frame **22** *we worked the same problem by the wage bracket method and got a different answer, $20.60."*

Yes, you *will* get different answers with each method. But if you use the amount in the middle (average) of the wage bracket interval, you will get approximately the same answer with both methods. (Try computing the income tax on a weekly wage of $285.00, married with 4 exemptions. The tax is $20.60 with the wage bracket method and $20.60 with the percentage method. The wage $285.00 is the middle of the interval "at least $280, but less than $290.") Try a few more practice problems. Find the income tax to be deducted for each of the following employees by the percentage method.

Name	Marital Status	Exemptions	Semimonthly Wage	Income Tax
Clark, Margie	Single	2	$498.76	
Henweigh, L. B.	Married	4	587.21	
Ross, Donald	Single	1	506.94	
Sturdivant, Reba	Married	3	610.25	

Check your work in **24**.

I'm married with two children, and both my husband and I work. Can we each declare four exemptions?

No. Careful there. The two of you should claim no more than a *total* of four exemptions.

24

Name	Marital Status	Exemptions	Semimonthly Wage	Income Tax
Clark, Margie	Single	2	$498.76	$51.00
Henweigh, L. B.	Married	4	587.21	39.50
Ross, Donald	Single	1	506.94	60.47
Sturdivant, Reba	Married	3	610.25	50.50

Clark, Margie: $498.76
                  -83.32 ⟵——— 2 exemptions × $41.66
                  $415.44

Tax = $47.31 + 19% × ($415.44 − 396)
    = $47.31 + 19% × 19.44
    = $47.31 + 3.69 rounded
    = $51.00

Henweigh, L. B.: $587.21
                    -166.64 ⟵——— 4 exemptions × $41.66
                    $420.57

Tax = $36.00 + 17% × ($420.57 − 400)
    = $36.00 + 17% × 20.57
    = $36.00 + 3.50
    = $39.50

Ross, Donald: $506.94
                 -41.66 ⟵——— 1 exemption
                 $465.28

Tax = $47.31 + 19% × ($465.28 − 396)
    = $47.31 + 19% × 69.28
    = $47.31 + 13.16
    = $60.47

Sturdivant, Reba: $610.25
                     -124.98 ⟵——— 3 exemptions × $41.66
                     $485.27

Tax = $36.00 + 17% × ($485.27 − 400)
    = $36.00 + 17% × 85.27
    = $36.00 + 14.50
    = $50.50

Turn to **25** for a discussion on FICA.

FICA

25

Another required deduction is FICA (Federal Insurance Contributions Act), commonly known as Social Security. When FICA was started in 1937, the original FICA rate was 1% on the first $3000 in wages. As the cost of living and Social Security expenses increased, the FICA rate and maximum amount was increased.

Year	FICA Rate	Wage Limitation	Maximum Amount of Tax
1969–1970	4.80%	$ 7,800	$ 374.40
1971	5.20%	7,800	405.60
1972	5.20%	9,000	468.00
1973	5.85%	10,800	631.80
1974	5.85%	13,200	772.20
1975	5.85%	14,100	824.85
1976	5.85%	15,300	895.05
1977	5.85%	16,500	965.25
1978	6.05%	17,700	1070.85
1979	6.13%	22,900	1403.77
1980	6.13%	25,900	1587.67
1981	6.65%	29,700	1975.05
1982	6.7%	32,400	2170.80
1983	6.7%	35,700	2391.90

The 1983 rate is 6.7% on the first $35,700 in wages. No FICA is withheld on earnings over $35,700. The maximum amount of tax is the product of the FICA rate times the wage limitation. For 1983,

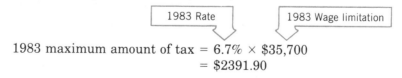

1983 maximum amount of tax = 6.7% × $35,700
                            = $2391.90

In addition to the amount withheld from the employee's wage, the employer is required to match the deduction. That means the federal government actually receives 2 × 6.7% or 13.4%; 6.7% from the employee and 6.7% from the employer.

There are two methods of calculating FICA, the table "look-up" and the percentage method. Both methods give the same result. The table look-up method uses the "Social Security Employee Tax Table" in the IRS Circular E. The percentage method is the easiest method, you simply multiply the gross pay by the FICA rate, and round when necessary.

> FICA = 6.7% × gross pay

Example: Calculate the amount of FICA withheld on a monthly wage of $865.

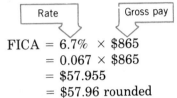

$$
\begin{aligned}
\text{FICA} &= 6.7\% \times \$865 \\
&= 0.067 \times \$865 \\
&= \$57.955 \\
&= \$57.96 \text{ rounded}
\end{aligned}
$$

The FICA withheld from the employee's wages is $57.96. The employer must also "contribute" an additional $57.96.

Your turn. Calculate the FICA withholding on the following wages.

(a) $576.42   (b) $956.85

Check your answers in 26.

Why do you computers use the percentage method instead of tax tables when you calculate federal income tax?

THE PERCENTAGE METHOD DOES NOT REQUIRE ME TO STORE ALL OF THOSE TAX TABLES, SO IT TAKES UP LESS OF MY MEMORY SPACE. SAVES MONEY.

26 (a) $$
\begin{aligned}
\text{FICA} &= 6.7\% \times \$576.42 \\
&= 0.067 \times \$576.42 \\
&= \$38.620 \ldots \\
&= \$38.62 \text{ rounded}
\end{aligned}
$$

(b) $$
\begin{aligned}
\text{FICA} &= 6.7\% \times \$956.85 \\
&= 0.067 \times \$956.85 \\
&= \$64.108 \ldots \\
&= \$64.11 \text{ rounded}
\end{aligned}
$$

In addition to federal income and FICA taxes, many other deductions are frequently made. Most states require a state income tax to be withheld. Although the state tax tables vary from state to state, the withholding methods are similar to those for the federal tax.

Other possible deductions include: health insurance, life insurance, union dues, annuities, bond purchases, retirement plans, and so on. The list seems endless.

A statement of deductions must be added to the payroll sheet. Starting with the gross pay, calculate the federal income tax, and FICA. The total deductions will be the sum of federal income tax, FICA, and the other deductions. The net pay will be the gross pay minus the total deductions.

Try it by completing the following weekly payroll sheet. Don't forget to total the gross pay, federal income tax, FICA, other deductions, total deductions, and net pay columns. Use the wage bracket method to calculate the federal income tax.

						Sum of tax, FICA, and Other Deductions	Gross Pay − Total Deductions

Name	Marital Status	No. of Exemptions	Gross Pay	Federal Income Tax	FICA	Other Deductions	Total Deductions	Net Pay
Ayarza, Silvia	S	1	247.86			22.50		
Cobb, Dennis	M	4	236.27			25.75		
Moomey, Roy	S	1	196.42			14.96		
Pham, Vinh	M	3	289.23			23.28		
TOTALS								

Check your work in **27**.

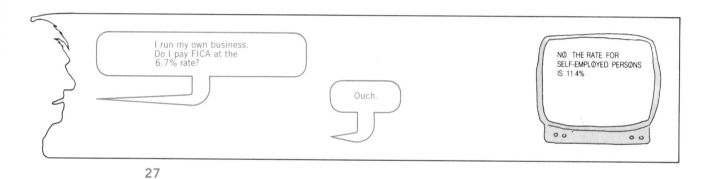

27

Name	Marital Status	No. of Exemptions	Gross Pay	Federal Income Tax	FICA	Other Deductions	Total Deductions	Net Pay
Ayarza, Silvia	S	1	247.86	30.00	16.61	22.50	69.11	178.75
Cobb, Dennis	M	4	236.27	13.40	15.83	25.75	54.98	181.29
Moomey, Roy	S	1	196.42	20.80	13.16	14.96	48.92	147.50
Pham, Vinh	M	3	289.23	23.90	19.38	23.28	66.56	222.67
TOTALS			969.78	88.10	64.98	86.49	239.57	730.21

The sum of these three numbers should equal the total deduction.

Total gross pay − total deduction

Note Be sure to check your work.

When this payroll sheet is completed, the totals of the federal income tax, FICA, and other deductions columns will equal the sum of the total deductions column.

After this amount has been found, the sum of the total deductions column is subtracted from the total of the gross pay column to verify the total of the net pay column.

For some practice problems calculating deductions, go to frame **28**.

6 Payroll

PROBLEM SET 3

28   The answers are on page 584.

6.3   Federal Income
      Tax and FICA

A. Complete the following weekly payroll sheet. Calculate the federal income tax by the wage bracket method.

Name	Marital Status	No. of Exemp- tions	Gross Pay	Federal Income Tax	FICA	Other Deduc- tions	Total Deduc- tions	Net Pay
1. Anderson, Lynda	M	2	$214.29			25.76		
2. Cusack, Mary	S	1	256.65			18.95		
3. Harris, Thomas	M	4	289.19			19.42		
4. Hendricks, Ray	M	3	198.57			23.57		
5. Lytle, Leslie	S	1	152.95			15.16		
6. McCormick, George	S	1	165.41			21.95		
7. Morgan, William	M	3	217.89			32.56		
8. Trivitt, Antoinette	S	2	187.42			25.18		
TOTALS								

B. Complete the following semimonthly payroll sheet. Calculate the federal income tax using the percentage method.

Name	Marital Status	No. of Exemp- tions	Gross Pay	Federal Income Tax	FICA	Other Deduc- tions	Total Deduc- tions	Net Pay
1. Churchman, Stephen	S	1	516.42			42.50		
2. Franklin, Gary	M	4	495.52			36.95		
3. Hall, Margaret	M	2	455.26			35.14		
4. Laws, Michael	S	1	314.57			29.36		
5. Miller, Coy	S	2	298.42			25.42		
6. Stevenson, Anthony	M	5	456.89			38.95		
7. Woodard, Mike	M	3	567.49			45.56		
8. Young, Judy	S	1	485.42			43.21		
TOTALS								

Date

Name

Course/Section

C. Complete the weekly payroll sheet. Use the wage bracket method to calculate the federal income tax.

Name	M	T	W	T	F	Total Hours	Regular Hours	Regular Rate	O.T. Hours	O.T. Rate	Regular Pay	O.T. Pay	Total Pay	Marital Status	No. of Exemptions	Federal Income Tax	FICA	Other Deductions	Total Deductions	Net Pay
1. Atkinson, Stephen	8	9	9	8	9			3.15						M	5			23.15		
2. Brasier, Scott	8	9	9	10	9			3.18						S	1			22.95		
3. Collins, Melinda	8	8	8	8	8			2.98						S	1			19.86		
4. Keith, Thomas	8	7	7	8	7			2.86						M	2			18.49		
5. Norman, Gary	8	9	10	10	10			3.28						M	4			25.62		
6. Smith,, Vanessa	9	10	10	10	8			3.45						S	2			24.86		
7. Wade, Douglas	9	9	9	9	8			3.50						M	6			23.14		
8. White, Ian	9	8	8	8	8			3.26						M	3			26.87		
TOTALS																				

D. The Stagg's Motorcycle Shop pays its sales people a salary plus a $2\frac{1}{2}\%$ commission on their total sales. Complete the monthly payroll sheet. Use the percentage method to calculate the federal income tax.

Name	Salary	Sales	Commission	Total Pay	Marital Status	No. of Exemptions	Federal Income Tax	FICA	Other Deductions	Total Deductions	Net Pay
1. Adams, Jerry	325.00	3,825.36			M	3			42.96		
2. Carr, Tim	350.00	4,156.75			S	1			39.51		
3. Curtis, Cheryl	330.00	3,526.33			S	1			35.50		
4. Fleming, Jimmie	300.00	2,856.89			S	2			29.98		
5. Hise, Leslie	337.50	2,546.71			M	4			26.75		
6. Landsberger, Dale	340.00	3,922.52			S	1			36.72		
7. Staggs, Patty	800.00	4,526.52			M	1			53.86		
8. Staggs, Virgil	800.00	2,851.27			M	1			52.49		
TOTALS											

When you have had the practice you need, either return to the Preview on page 203 or continue in 29 with Payroll records.

SECTION 4: PAYROLL RECORDS

29 | **OBJECTIVE**

Complete a given employee's earning record and quarterly payroll summary.

The payroll sheet is used to calculate the employee's gross pay, various deductions, and finally the net pay. In addition to the payroll sheet, several other records are kept by employers.

Employers are required to file certain quarterly (every three months) and yearly summaries. This information is usually obtained from the *employee's earning record*. The record contains the employee's name, Social Security number, address, marital status, number of exemptions, and a summary of each pay period. A pay period summary includes gross pay, federal income tax, FICA, other deductions, total deductions, and net pay. Information from the summary is totaled quarterly and yearly for the required federal reports.

A typical employee earning record is shown below. Calculate the income tax, FICA, total deductions, net pay, and quarterly totals for this employee. Use the percentage method to calculate the income tax.

Name	Bird, Susan		Social Security No.		547-76-6525	
Address	600 S.W. 52		Marital Status M		No. of Exemptions	4

Pay Period	Gross Pay	Federal Income Tax	FICA	Other Deductions	Total Deductions	Net Pay
Jan. Feb. March	845.00 855.00 860.00			34.60 35.57 36.15		
First Quarter						

Check your work in **30**.

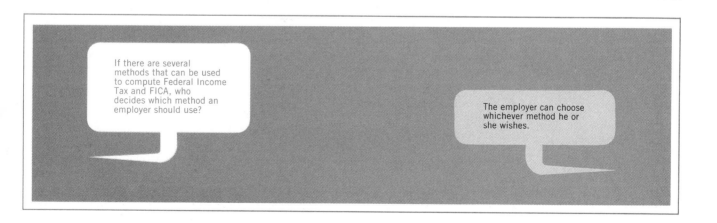

If there are several methods that can be used to compute Federal Income Tax and FICA, who decides which method an employer should use?

The employer can choose whichever method he or she wishes.

Pay Period	Gross Pay	Federal Income Tax	FICA	Other Deductions	Total Deductions	Net Pay
Jan.	845.00	37.40	56.62	34.60	128.62	716.38
Feb.	855.00	38.60	57.29	35.57	131.46	723.54
March	860.00	39.20	57.62	36.15	132.97	727.03
First Quarter	2560.00	115.20	171.53	106.32	393.05	2166.95

Whew! That's a lot of work. A calculator is a big help, but it is very important that you always check your answers.

Notice that we added to find the first quarter totals for the federal income tax, FICA, and other deductions. The sum of these totals should equal the sum of the total deductions column. We also subtracted the sum of the total deductions from the total gross pay to find the total net pay.

Using the employee's earning record, the employer must construct a quarterly payroll summary. The *quarterly payroll summary* will include the employees' names, Social Security numbers, gross earnings, federal income taxes, FICAs, and the totals.

The following quarterly payroll summary is a typical example. Complete the summary by calculating the totals.

QUARTERLY PAYROLL SUMMARY
Quarter Ending March 31, 19—

Employee	Social Security No.	Quarterly Earnings	Federal Tax Withheld	FICA Withheld
Bird, Susan	547-76-6525	2560.60	218.39	171.56
Frazier, Mary	423-68-1100	2156.16	216.28	144.46
Hughes, Linda	587-19-4215	2615.42	225.17	175.23
Miller, Carrie	321-54-8176	1956.45	195.14	131.08
Smedley, Gary	456-78-9156	1856.78	192.87	124.40
TOTALS				

Check your answers in 31.

Employee	Social Security No.	Quarterly Earnings	Federal Tax Withheld	FICA Withheld
Bird, Susan	547-76-6525	2560.60	218.39	171.56
Frazier, Mary	423-68-1100	2156.16	216.28	144.46
Hughes, Linda	587-19-4215	2615.42	225.17	175.23
Miller, Carrie	321-54-8176	1956.45	195.14	131.08
Smedley, Gary	456-78-9156	1856.78	192.87	124.40
TOTALS		11145.41	1047.85	746.73

The information from the quarterly payroll summary above is reported to the IRS (Form 941) at the end of each quarter. The employer must deposit the total federal income tax and FICA withholdings in addition to the employer's FICA in a qualified commercial or Federal Reserve bank each month.

One additional report required by the IRS is the W-2 form. The W-2 contains an annual summary of earnings and taxes withheld. Several copies are sent to each employee to be used when filing individual tax forms.

Wage and Tax Statement 1984

Copy B   To be filed with employee's FEDERAL tax return

Type or print EMPLOYER'S Federal identifying number, name, address and ZIP code above.

Employer's State identifying number

Federal Income Tax Withheld	Total Wages Paid	Other Compensation (1)	Deferred Annuity (2)	Total FICA Wages Paid	FICA Employee Tax Withheld	State Income Tax Withheld

EMPLOYEE'S social security number ▶

(1) Other Compensation certified to the Oklahoma Director of State Finance as non-taxable income per IRS Regulations. Not included in Total Wages Paid.
(2) Not subject to Income Tax. Not included in Total Wages Paid.

NOTICE TO EMPLOYEE:

A. INCOME TAX WAGES — This statement is important. Copy B must be filed with your Federal Income Tax Return and Copy 2 must be filed with your State Income Tax Return. If your social security number, name, or address is stated incorrectly, notify your employer.

B. CREDIT FOR FICA TAX — If more than the maximum of FICA (social security and hospital insurance) employee tax was withheld during 1976 because you received wages from more than one employer, the excess should be claimed as a credit against your Federal income tax. See instructions for your Federal income tax return. The social security (FICA) rate of 5.85% includes 0.90% for Hospital Insurance Benefits and 4.95% for old-age, survivors, and disability insurance.

Type or print EMPLOYEE'S name, address and ZIP code above.

Form **W-2**   If this is a corrected form check the box in the left margin and type the words CORRECTED RETURN in all caps directly above the title "Wage and Tax Statement." This information is being furnished the Internal Revenue Service, and appropriate State Officials.

Go to **32** for a set of practice problems.

6 Payroll

PROBLEM SET 4

32   The answers are on page 586.

6.4   Payroll Records

A.   Complete the following employee earning record. Calculate the federal income tax (percentage method), FICA, total deductions, net pay, and quarterly totals.

Name	Mauldin, Mack		Social Security No.		423-67-1894		
Address	427 N.W. 89		Marital Status M	No. of Exemptions 2			

Pay Period	Gross Pay	Federal Income Tax	FICA	Other Deductions	Total Deductions	Net Pay
Jan. Feb. March	427.86 435.95 431.56			25.57 26.89 26.14		
First Quarter						

B.   Complete the following employee earnings record. Calculate the federal income tax (wage bracket method), FICA, total deductions, net pay, and quarterly, totals.

Name	McDaniel, Mark		Social Security No.		548-89-1037		
Address	325 S. Dump St.		Marital Status S	No. of Exemptions 1			

Week Ending	Gross Pay	Federal Income Tax	FICA	Other Deductions	Total Deductions	Net Pay
1/6	178.50			15.40		
1/13	180.00			16.72		
1/20	180.00			16.72		
1/27	179.00			16.55		
2/3	195.75			17.14		
2/10	210.50			18.57		
2/17	205.00			18.13		
2/24	195.00			17.09		
3/3	180.00			16.72		
3/10	178.25			15.36		
3/17	170.50			14.56		
3/24	195.00			17.09		
3/31	198.75			17.53		
QUARTER TOTALS						

Date _____

Name _____

Course/Section _____

C. Complete the following quarterly payroll summary.

QUARTERLY PAYROLL SUMMARY
Quarter Ending June 30, 19—

Employee	Social Security No.	Quarterly Earnings	Federal Tax Withheld	FICA Withheld
Ashmore, Harvey	423-58-1096	1,954.37	145.29	130.94
Buckner, Billy	564-48-2037	2,043.89	152.53	136.94
Davis, Leslie	209-47-8943	2,516.47	205.27	168.60
Hatcher, Deborah	325-86-8132	2,056.82	153.48	137.81
Humphries, Tina	558-31-8942	2,956.42	245.37	198.08
Jones, Judy	356-68-3529	1,836.25	142.35	123.03
Kinchion, John	321-54-4687	2,156.83	155.27	144.51
McDonald, Mark	456-31-5101	3,215.67	289.36	215.45
Wallis, Charlotte	105-25-4015	2,259.21	195.47	151.37
Whitworth, Brenda	538-86-3925	1,997.43	146.38	133.83
TOTALS				

When you have had the practice you need, turn to **33** for the Self-Test on payroll.

6 Payroll

The answers are on page 587.

1. Susan Kortemeier earns an annual salary of $13,548. What is her semi-monthly salary?

2. The Limon Box Company pays its workers on the following differential piecework schedule:

   $$\begin{array}{lll} 1\text{–}350 \text{ boxes} & @ & \$0.06 \\ 351\text{–}450 \text{ boxes} & @ & 0.08 \\ \text{Over } 450 \text{ boxes} & @ & 0.105 \end{array}$$

   What is the gross pay for a worker who constructed 503 boxes?

3. Juwana Oliver earns a weekly salary of $165, plus a commission of $2\frac{1}{2}\%$ of her total sales. What is her gross pay for a week in which she sold $2956.72 of merchandise?

4. Paul Harralson is paid based on the following sliding scale commission schedule:

   $$\begin{array}{lll} \text{Up to } \$2000 & @ & 9\% \\ \$2000 \text{ to } \$2750 & @ & 10\frac{1}{4}\% \\ \text{Over } \$2750 & @ & 11\frac{1}{2}\% \end{array}$$

   What is his gross pay for a total of $3156.78 in sales?

5. Complete the following weekly payroll sheet.

Name	M T W T F	Total Hours	Regular Hours	Regular Rate	O.T. Hours	O.T. Rate	Regular Pay	O.T. Pay	Gross Pay
Andree, Denise	8 8 7 8 7			4.14					
Cook, Chris	8 9 9 9 9			4.25					
Gatlin, Dennis	8 9 7 9 10			3.86					
Hernandez, Nasario	9 9 10 9 9			4.15					
Walker, Robert	9 8 8 8 8			4.02					
TOTALS									

6. Mary Lee is single and declares one exemption. Compute her federal income tax by the wage bracket method, and FICA on her weekly wage of $267.42.

7. Complete the following employee's earning record. Use the percentage method to calculate the federal income tax.

Date _____

Name _____

Course/Section _____

Name	Hill, Darrell		Social Security No.	443-56-7513			
Address	302 Bumpy Road		Marital Status M	No. of Exemptions 3			

Pay Period	Gross Pay	Federal Income Tax	FICA	Other Deduc-tions	Total Deduc-tions	Net Pay
Jan.	657.50			34.96		
Feb.	721.56			38.37		
March	694.28			36.92		
First Quarter						

8. Complete the following quarterly payroll summary.

QUARTERLY PAYROLL SUMMARY
Quarter Ending March 31, 19—

Employee	Social Security No.	Quarterly Earnings	Federal Income Tax	FICA
DeLong, Pamela	445-66-1023	2,857.86	313.96	191.48
Griggs, Lesa	443-64-7270	3,062.57	328.76	205.19
Holleman, William	054-36-0250	2,525.78	288.45	169.23
McManus, Michael	526-11-3538	2,896.42	298.76	194.06
Swyden, Randy	548-32-4156	3,158.59	326.42	211.63
TOTALS				

7 Simple Interest

WHAT'S SO INTERESTING ABOUT SIMPLE INTEREST?

COLLECTING IT!

Many times an individual or business needs additional capital for improvements. It's "interesting" to note how many people are borrowing money these days to make improvements.

Banks and lending institutions are often eager to lend money to individuals. In return for the loan of the money, lending institutions receive a payment called "interest" for the amount borrowed.

This unit is designed to aid individuals in computing the amount of interest charged for the use of a bank loan according to the amount of the loan, time, and interest rate.

Is your pencil handy? Good! This unit is simple but interesting, so let's get started.

UNIT OBJECTIVE

After successfully completing this unit, you will be able to calculate the number of days between two dates using exact time and 30-day-month time. You will be able to calculate simple interest using the following methods: (a) accurate interest, (b) ordinary interest at 30-day-month time, and (c) ordinary interest at exact time. You will be also able to solve a simple interest problem for principal, interest rate, or time.

This is a sample test for Unit 7. All the answers are placed immediately after the test. Work as many of the problems as you can, then check your answers and look after the answers for further directions.

		Where to Go for Help	
		Page	**Frame**
1.	Calculate the number of days from June 16 to October 9 using:		
	(a) exact time	251	1
	(b) 30-day-month time	251	1
2.	What is the interest on a loan of $2500 at 16% starting March 10 and due on October 15, 1983 using accurate simple interest?	259	8
3.	What is the interest on a loan of $800 at 12% starting April 9 and due on May 24, 1983 using ordinary simple interest at 30-day-month time?	259	8
4.	What is the principal on a loan at 18% for 73 days with accurate interest $12.60 (not a leap year)?	271	17
5.	What is the time in days on a loan of $500 at $12\frac{1}{2}$% with ordinary simple interest at 30-day-month time of $18.75?	271	17

* If you cannot work any of these problems, start with frame 1 on page 251.
* If you missed some of the problems, turn to the page numbers indicated.
* If all of your answers were correct, you are probably ready to proceed to Unit 8. If you would like more practice on Unit 7 before turning to Unit 8, try the Self-Test on page 285.
* Super-students—those who want to be certain they learn all of this—will turn to frame 1 and begin work there.

5. 108 days

4. $350

3. $12.00

2. $240.00

1. (a) 115 days
   (b) 113 days

7 Simple Interest

WHAT'S SO INTERESTING ABOUT SIMPLE INTEREST?

COLLECTING IT!

Many times an individual or business needs additional capital for improvements. It's "interesting" to note how many people are borrowing money these days to make improvements.

Banks and lending institutions are often eager to lend money to individuals. In return for the loan of the money, lending institutions receive a payment called "interest" for the amount borrowed.

This unit is designed to aid individuals in computing the amount of interest charged for the use of a bank loan according to the amount of the loan, time, and interest rate.

Is your pencil handy? Good! This unit is simple but interesting, so let's get started.

UNIT OBJECTIVE

After successfully completing this unit, you will be able to calculate the number of days between two dates using exact time and 30-day-month time. You will be able to calculate simple interest using the following methods: (a) accurate interest, (b) ordinary interest at 30-day-month time, and (c) ordinary interest at exact time. You will be also able to solve a simple interest problem for principal, interest rate, or time.

This is a sample test for Unit 7. All the answers are placed immediately after the test. Work as many of the problems as you can, then check your answers and look after the answers for further directions.

Where to Go for Help

	Page	Frame

1. Calculate the number of days from June 16 to October 9 using:

   (a) exact time   251   1

   (b) 30-day-month time   251   1

2. What is the interest on a loan of $2500 at 16% starting March 10 and due on October 15, 1983 using accurate simple interest?   259   8

3. What is the interest on a loan of $800 at 12% starting April 9 and due on May 24, 1983 using ordinary simple interest at 30-day-month time?   259   8

4. What is the principal on a loan at 18% for 73 days with accurate interest $12.60 (not a leap year)?   271   17

5. What is the time in days on a loan of $500 at $12\frac{1}{2}$% with ordinary simple interest at 30-day-month time of $18.75?   271   17

to frame 1 and begin work there.

★ Super-students—those who want to be certain they learn all of this—will turn Test on page 285.

If you would like more practice on Unit 7 before turning to Unit 8, try the Self-

★ If all of your answers were correct, you are probably ready to proceed to Unit 8.

★ If you missed some of the problems, turn to the page numbers indicated.

★ If you cannot work any of these problems, start with frame 1 on page 251.

5. 108 days

4. $350

3. $12.00

2. $240.00

1. (b) 113 days
   (a) 115 days

Answers

Preview 7

SECTION 1: TIME

1 | **OBJECTIVE**
|
| Calculate the number of days between two given dates using exact time and 30-day-month time.

"Time? I can count days, this sounds easy."

Exact Time

It is easy, but there are a few details to cover. In business transactions, there are two methods of counting days: exact time, and 30-day-month time. In *30-day-month time,* we count each month as having 30 days. With *exact* time, we count the actual number of days in the month.

Important ⟩

With both methods we do not count the starting day. For example, the number of days from October 12 to October 15 will be 3 days. One way to determine the number of days is to count them individually.

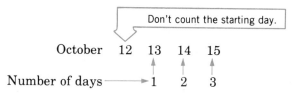

Don't count the starting day.

October   12   13   14   15

Number of days ────→ 1   2   3

Find the number of days from September 5 to September 19.

Check your answer in 2.

2   14 days. Was that your answer?

Count them: September 6, 7, 8, 9, 10, 11, 12, 13, 14, 15, 16, 17, 18, 19. A total of 14 days. Remember, don't count the starting day (in this case, September 5). Did you notice an easier way to do this problem? The best method is to subtract the starting day from the last day. This gives the difference between the two dates or the number of days between them. As required, this method doesn't count the starting day. In the above problem, the number of days will be $19 - 5 = 14$ days.

Use the short cut to find the number of days from June 9 to June 27.

Check your work in 3.

3   27 ◄─────── June 27, the ending date
   $\underline{-9}$ ◄─────── June 9, the starting date
   18 days

To calculate the number of days between *two different months* using exact time, you must first remember the number of days in each month.

January	31		July	31
February	28	(29 in leap year)	August	31
March	31		September	30
April	30		October	31
May	31		November	30
June	30		December	31

A leap year is one in which the last two digits of the year are evenly divisible by 4. The year 1984 is a leap year, since 84 is evenly divisible by 4: 84 ÷ 4 = 21. The year 1983 is not a leap year: 83 ÷ 4 = 20¾. Four does not evenly divide into 83.

To calculate the number of days between two dates using exact time, find the number of days for each individual month, then add. For example, to calculate the number of days from March 16 to July 4, find the number of days in March, April, May, June, and July.

$$
\begin{array}{lll}
\text{March 16 to March 31:} & 15 \text{ days} & \longleftarrow \; 31 - 16 = 15 \text{ days} \\
\text{All April:} & 30 \\
\text{All May:} & 31 \\
\text{All June:} & 30 \\
\text{to July 4:} & \underline{4} \\
\text{Total} \longrightarrow & 110 \text{ days}
\end{array}
$$

In July, we count all 4 days, but in March we do not count March 16. Remember, never count the starting day.

Find the number of days between the two dates using exact time. Remember to count February as 29 days in leap year.

1. May 9 to July 18, 1982
2. April 14 to November 5, 1983
3. January 27 to August 17, 1984

Check your solutions in 4.

4   1. May 9 to May 31: 22 days ◄——— 31 − 9 = 22 days
        All June: 30
        to July 18: 18
        TOTAL ——► 70 days

    2. April 14 to April 30:   16 days ◄——— 30 − 14 = 16 days
                     May: 31
                     June: 30
                     July: 31
                   August: 31
                September: 30
                  October: 31
            to November 5:   5
            TOTAL ——► 205 days

3. January 27 to January 31: 4 days ◄——— 31 − 27 = 4 days
February: 29 ◄——— 1984 is a leap year
March: 31
April: 30
May: 31
June: 30
July: 31
August: <u>17</u>
TOTAL ——► 203 days

30-day-month time

The second method of counting days is *30-day-month time*. This method assumes each month has 30 days and the year has 360 days. Thirty-day-month time is commonly used by businesses.

Use the following procedure for 30-day-month time:

Step 1. Find the number of months between the starting date and the same day in the ending month. Multiply the number of months by 30.

  Example: Find the number of days from March 13 to August 27 using 30-day-month time.

March 13 to August 13 = 5 months

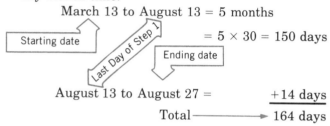

= 5 × 30 = 150 days

August 13 to August 27 = <u>+14 days</u>
Total ————► 164 days

Step 2. Identify the ending date and the last day in Step 1. Generally, these are different dates. Now, find the difference between these dates. With any additional days, add them to the total.

Now, it's your turn.

Find the number of days between the following two dates using 30-day-month time.

1. May 9 to July 23
2. January 6 to December 18
3. February 21 to September 30.

Turn to **5** to check your solutions.

5

1. May 9 to July 9 = 2 months = 2 × 30 days = 60 days
   July 9 to July 23 = <u>+14 days</u>
   TOTAL ——► 74 days

2. January 6 to December 6 = 11 months = 11 × 30 days = 330 days
   December 6 to December 18 = <u>+12 days</u>
   TOTAL ——► 342 days

3. February 21 to September 21 = 7 months = 7 × 30 = 210 days
   September 21 to September 30 = <u>+9 days</u>
   TOTAL ——► 219 days

When using *30-day-month time*, we assume each year has 360 days. It doesn't matter whether it's a leap year.

In the preceding problems, the day in the ending month was always "later" than the day in the first month. In this case, we always added the additional days (Step 2). Let's look at an example where there are "extra" or "left-over" days.

Find the number of days from February 9 to July 2 using 30-day-month time.

Step 1. February 9 to July 9 = 5 months = 5 × 30 days = 150 days
Step 2. July 2 to July 9 = $\qquad\qquad\qquad\qquad$ −7 days

$\qquad\qquad\qquad\qquad\qquad\qquad$ TOTAL ⟶ 143 days

Note that in Step 1 we have gone past the ending date. Hence, we subtract the *extra* days in Step 2.

Find the number of days between the two dates using 30-day-month time.

1. May 15 to October 4 $\qquad\qquad$ 3. October 25 to December 9
2. July 27 to September 27 $\qquad$ 4. April 24 to October 25

Check your answers in **6**.

6
1. May 15 to October 15 = 5 months = 5 × 30 days = 150 days
   October 4 to October 15 = $\qquad\qquad\qquad\qquad$ −11 days

   $\qquad\qquad\qquad\qquad\qquad$ TOTAL ⟶ 139 days

2. July 27 to September 27 = 2 months = 2 × 30 days = 60 days

   $\qquad\qquad\qquad\qquad\qquad$ TOTAL ⟶ 60 days

3. October 25 to December 25 = 2 months = 2 × 30 days = 60 days
   December 9 to December 25 = $\qquad\qquad\qquad\qquad$ −16 days

   $\qquad\qquad\qquad\qquad\qquad$ TOTAL ⟶ 44 days

4. April 24 to October 24 = 6 months = 6 × 30 days = 180 days
   October 24 to October 25 = $\qquad\qquad\qquad\qquad$ +1 day

   $\qquad\qquad\qquad\qquad\qquad$ TOTAL ⟶ 181 days

In Problem 4, the ending date, 25th, is later in the month than the beginning date, 24th. Since we have an additional day, we add. In Problem 3, the ending date, 9th, is earlier in the month than the beginning date, 25th. Here we subtract the extra days.

Be Careful ⟩

Remember, add for additional days and subtract for extra or left-over days. Now, turn to **7** for some practice problems on calculating the number of days.

---

7 Simple Interest

PROBLEM SET 1

7   The answers are on page 588.

7.1 Time   **A. Calculate the exact number of days between the two dates. Assume it is not a leap year.**

1. April 17 to July 23   _____

2. July 3 to September 14   _____

3. October 15 to December 23   _____

4. May 10 to July 28   _____

5. January 12 to November 18   _____

6. June 1 to October 30   _____

7. November 17 to December 18   _____

8. February 3 to November 5   _____

9. April 4 to July 18   _____

10. August 2 to September 23   _____

11. March 23 to June 5   _____

12. May 8 to June 17   _____

13. September 22 to December 29   _____

14. August 8 to October 15   _____

15. June 5 to November 8   _____

16. July 23 to December 27   _____

17. February 17 to November 5   _____

18. May 12 to September 7   _____

19. August 2 to November 12   _____

20. January 15 to April 14   _____

21. March 25 to August 3   _____

22. September 9 to December 18   _____

Date

23. November 14 to December 13   _____

Name

24. April 3 to October 5   _____

Course/Section

25. October 7 to December 13   _____

26. February 12 to March 2   —————————

27. June 15 to December 23   —————————

28. March 3 to July 14   —————————

29. May 17 to August 3   —————————

30. January 27 to November 14   —————————

B. Calculate the number of days between the two dates using 30-day-month time.

1. March 5 to April 7   —————————

2. August 17 to October 25   —————————

3. April 21 to June 29   —————————

4. September 18 to December 23   —————————

5. March 25 to June 29   —————————

6. May 7 to August 16   —————————

7. June 18 to September 7   —————————

8. November 13 to December 6   —————————

9. August 21 to October 18   —————————

10. July 25 to September 23   —————————

11. October 18 to December 4   —————————

12. April 30 to July 27   —————————

13. October 15 to December 17   —————————

14. January 17 to October 25   —————————

15. June 3 to November 14   —————————

16. February 12 to December 2   —————————

17. July 18 to September 5   —————————

18. September 27 to November 4   —————————

19. February 2 to July 27   —————————

20. June 6 to October 3   —————————

21. January 23 to December 15   —————————

22. April 5 to November 6   —————————

23. August 18 to October 18   —————————

24. May 29 to August 18 _____

25. March 25 to October 29 _____

26. April 12 to July 23 _____

27. July 17 to August 4 _____

28. February 27 to March 15 _____

29. May 23 to September 18 _____

30. January 18 to November 23 _____

C. **Calculate the exact number of days between the two dates.**

1. January 7 to March 13, 1984 _____

2. February 14 to July 15, 1982 _____

3. March 27 to August 23, 1983 _____

4. January 15 to August 12, 1984 _____

5. January 13 to July 4, 1983 _____

6. April 15 to August 7, 1984 _____

7. February 3 to June 18, 1984 _____

8. March 5 to May 19, 1983 _____

9. August 2 to October 15, 1983 _____

10. February 3 to October 13, 1983 _____

11. January 23 to March 15, 1984 _____

12. September 5 to November 23, 1982 _____

13. July 19 to December 18, 1982 _____

14. February 23 to June 5, 1984 _____

15. January 17 to March 8, 1982 _____

16. January 7 to April 15, 1984 _____

17. February 12 to June 3, 1984 _____

18. March 3 to July 7, 1983 _____

19. February 6 to April 15, 1984 _____

20. April 15 to December 3, 1983 _____

21. January 23 to March 3, 1982 _____

22. June 27 to September 18, 1984 —————————

23. February 4 to December 14, 1984 —————————

24. January 6 to March 3, 1982 —————————

25. May 13 to August 5, 1983 —————————

26. February 2 to June 12, 1984 —————————

27. January 13 to August 5, 1982 —————————

28. March 25 to December 3, 1983 —————————

29. January 3 to September 12, 1984 —————————

30. February 23 to April 5, 1984 —————————

When you have had the practice you need, either return to the Preview on page 250 or continue in 8 with the study of the three methods to calculate simple interest.

SECTION 2: CALCULATING SIMPLE INTEREST

8

> **OBJECTIVE**
>
> Calculate simple interest and maturity value using the following three methods: accurate simple interest, ordinary simple interest at 30-day-month time, and ordinary simple interest at exact time.

Simple interest is always calculated with the following formula:

Important

$$I = Prt$$

where

I = interest
P = principal
r = interest rate
t = time (as a fractional portion of a year)

Simple interest is always the product of principal, interest rate, and time, $I = Prt$. The interest rate r must be converted to a decimal number before multiplying.

There are three commonly used methods of calculating simple interest. All three methods use the same formula, $I = Prt$. The difference between these methods lies in the procedure for calculating the time. Two of the forms use exact time and the third uses 30-day-month time. In this section we will study all three methods.

ACCURATE SIMPLE INTEREST

Accurate simple interest (or *exact interest*) uses exact time. The time in the formula is the exact number of days between the starting and ending dates divided by the exact number of days in the year, 365 or 366 in a leap year.

For example, to calculate a loan for 132 days in 1983, you would use time as $132 \div 365$.

Accurate Simple Interest

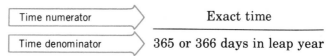

Time numerator

$$\frac{\text{Exact time}}{365 \text{ or } 366 \text{ days in leap year}}$$

Time denominator

Now, let's work an example.

Calculate the accurate simple interest on a loan of $750 at 12% made on May 13 and due on October 6, 1983.

First, calculate the exact number of days:

May 13 to May 31:	18 days
June:	30
July:	31
August:	31
September:	30
to October 6:	6
Total ⟶	146 days

Next, put the principal, rate, and time in the simple interest formula.

$P = \$750$

$r = 12\%$

$t = \dfrac{146 \longleftarrow \text{Exact number of days}}{365 \longleftarrow \text{Exact number of days in 1983}}$

Complete this problem and check your work in **9**.

I keep forgetting how to spell "principal" ...or is it "principle"? Which one is money? The boss at the high school? A rule?

Your pals at the high school and at the bank both spell it "princi**pal**." Remember, money is a business person's best **pal**.

9 $I = Prt = \$750 \times 12\% \times \dfrac{146}{365}$

$\quad\quad\quad = \$750 \times 0.12 \times \dfrac{146}{365} = \$36.00$

Always change the percent to a decimal number before multiplying. If you need a reminder of how to change a percent to a decimal number, return to page 122.

Whether using a calculator or doing the work by hand, the easiest method of calculating the interest in the above problem is to multiply \$750 times 0.12 times 146, and then divide this result by 365. This gives the interest, \$36. For help in using your calculator to compute simple interest, turn to the box on page 261.

This calculation can also be done using fractions:

$$I = \$750 \times 12\% \times \frac{146}{365} = \frac{\overset{6}{\cancel{750}}}{1} \times \frac{\overset{3}{\cancel{12}}}{\underset{25}{\cancel{100}}} \times \frac{\overset{2}{146}}{\underset{1}{\cancel{365}}}$$

$$= \frac{6 \times 3 \times 2}{1 \times 1 \times 1} = \frac{36}{1} = \$36$$

Many problems are not as easy to calculate as the preceding one. Generally, using decimal numbers (and a calculator, if possible) is the best bet. If the answer is not a simple number, round to the nearest cents position. For a quick review of rounding, turn to page 102.

When a loan is due, the borrower must pay back the principal plus the interest. This amount is called the *maturity value*. The maturity value is the total amount due when the loan is due or matures. Maturity Value = Principal + Interest. The letter S is generally used to represent maturity value.

$$\boxed{S = P + I}$$

S = maturity value
P = principal
I = interest

Calculate the maturity value of the preceding problem.

Check your answer in **10**.

COMPUTING INTEREST WITH A CALCULATOR

Once you have set them up, simple interest problems on the calculator are indeed simple. For example, to complete the following problem:

$$I = Prt = \$750 \times 12\% \times \frac{146}{365} = \$750 \times 0.12 \times \frac{146}{365}$$

Use the following keystroke sequence (for an algebraic-type calculator)

$$\boxed{7}\,\boxed{5}\,\boxed{0}\,\boxed{\times}\,\boxed{0}.\boxed{1}\,\boxed{2}\,\boxed{\times}\,\boxed{1}\,\boxed{4}\,\boxed{6}\,\boxed{\div}\,\boxed{3}\,\boxed{6}\,\boxed{5}\,\boxed{=}\qquad 36.$$

The answer is \$36.00. It's simple (interest)!

10   Maturity value $= S = P + I$
$$= \$750 + 36 = \$786$$

Easy, isn't it?

Try the following problems using accurate simple interest. Be careful when calculating the time.

1. Calculate the interest and maturity value on a loan of \$450 at $11\frac{1}{2}\%$ made on March 12 and due on October 17, 1983.
2. What is the interest and maturity value on a loan of \$375 at 17.3% made on April 5 and due on September 17, 1983?
3. Calculate the interest and maturity value on a loan of \$1200 at 19% made on January 16 and due on April 5, 1984.

Work each problem carefully and check your work in 11.

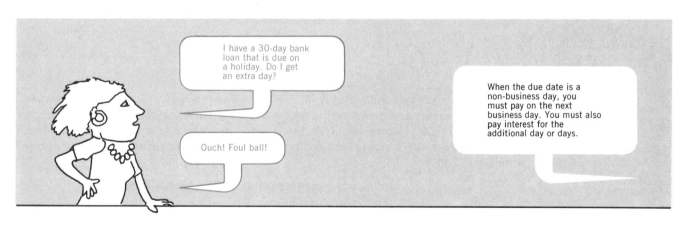

11   1. March 12 to March 31:   19
                        April: 30
                          May: 31
                         June: 30
                         July: 31
                       August: 31
                    September: 30
                to October 17:   17
                                 ———
                                 219

$$I = Prt = \$450 \times 11\frac{1}{2}\% \times \frac{219}{365} = \$450 \times 0.115 \times \frac{219}{365}$$

$$= \$31.05$$

$$S = P + I = \$450 + 31.05 = \$481.05$$

2. April 5 to April 30: 25
                  May: 31
              June: 30
              July: 31
         August: 31
  to September 17: <u>17</u>
                    165

$$I = Prt = \$375 \times 17.3\% \times \frac{165}{365} = \$375 \times 0.173 \times \frac{165}{365}$$

$$= \$29.327 \ldots = \$29.33 \text{ rounded}$$

$$S = P + I = \$375 + 29.33 = \$404.33$$

3. January 16 to January 31: 15
             February: 29 ←——1984 is a leap year.
              March: 31
       to April 5: <u>5</u>
                  80

$$I = Prt = \$1200 \times 19\% \times \frac{80}{366} \leftarrow\!\!-\!\!-1980 \text{ is a leap year.}$$

$$= \$1200 \times 0.19 \times \frac{80}{366} = \$49.836 \ldots = \$49.84 \text{ rounded}$$

$$S = P + I = \$1200 + 49.84 = \$1249.84$$

Always remember to check for leap year. With accurate simple interest, the denominator of the time fraction for a leap year is 366 (as with Problem 3 above).

ORDINARY SIMPLE
INTEREST AT
30-DAY-MONTH

Another way to calculate simple interest is *ordinary simple interest at 30-day-month time.* The difference between this method and accurate interest is the time fraction. In ordinary simple interest at 30-day-month time, the numerator is the number of days between the starting and ending dates using 30-day-month time. The denominator is 360.

Ordinary Simple Interest at 30-Day-Month Time

$$\frac{\text{Time numerator}}{\text{Time denominator}} \quad \frac{\text{30-day month time}}{360}$$

Work the following problem using ordinary simple interest at 30-day-month time.

Calculate the simple interest and maturity value on a loan of $750 at 12% made on May 13 and due on October 6, 1983.

Set it up using the simple interest formula, $I = Prt$, and the time fraction above.

Our work is in **12**.

12   May 13 to October 13 = 5 months = 5 × 30 days = 150 days
October 6 to October 13 =   −7 days

Total days ⟶ 143 days

$$I = Prt = \$750 \times 12\% \times \frac{143}{360}$$

$$= \$750 \times 0.12 \times \frac{143}{360} = \$35.75$$

$$S = P + I = \$750 + 35.75 = \$785.75$$

Remember, with ordinary simple interest at 30-day-month time, always use 30-day-month time in the numerator and 360 in the denominator of the time fraction.

Did you notice that this is the same problem we solved using accurate simple interest in frames **8** and **9**? The accurate simple interest is $36.00. The ordinary simple interest at 30-day-month time is $35.75.

Work the following problems using ordinary simple interest at 30-day-month time.

1. What is the interest and maturity value on a loan of $450 at $11\frac{1}{2}\%$ made on March 12 and due on October 17, 1983?
2. What is the interest and maturity value on a loan of $375 at 17.3% made on April 5 and due on September 17, 1983?
3. Calculate the interest and maturity value on a loan of $1200 at 19% made on January 16 and due on April 5, 1984.

Check your answers in **13**.

---

13   1. March 12 to October 12 = 7 months = 7 × 30 days = 210 days
        October 12 to October 17 =   +5 days

TOTAL ⟶ 215 days

$$I = Prt = \$450 \times 11\tfrac{1}{2}\% \times \frac{215}{360} = \$450 \times 0.115 \times \frac{215}{360}$$

$$= \$30.90625 = \$30.91 \text{ rounded}$$

$$S = P + I = \$450 + 30.91 = \$480.91$$

2. April 5 to September 5 = 5 months = 5 × 30 days = 150 days
   September 5 to September 17 =   +12 days

TOTAL ⟶ 162 days

$$I = Prt = \$375 \times 17.3\% \times \frac{162}{360} = \$375 \times 0.173 \times \frac{162}{360}$$

$$= \$29.19375 = \$29.19 \text{ rounded}$$

$$S = P + I = \$375 + 29.19 = \$404.19$$

3. January 16 to April 16 = 3 months = 3 × 30 days = 90 days
   April 5 to April 16 =   −11 days

TOTAL ⟶ 79 days

$$I = Prt = \$1200 \times 19\% \times \frac{79}{360} = \$1200 \times 0.19 \times \frac{79}{360}$$

$$= \$50.033 \ldots = \$50.03 \text{ rounded}$$

$$S = P + I = \$1200 + 50.03 = \$1250.03$$

In Problem 3, the fact that 1984 is a leap year does not affect the calculation. In fact, the year can be totally ignored when using ordinary simple interest at 30-day-month time, since we assume each month has 30 days and each year has 360 days.

The above three problems were the same ones we worked in frame 11 using accurate simple interest. The answers are different and in some cases, the interest is greater with accurate interest. In others, it is less.

ORDINARY SIMPLE
INTEREST AT EXACT
TIME

In the third method, ordinary simple interest at exact time (or banker's interest), the interest is always greater than or equal to the other two methods.

Ordinary simple interest at exact time is a combination of the other methods. The numerator of the time fraction uses exact time. The denominator is 360. In both types of "ordinary" interest, the denominator is *always* 360.

Ordinary Simple Interest at Exact Time

$$\frac{\text{Time numerator}\rangle \; \text{exact time}}{\text{Time denominator}\rangle \; 360}$$

Try the following problem.

Calculate the interest and maturity value on a loan of $750 at 12% made on May 13 and due on October 6, 1983. Use ordinary simple interest at exact time.

Set the problem up using the simple interest formula, $I = Prt$, and the time fraction given above.

Check your answer in 14.

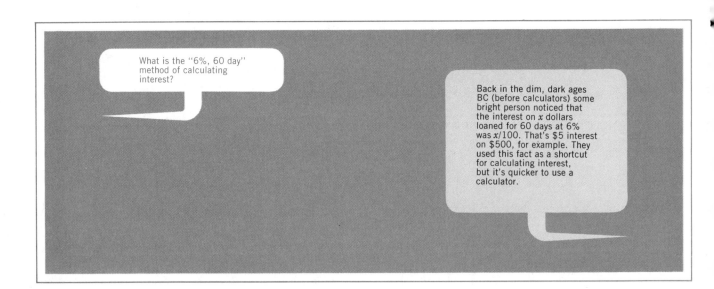

What is the "6%, 60 day" method of calculating interest?

Back in the dim, dark ages BC (before calculators) some bright person noticed that the interest on x dollars loaned for 60 days at 6% was $x/100$. That's $5 interest on $500, for example. They used this fact as a shortcut for calculating interest, but it's quicker to use a calculator.

14   May 13 to May 31: 18 days
June: 30
July: 31
August: 31
September: 30
to October 6: 6
TOTAL ⟶ 146 days

$$I = Prt = \$750 \times 12\% \times \frac{146}{360} = \$750 \times 0.12 \times \frac{146}{360} = \$36.50$$

$$S = P + I = \$750 + 36.50 = \$786.50$$

Remember, with ordinary simple interest at exact time, the numerator uses exact time; the denominator is 360.

The answer to the above problem is $36.00 using accurate interest and $35.75 with ordinary simple interest at 30-day-month time. When using ordinary simple interest at exact time, the interest is always more.

Try a few more problems. Use ordinary simple interest at exact time.

1. What is the interest and maturity value on a loan of $150 at 16% starting April 14 and due on September 5, 1983?
2. Calculate the interest and maturity value on a loan of $1275 at 13.8% made on March 17 and due on May 5, 1983.
3. What is the interest and maturity value on a loan of $825 at 19% made on January 2 and due on April 15, 1984.

Our work is in **15**. Check your work when you have completed these problems.

Hey, you did this problem incorrectly. Looks like you were using inaccurate, unordinary hard interest at inexact time.

SØRRY. I WAS TAKING A CIRCUIT BREAK.

15   1.   April 14 to April 30:   16 days
                        May: 31
                       June: 30
                       July: 31
                     August: 31
        to September 5:   5
                    TOTAL ———▶ 144 days

$$I = Prt = \$150 \times 16\% \times \frac{144}{360} = \$150 \times 0.16 \times \frac{144}{360} = \$9.60$$

$$S = P + I = \$150 + 9.60 = \$159.60$$

2.   March 17 to March 31:   14 days
                     April: 30
             to May 5:   5
              TOTAL ———▶ 49 days

$$I = Prt = \$1275 \times 13.8\% \times \frac{49}{360} = \$1275 \times 0.138 \times \frac{49}{360}$$

$$= \$23.94875 = \$23.95 \text{ rounded}$$

$$S = P + I = \$1275 + 23.95 = \$1298.95$$

3. January 2 to January 31: 29 days
                   February: 29 ←————————1984 is a leap year
                      March: 31
                to April 15: 15
                     TOTAL ———→ 104

$$I = Prt = \$825 \times 19\% \times \frac{104}{360} = \$825 \times 0.19 \times \frac{104}{360}$$

$$= \$45.283 \ldots = \$45.28 \text{ rounded}$$

$$S = P + I = \$825 + 45.28 = \$870.28$$

In Problem 3, 1984 is a leap year. When using exact time, you may find that leap year makes a difference.

All three simple interest methods use the same basic formula, $I = Prt$, but each method uses a different time fraction. If is very important for you to remember the numerator and denominator for each method.

Summary >

	Accurate Simple Interest	Ordinary Simple Interest at 30-day-month time	Ordinary Simple Interest at Exact time
Time numerator	Exact time	30-day-month time	Exact time
Time denominator	365 or 366 in leap year	360	360

Turn to **16** for a set of practice problems on calculating simple interest.

7 Simple Interest

PROBLEM SET 2

16   The answers are on page 588.

7.2 Calculating
Simple Interest

A. In the following problems, calculate the time, accurate simple interest, and maturity value.

	Principal	Rate	Beginning Date	Due Date	Time	Interest	Maturity Value
1.	$750	19%	March 16	October 21, 1983	___	___	___
2.	200	12%	August 12	October 24, 1982	___	___	___
3.	500	13%	May 5	September 28, 1983	___	___	___
4.	825	16%	July 28	October 9, 1982	___	___	___
5.	470	17%	January 5	November 3, 1983	___	___	___
6.	855	18%	May 17	July 4, 1983	___	___	___
7.	325	$12\frac{1}{2}$%	March 23	June 18, 1983	___	___	___
8.	450	$11\frac{3}{4}$%	February 8	August 23, 1982	___	___	___
9.	329	$18\frac{1}{4}$%	June 25	September 14, 1983	___	___	___
10.	856	$10\frac{1}{2}$%	April 17	November 3, 1983	___	___	___
11.	475	19%	January 7	April 14, 1984	___	___	___
12.	587	11%	February 18	December 18, 1984	___	___	___
13.	750	18%	February 6	October 15, 1984	___	___	___
14.	1270	15%	January 15	September 12, 1984	___	___	___
15.	2500	16%	March 3	May 25, 1983	___	___	___
16.	1300	$17\frac{1}{2}$%	April 27	June 5, 1983	___	___	___
17.	650	12%	February 17	August 15, 1983	___	___	___
18.	875	$19\frac{1}{4}$%	January 12	September 4, 1983	___	___	___
19.	6200	$16\frac{3}{4}$%	November 13	December 12, 1984	___	___	___
20.	9800	$17\frac{1}{2}$%	April 15	July 5, 1984	___	___	___
21.	7650	16%	January 4	August 17, 1984	___	___	___
22.	6200	$16\frac{1}{4}$%	February 17	November 23, 1984	___	___	___
23.	8500	$15\frac{1}{2}$%	April 3	July 21, 1984	___	___	___
24.	7275	$12\frac{3}{4}$%	February 23	August 29, 1984	___	___	___
25.	8350	$13\frac{1}{2}$%	January 14	March 19, 1984	___	___	___

Date

Name

Course/Section

B. In the following problems, calculate the time, ordinary simple interest at 30-day-month time, and maturity value.

	Principal	Rate	Beginning Date	Due Date	Time	Interest	Maturity Value
1.	$900	18%	March 3	April 13	——	——	——
2.	250	12%	August 5	November 17	——	——	——
3.	400	19%	June 14	July 29	——	——	——
4.	600	18%	April 6	August 21	——	——	——
5.	500	19%	May 14	September 23	——	——	——
6.	700	11%	January 27	March 29	——	——	——
7.	530	$12\frac{1}{4}$%	June 18	September 7	——	——	——
8.	675	$11\frac{3}{4}$%	February 23	May 4	——	——	——
9.	325	18%	May 19	October 26	——	——	——
10.	850	10%	July 25	December 29	——	——	——
11.	427	$12\frac{1}{2}$%	April 13	July 7	——	——	——
12.	836	$18\frac{3}{4}$%	March 11	November 4	——	——	——
13.	750	18%	February 6	October 15	——	——	——
14.	1270	15%	January 15	September 12	——	——	——
15.	2500	16%	March 3	May 25	——	——	——
16.	1300	$17\frac{1}{2}$%	April 27	June 5	——	——	——
17.	650	12%	February 17	August 15	——	——	——
18.	875	$19\frac{1}{4}$%	January 12	September 4	——	——	——
19.	6200	$16\frac{3}{4}$%	November 13	December 12	——	——	——
20.	9800	$17\frac{1}{2}$%	April 15	July 5	——	——	——
21.	7650	16%	January 4	August 17	——	——	——
22.	6200	$16\frac{1}{4}$%	February 17	November 23	——	——	——
23.	8500	$15\frac{1}{2}$%	April 3	July 21	——	——	——
24.	7275	$12\frac{3}{4}$%	February 23	August 29	——	——	——
25.	8350	$13\frac{1}{2}$%	January 14	March 19	——	——	——

C. In the following problems, calculate the time, ordinary simple interest at exact time, and maturity value.

	Principal	Rate	Beginning Date	Due Date	Time	Interest	Maturity Value
1.	$600	12%	March 4	April 13, 1983	____	____	____
2.	500	19%	June 17	August 28, 1983	____	____	____
3.	250	18%	April 14	May 29, 1983	____	____	____
4.	300	12%	May 23	October 5, 1984	____	____	____
5.	500	18%	June 15	September 8, 1984	____	____	____
6.	700	16%	May 7	December 23, 1983	____	____	____
7.	450	$19\frac{1}{2}$%	July 8	October 14, 1983	____	____	____
8.	375	$11\frac{3}{4}$%	March 17	November 21, 1983	____	____	____
9.	825	11%	April 14	June 19, 1982	____	____	____
10.	480	19%	August 3	October 7, 1983	____	____	____
11.	256	$18\frac{1}{4}$%	January 23	March 15, 1982	____	____	____
12.	438	$12\frac{1}{2}$%	February 12	April 3, 1984	____	____	____
13.	750	18%	February 6	October 15	____	____	____
14.	1270	15%	January 15	September 12	____	____	____
15.	2500	16%	March 3	May 25	____	____	____
16.	1300	$17\frac{1}{2}$%	April 27	June 5	____	____	____
17.	650	12%	February 17	August 15	____	____	____
18.	875	$19\frac{1}{4}$%	January 12	September 4	____	____	____
19.	6200	$16\frac{3}{4}$%	November 13	December 12	____	____	____
20.	9800	$17\frac{1}{2}$%	April 15	July 5	____	____	____
21.	7650	16%	January 4	August 17	____	____	____
22.	6200	$16\frac{1}{4}$%	February 17	November 23	____	____	____
23.	8500	$15\frac{1}{2}$%	April 3	July 21	____	____	____
24.	7275	$12\frac{3}{4}$%	February 23	August 29	____	____	____
25.	8350	$13\frac{1}{2}$%	January 14	March 19	____	____	____

When you have had the practice you need, either return to the Preview on page 250 or continue in **17** where we will learn to solve for other interest variables.

SECTION 3: SOLVING FOR OTHER INTEREST VARIABLES

17

OBJECTIVE

In a simple interest problem, calculate the principal, interest, rate, or time.

In the basic formula for simple interest, $I = Prt$, if three of the items are known, the fourth can be found by solving. In the previous section, I was calculated when P, r, and t were known. In this section, you will learn to solve for the other three quantities. The procedure is similar to that used in solving the percent problems in Unit 4.

If you are good at doing algebra, solving the interest formula for each of the variables will be easy. If you are not good at algebra you must either memorize the following formulas or learn how to use the visual helper shown below.

INTEREST
EQUATION FINDER

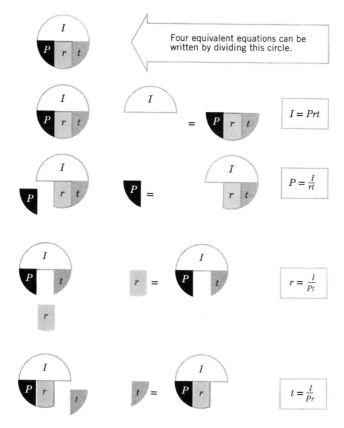

All four equations, $I = Prt$, $P = \dfrac{I}{rt}$, $r = \dfrac{I}{Pt}$, and $t = \dfrac{I}{Pr}$ are equivalent.

Now, turn to **18** to see how the equation for the principal is used.

Solving For
Principal

18

If the interest, interest rate, and time are known, the principal can be calculated. Using the *Interest Equation Finder*, we know $P = \dfrac{I}{rt}$.

To calculate the principal, substitute the interest, rate, and time in the formula.

Example. What is the principal on a loan with an interest rate of 12%, time of 146 days, and accurate interest of $36 (for a 365-day year).

$$I = \$36$$

$$r = 12\% = 0.12$$

$$t = \frac{146}{365} \longleftarrow 365 \text{ for accurate interest}$$

$$P = \frac{I}{rt} = \frac{\$36}{0.12 \times \frac{146}{365}}$$

Complete the calculation, then go to **19**.

SOLVING FOR OTHER INTEREST VARIABLES WITH A CALCULATOR

Calculators are very useful in solving for other interest variables. To complete the following problem,

$$P = \frac{I}{rt} = \frac{\$36}{12\% \times \frac{146}{365}} = \frac{\$36}{0.12 \times \frac{146}{365}}$$

Step 1. First, calculate the denominator. Use the following keystroke sequence (for an algebraic type calculator)

| 0 | . | 1 | 2 | × | 1 | 4 | 6 | ÷ | 3 | 6 | 5 | = | 0.048

The result will be 0.048.

Step 2. Divide $36 by 0.048. If you do not have an inverse key (1/x), you must remember, write down, or store 0.048 in your calculator memory. Then divide.

| 3 | 6 | ÷ | 0 | . | 0 | 4 | 8 | = | 750.

The answer is 750.

Alternate step 2. If you have an inverse key, you can complete the calculation without storing or remembering the denominator. Use the following sequence:

| 0 | . | 0 | 4 | 8 | 1/x | × | 3 | 6 | = | 750.

from previous calculation

Calculators are very helpful in problems that do not work out evenly. Try the following,

$$P = \frac{I}{rt} = \frac{22.72}{9\% \times \frac{127}{365}} = \frac{22.72}{0.09 \times \frac{127}{365}}$$

Step 1. First, calculate the denominator.

| 0 | . | 0 | 9 | × | 1 | 2 | 7 | ÷ | 3 | 6 | 5 | = | .03131506 |

The result is 0.03131506.

Step 2. Next, divide the numerator by the denominator and round.

| 2 | 2 | . | 7 | 2 | ÷ | 0 | . | 0 | 3 | 1 | 3 | 1 | 5 | 0 | 6 | = | 725.52950 |

The answer is $725.52950 = $725.53 rounded. This type of problem can be worked out by hand, but using a calculator is much, much easier.

Try a few problems on your calculator.

1. What principal is necessary to yield accurate interest of $19.02 at 11.7% in 213 days? (Assume it's a 365-day year.)
2. Find the principal on a loan with an interest rate of $9\frac{3}{8}\%$; the time is 41 days, and the ordinary interest is $14.92.
3. What interest rate, rounded to the nearest tenth of a percent, will earn accurate interest of $16.91 on $761.52 in 67 days? (Not a leap year.)
4. Find the interest rate, rounded to the nearest tenth of a percent, on a loan of $247.86 made on April 13 and due on June 18, with ordinary interest at 30-day-month time of $4.12.
5. How much time (in days) will it take a principal of $476.53 to earn $20.13 interest at $9\frac{5}{8}\%$ (ordinary interest at 30-day-month time)?
6. What is the time (in days) on a loan of $125.67 at $14\frac{3}{4}\%$ with $4.62 interest? (Accurate interest—not a leap year.)

The answers are on page 589.

19
$$P = \frac{I}{rt} = \frac{\$36}{12\% \times \frac{146}{365}} = \frac{\$36}{0.12 \times \frac{146}{365}} = \frac{\$36}{0.048} = \$750$$

Simplify this first.

Always change the percent to a decimal number before multiplying. The easiest way to complete the calculation is to simplify the denominator first. Then, divide the numerator by the denominator. Remember, the time fraction will always depend on the interest method used.

Work the following problems:

1. What principal is necessary to yield ordinary interest of $3.75 at 12% in 45 days?
2. Find the principal on a loan with interest rate 18%, time 73 days, and accurate interest $26.10? (Assume a 365-day year.).

Be very careful with the time fraction.

Turn to **20** to check your work.

20

1. $P = \dfrac{I}{rt} = \dfrac{\$3.75}{12\% \times \dfrac{45}{360}} = \dfrac{\$3.75}{0.12 \times \dfrac{45}{360}} = \dfrac{\$3.75}{0.015} = \$250$

Ordinary interest—use 360 days.

2. $P = \dfrac{I}{rt} = \dfrac{\$26.10}{18\% \times \dfrac{73}{365}} = \dfrac{\$26.10}{0.18 \times \dfrac{73}{365}} = \dfrac{\$26.10}{0.036} = \$725$

Accurate interest—use 365 days.

Solving for interest rate

If the interest, principal, and time are known, the interest rate can be computed. Using the Interest Equation Finder, we have the following formula for interest rate:

$$r = \frac{I}{Pt}$$

To calculate the interest rate, just substitute the interest, principal, and time into the formula and simplify. As before, the time fraction will depend on the interest method. The answer will be a decimal number which must be changed to a percent.

Example: What is the interest rate on a loan of $900 for 65 days with ordinary interest $19.50?

$I = \$19.50$

$P = \$900$

$t = \dfrac{65}{360}$ ◄———— 360-day year for ordinary interest

$r = \dfrac{I}{Pt} = \dfrac{19.50}{900 \times \dfrac{65}{360}}$

Complete the calculation and check your work in **21**.

21 $$r = \frac{I}{Pt} = \frac{19.50}{900 \times \frac{65}{360}} = \frac{19.50}{162.5} = 0.12 = 12\%$$

First, simplify the denominator. Then divide the numerator by the denominator. The result will be a decimal number that must be changed to a percent. (Remember, we are solving for interest *rate*). If you need a quick review of how to change a decimal number to a percent, turn to page 122.

Try a few more interest rate problems.

1. What interest rate will earn accurate interest of $37.50 on $500 in 219 days? (Not a leap year.)
2. Find the interest rate on a loan of $250 made on March 6 and due on July 21, with ordinary interest at 30-day-month time of $15.00.

Check your calculations in **22**.

---

22   1. $$r = \frac{I}{Pt} = \frac{37.50}{500 \times \frac{219}{365}} = \frac{37.50}{300} = 0.125 = 12.5\% \qquad \text{or} \qquad 12\tfrac{1}{2}\%$$

Accurate interest

2.   March 6 to July 6 = 4 months = 120 days
July 6 to July 21 =   +15 days
           TOTAL ⟶ 135 days

$$r = \frac{I}{Pt} = \frac{15.00}{250 \times \frac{135}{360}} = \frac{15.00}{93.75} = 0.16 = 16\%$$

Ordinary interest

When solving for the interest rate, *always* convert the answer to a percent.

Solving for time
The final type of interest problem involves solving for the time. If the interest, principal, and interest rate are known, you can calculate the time. The formula can easily be obtained from our Interest Equation Finder.

To calculate the time, first substitute the interest, principal, and interest rate in the formula and simplify. The result will be a (common or decimal) fraction that represents years. To arrive at the number of days, multiply the fraction by the number of days in the year. Multiply by 365 or 366 for exact time, and by 360 for 30-day-month time.

Example: What is the time (in days) on a loan of $400 at 11.7% if the interest is $5.85 (ordinary simple interest at 30-day-month time).

$$t = \frac{I}{Pr} = \frac{5.85}{400 \times 11.7\%}$$

Finish the example and find the time.

Time =   _____?_____

Choose an answer:

(a) 0.125 days   Go to 23.
(b) 45 days   Go to 24.
(c) What? I don't know.   Go to 25.

---

23   No. The time is 0.125 *years,* not 0.125 days. The answer is found in years and must be converted to days. Multiply 0.125 by the number of days in the year using 30-day-month time. Complete the problem and check your answer in 24.

---

24   Right you are!

The correct answer is 45 days.

Multiply: $0.125 \times 360 = 45$ days

30-day-month time

Always remember to multiply the time fraction by the number of days in the year.

Work the following problems:

1. How much time (in days) will it take a principal of $250 to earn $5.00 interest at 18% (ordinary interest at 30-day-month time)?
2. What is the time (in days) on a loan of $525 at 18% with $18.90 interest? (Accurate interest—not a leap year.)

Check your work in 26.

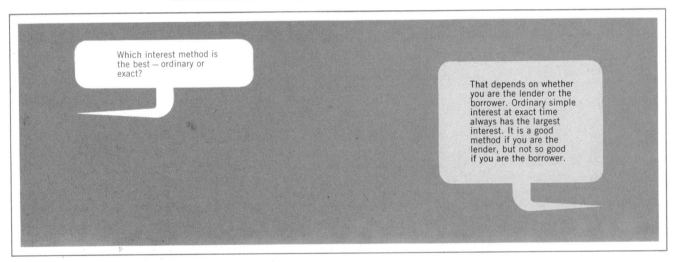

Which interest method is the best — ordinary or exact?

That depends on whether you are the lender or the borrower. Ordinary simple interest at exact time always has the largest interest. It is a good method if you are the lender, but not so good if you are the borrower.

25   Don't worry, we'll explain. First, simplify the fraction just as you did in the previous problems.

$$t = \frac{I}{Pr} = \frac{5.85}{400 \times 11.7\%} = \frac{5.85}{400 \times 0.117} = \frac{5.85}{46.8} = 0.125$$

Remember . . .

1. Change the percent to a decimal number.
2. Then, simplify the denominator.
3. Finally, divide the numerator, 5.85 by the denominator, 46.8.
4. The answer is 0.125 *years.* All answers should be written in days.

Multiply 0.125 by the number of days in the year using 30-day-month time. Complete the problem and check your answer in 24.

---

1. $t = \dfrac{I}{Pr} = \dfrac{5.00}{250 \times 18\%} = \dfrac{5.00}{250 \times 0.18} = \dfrac{5.00}{45} = 0.111 \ldots 0.111$ rounded

   Days $= 0.111 \times 360 = 39.96 = 40$ days rounded

   > 30-day-month time

Getting a repeated decimal number like 0.111 . . . may seem a little strange. But in this case, just round. Three decimal places are sufficient even for bankers.

If you worked this problem using fractions, you would get $t = \frac{1}{9}$. Then the number of days would be $\frac{1}{9} \times 360 = 40$ days. In this case, working with fractions yields an exact answer.

If you're using a calculator, use all the decimal places your calculator has. To obtain the number of days multiply $0.1111111 \times 360 = 39.999996 = 40$ rounded.

2. $t = \dfrac{I}{Pr} = \dfrac{18.90}{525 \times 0.18} = \dfrac{18.90}{94.5} = 0.2$

   Days $= 0.2 \times 365 = 73$ days

   > Exact time

In Problem 2 remember to multiply by 365 for exact time.

Be sure to remember the three formulas used to solve for principal, interest rate, and time.

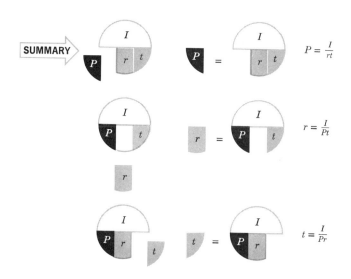

SUMMARY

$P = \dfrac{I}{rt}$

$r = \dfrac{I}{Pt}$

$t = \dfrac{I}{Pr}$

Summary ⟩ Are you ready for a set of practice problems? Then, turn to **27**.

7 Simple Interest

The answers are on page 589.

7.3 Solving for Other Interest Variables

A. In the following problems, calculate the principal. Use ordinary simple interest at 30-day-month time.

	Interest	Rate	Time	Principal
1.	$10.00	16%	45 days	_____
2.	33.00	11%	72 days	_____
3.	40.00	12%	40 days	_____
4.	30.00	10%	90 days	_____
5.	7.00	7%	30 days	_____
6.	88.00	11%	144 days	_____
7.	20.25	6%	180 days	_____
8.	7.00	14%	40 days	_____
9.	22.00	11%	72 days	_____
10.	18.00	12%	60 days	_____
11.	11.20	8%	72 days	_____
12.	9.00	12%	45 days	_____
13.	6.25	12.5%	45 days	_____
14.	4.50	9%	30 days	_____
15.	14.00	12%	60 days	_____
16.	112.00	18%	40 days	_____
17.	40.00	16%	36 days	_____
18.	100.00	15%	30 days	_____
19.	128.00	12%	60 days	_____
20.	231.20	17%	72 days	_____
21.	29.25	18%	90 days	_____
22.	60.00	20%	30 days	_____

Date

Name

Course/Section

B. In the following problems, calculate the interest rate. Use ordinary simple interest at 30-day-month time.

	Interest	Principal	Time	Rate
1.	$12.00	$600	60 days	_____
2.	33.00	1200	90 days	_____
3.	28.00	500	144 days	_____
4.	36.00	1800	80 days	_____
5.	14.00	800	45 days	_____
6.	6.00	600	40 days	_____
7.	10.00	500	120 days	_____
8.	2.50	250	36 days	_____
9.	6.00	800	45 days	_____
10.	14.00	900	20 days	_____
11.	7.00	450	40 days	_____
12.	27.00	750	72 days	_____
13.	6.25	400	45 days	_____
14.	4.75	600	30 days	_____
15.	8.50	600	60 days	_____
16.	50.00	2000	60 days	_____
17.	119.00	6300	40 days	_____
18.	95.00	5000	36 days	_____
19.	15.00	1500	30 days	_____
20.	50.40	1800	72 days	_____
21.	100.00	2500	90 days	_____
22.	185.00	9000	40 days	_____

C. In the following problems, calculate the time in days. Use ordinary simple interest at 30-day-month time.

	Interest	Principal	Rate	Time
1.	$12.00	$800	12%	_____
2.	5.00	400	10%	_____
3.	90.00	1500	18%	_____
4.	6.00	500	12%	_____
5.	24.00	1000	12%	_____
6.	27.00	900	18%	_____
7.	4.50	600	9%	_____
8.	8.00	400	12%	_____
9.	2.50	250	10%	_____
10.	3.00	300	9%	_____
11.	5.00	500	10%	_____
12.	2.55	300	$12\frac{3}{4}\%$	_____
13.	24.50	800	$12\frac{1}{4}\%$	_____
14.	20.25	675	6%	_____
15.	13.20	1100	12%	_____
16.	292.50	6500	18%	_____
17.	62.50	5000	15%	_____
18.	323.00	9500	17%	_____
19.	105.00	4500	14%	_____
20.	31.00	2000	$15\frac{1}{2}\%$	_____
21.	60.00	2400	15%	_____
22.	114.00	3000	19%	_____

D. Solve the following problems. (Use a 365-day year for accurate interest.)

1. What is the principal on a loan at 18% for 73 days with accurate interest of $9.00?

2. What is the principal on a loan at 11% for 45 days with ordinary interest of $5.50?

   _____

3. What is the interest rate on a loan of $657 for 40 days with accurate interest of $6.48?

   _____

4. At what interest rate must $350 be invested for 40 days to yield ordinary interest of $15.00?

   _____

5. For what length of time (in days) must $725 be invested at $12\frac{1}{2}$% to yield accurate interest of $36.25?

   _____

6. What is the time (in days) on a loan of $400 at $12\frac{3}{4}$% with ordinary interest of $12.75?

   _____

7. What is the principal on a loan at $10\frac{3}{4}$% for 219 days with accurate interest of $51.60?

   _____

8. What is the time (in days) on a loan of $475 at 18% with ordinary interest of $42.75?

   _____

9. At what rate must $300 be invested for 60 days to earn ordinary simple interest of $3.00?

   _____

10. What is the interest rate on a loan of $1825 for 50 days with accurate interest of $27.50?

    _____

11. What is the time on a loan of $500 at $12\frac{3}{4}$% with accurate interest of $25.50?

    _____

12. What amount must be invested at 12% for 120 days to yield ordinary interest of $36.00?

    _____

13. What is the principal on a loan starting May 9 and due July 21, 1983, with interest rate 7% and accurate interest of $5.60?

    _____

14. What is the principal on a loan starting July 15 and due August 20, 1982 with interest rate 9% and ordinary interest at exact time of $4.50?

    _____

15. What is the interest rate on a loan of $500, starting March 18 and due August 11, 1983 with accurate interest of $24.00?

    _____

16. What is the interest rate on a loan of $450 starting November 3 and due on December 13, 1982 with ordinary interest of $4.50?

    _____

17. What is the time (in days) on a loan of $950 at 10% with accurate interest of $57.00?

    _____

18. What is the time (in days) on a loan of $600 at $12\frac{1}{2}$% with ordinary interest of $15.00?

    _____

19. What amount must be invested at $10\frac{1}{2}$% for 73 days to yield accurate interest of $52.50?

    _____

20. What is the time (in days) on a loan of $600 at 11% with ordinary interest of $44.00?

    _____

21. What is the interest rate on a loan of $5000 for 146 days with accurate interest of $380?

    _____

22. Calculate the principal on a loan for 40 days, with interest rate 15% and accurate interest of $96.

    _____

When you have had the practice you need, turn to 28 for the Self-Test on simple interest.

The answers are on page 589.

1. Calculate the number of days from April 17 to August 9 using: (a) exact time, (b) 30-day-month time.

    (a) _____

    (b) _____

2. Calculate the (a) interest and (b) maturity value on a loan of $750 at 12% starting March 6 and due on May 16, 1983 using accurate simple interest.

    (a) _____

    (b) _____

3. Calculate the (a) interest and (b) maturity value on a loan of $500 at 9% starting March 26 and due on May 2, 1984 using ordinary simple interest at 30-day-month time.

    (a) _____

    (b) _____

4. Calculate the (a) interest and (b) maturity value on a loan of $450 at $12\frac{1}{2}$% starting October 7 and due on November 16, 1982 using ordinary simple interest at exact time.

    (a) _____

    (b) _____

5. What is the principal on a loan with interest rate 18%, time 60 days, and ordinary interest $21.00?

    _____

6. What is the interest rate on a loan of $200, starting March 16 and due August 9, 1983, with accurate interest of $7.60?

    _____

7. What is the time in days on a loan of $600 at 13% with ordinary interest at 30-day-month time of $19.50?

    _____

Date _____

Name _____

Course/Section _____

8 Bank Discount, Compound Interest, and Present Value

Did you get that one? If your answer is yes, then you'll have no trouble going through this unit. If you're still sitting there trying to figure it out—you need to pay close attention to the material covered in this unit.

This unit is designed to help you understand bank discount and how compound interest and present value are computed.

Let's get started! The "present value" of this chapter should "compound your interest" by letting you know how to get more money for your money.

UNIT OBJECTIVE

After successfully completing this unit, you will be able to calculate compound interest, present value, and bank discount on non-interest-bearing and interest-bearing notes.

PREVIEW 8   This is a sample test for Unit 8. All the answers are placed immediately after the test. Work as many of the problems as you can, then check your answers and look after the answers for further directions.

Where to Go for Help

	Page	Frame

1. Calculate the (a) maturity value, (b) bank discount, and (c) net proceeds for a $500, 12%, 90-day note, dated June 18 and discounted July 18 at 9%. Use ordinary simple interest at 30-day-month time.

(a) _____

(b) _____

(c) _____   289   1

2. Calculate the maturity value of $275 invested at 6% compounded monthly for five years. (Use the table on page 299).   _____   297   8

3. What amount must be invested at 8% compounded quarterly for 10 years to yield $1500? (Use the table on page 306).   _____   305   13

turn to frame 1 and begin work there.

★ Super-students—those who want to be certain to learn all of this material—will (If you would like more practice on Unit 8, try the Self-Test on page 311.)

★ If all of your answers were correct, you are probably ready to proceed to Unit 9.

★ If you missed some of the problems, turn to the page numbers indicated.

★ If you cannot work any of these problems, start with frame 1 on page 289.

3. $679.34

2. $370.93

1. (a) $515
   (b) $7.73
   (c) $507.27

SECTION 1: BANK DISCOUNT

1

> **OBJECTIVE**
>
> Calculate the bank discount and the net proceeds of non-interest-bearing and interest-bearing notes.

Bank discount is a method that was originally used to lend money to businesses. With bank discount, the interest is subtracted from the principal. The actual amount borrowed, called the *net proceeds,* will be the original principal "discounted" or decreased by the amount of interest.

An example will make it clear. A loan of $500 at 12% for 60 days will cost $10 interest (ordinary simple interest at 30-day-month time). Under a bank discount scheme, the lender would subtract the interest from the amount loaned and the borrower would actually receive the net proceeds of $500 − $10 = $490.

Although bank discount is not widely utilized for ordinary loans, it is used for businesses accepting promissory notes as collateral. A business will often accept a note from a customer in lieu of a cash payment for merchandise. If the business needs cash, at some time before the due date of the note, it can borrow the money from a bank using the note as collateral or by signing the note over to the bank. The bank will lend the money using the bank discount.

Although a business may accept a note on a particular date, there may be a significant time lapse before that business borrows the cash from a bank. In order to calculate the bank discount, you must first figure the amount of time for which the money will be borrowed.

Generally, the length of the note and the amount of time the money is borrowed from the bank is not the same.

Example: Find the number of days for the bank loan in the following problem.

Freddy the barber bought some equipment for his shop using a promissory note. Sonny's Barber Supply Company holds his 90-day note for $900 made on August 13. The company needed cash to pay current bills so, on September 13, the note was discounted by Sonny's bank at 9% (ordinary simple interest at 30-day-month time).

The note was made on August 13 but wasn't taken to the bank until September 13. The bank will therefore lend money for the time remaining on the note as of September 13. Find the number of days of time remaining.

Check your work in **2.**

---

2   The correct time is 60 days.

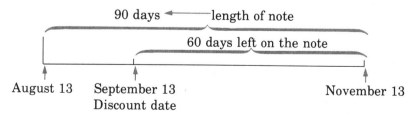

The total time of the note was 90 days. The note was made on August 13 but not discounted until September 13, 30 days later. This left 60 days on the note. Remember, the money was borrowed from the bank only for the time remaining on the note.

Next, you must calculate the bank discount. **Bank discount is always equal to the interest on the note.**

$$\text{Discount} = I = Prt = \$900 \times 9\% \times \frac{60}{360} = \$13.50$$

Finally, **the *net proceeds* will be the original principal less the discount.**

Principal   Discount

$$\text{Net proceeds} = \$900 - \$13.50 = \$886.50$$

The bank buys the note from the company for $886.50. Sixty days later, Freddie pays the bank $900. Freddie has his equipment, the company has the cash when it was needed, and the bank made a profit by lending out some of its cash.

The note in the preceding problem is a *non-interest-bearing* note. The note was for $900 but Sonny's Barber Supply Company didn't charge its customer interest. The 9% was the interest rate charged by the bank.

Find the bank discount and net proceeds for the following problems. Use ordinary simple interest at 30-day-month time.

1. What is the bank discount and net proceeds on a $500, 60-day note dated May 18 and discounted June 3 at 12%?
2. Arnold's Auto Supply holds a 60-day note for $217.86 made on March 23. On April 5 the note was discounted at the bank at 11.7%. What is the bank discount and net proceeds?

Check your work in 3.

The phrase "bank discount" makes it sound like the bank is having a sale on money! I wouldn't mind buying a hundred or so at 10% off.

Dreamer.
The only bargain you get with a bank discount loan is that you get to use the cash now instead of having to wait until the note is paid.

3   1. May 18 to June 3 = 15 days

Time remaining on note is 60 − 15 or 45 days.

$$\text{Discount} = I = Prt = \$500 \times 12\% \times \frac{45}{360} = \$7.50$$

Net proceeds = $500 − 7.50 = $492.50

2. March 23 to April 5 = 12 days

Time remaining on note is 60 − 12 or 48 days.

$$\text{Discount} = I = Prt = \$217.86 \times 11.7\% \times \frac{48}{360} = \$3.3986158$$

$$= \$3.40 \text{ rounded}$$

Net proceeds = $217.86 − 3.40 = $214.46

The examples we have just discussed involve non-interest-bearing notes—no interest on the original note.

Interest-Bearing Notes

Discounting non-interest-bearing notes is simple, but not widely practiced. Most businesses can't afford to lend money to their customers and not charge interest. Generally, they will accept only *interest-bearing* notes.

To compute the bank discount on an interest-bearing note, you must first calculate the maturity value of the note. Remember, the maturity value of a loan is the total amount paid back, principal plus interest.

Calculate the maturity value in the following problem. Use ordinary simple interest at 30-day-month time.

The Carlile Manufacturing Company holds a 60-day 18% note for $400 made on June 5 and discounted at the bank on June 20 at 12%.

Don't calculate the discount yet, just the maturity value of the note.

Turn to 4 to check your work.

---

4    $I = Prt = \$400 \times 18\% \times \dfrac{60}{360} = \$12$

Maturity value, $S = P + I = \$400 + \$12 = \$412$ | This maturity value is due in 60 days from the date of the note.

Notice that to compute the interest on the note, we must use the original length of the time of the note.

Important

After calculating the maturity value, you must next find the bank discount and net proceeds. **When finding the discount and proceeds, use the *maturity value* of the note, not the principal.**

Complete the Carlile Manufacturing Company problem by calculating the bank discount and net proceeds. Remember to use the maturity value of $412.

Check your calculations in 5.

What happens if the person making the original note doesn't pay? Who loses? The bank?

You can bet it's *not* the bank! The bank can collect the money from the payee, middleman who received the cash. Remember, the bank *lent* the payee money. *Bank discount* is simply the loan fee charged by the bank.

5   June 5 to June 20 = 15 days

Time remaining on note is 60 − 15 or 45 days

$r = 12\%$ ◄——— The bank rate

Maturity value

Discount = $I = Srt = \$412 \times 0.12 \times \dfrac{45}{360} = \$6.18$

Net proceeds = maturity value − discount
= $412 − 6.18 = $405.82

The bank pays $405.82 to the company. The maker of the note pays $412 to the bank by the due date.

For calculating the bank discount and net proceeds of interest-bearing notes, use the following steps:

Step 1. Calculate the *maturity value* of the note,

$$I = Prt$$

Maturity value $\longrightarrow S = P + I$

Remember to use the principal of the note, the entire time of the note, and the interest rate charged by the business to the customer.

Step 2. Calculate the *bank discount*

> Maturity value

$$\text{Discount} = I = Srt$$

Remember to use the maturity value of the note, the time *remaining* on the note, and the bank's interest rate.

Step 3. Calculate the *net proceeds*

Net proceeds = maturity value − discount

Now it's your turn. In the following problems, calculate the maturity value, bank discount, and net proceeds. Use ordinary simple interest at 30-day-month time.

1. Bob's Jog-Along Shop holds a 90-day, 15% note for $200 made on April 17 and discounted May 2 at 9%. What is the maturity value, bank discount, and net proceeds?
2. What is the maturity value, bank discount, and net proceeds for a $500, 60-day, 18% note made on July 5 and discounted August 13 at $12\frac{1}{2}$%?

Our solutions are in 6.

---

6   1. **Step 1.** $I = Prt = \$200 \times 15\% \times \dfrac{90}{360} = \$7.50$

> Entire time of the note

> Interest rate charged by the business to the customer

Maturity value $= S = P + I = \$200 + 7.50 = \$207.50$

> Time remaining on note

Step 2.   Discount $= I = Srt = \$207.50 \times 9\% \times \dfrac{75}{360}$

> Interest rate charged by the bank

$$= \$3.890625 = \$3.89 \text{ rounded}$$

Step 3.   Net proceeds = maturity value − discount
$$= \$207.50 - 3.89 = \$203.61$$

2. **Step 1.** $I = Prt = \$500 \times 18\% \times \dfrac{60}{360} = \$15$

Maturity value $= S = P + I = \$500 + 15 = \$515$

$$\text{Time remaining on note}$$

Step 2. $\text{Discount} = I = Srt = \$515 \times 12\frac{1}{2}\% \times \dfrac{22}{360}$

$$= \$515 \times 0.125 \times \frac{22}{360} = \$3.934 \ldots$$

$$= \$3.93 \text{ rounded}$$

Step 3. Proceeds = maturity value − discount
$$= \$515 - 3.93 = \$511.07$$

It's easy, as long as you follow the three steps for interest-bearing notes. For non-interest-bearing notes just skip **Step 1** (since there is no interest on the note) and use the principal in **Step 2**.

For a set of practice problems on bank discount, turn to 7.

BANK DISCOUNT USING BANKERS' INTEREST

Although we used ordinary interest at 30-day-month time to calculate bank discounts, any of the three simple interest methods can be used. Another commonly used method is ordinary interest at exact time, which is usually called bankers' interest. Bankers' interest uses the time fraction

$$\frac{\text{exact time}}{360}$$

The following examples use bankers' interest.

Example: A \$500 non-interest-bearing note dated May 18 and due on August 18 is discounted June 2 at 12%. Calculate the bank discount and net proceeds using bankers' interest.

The time left on the note is from June 2 to August 18.

$$\begin{aligned}
\text{June 2 to June 30} = 30 - 2 &= 28 \text{ days} \\
\text{July} &= 31 \text{ days} \\
\text{to August 18} &= \underline{18 \text{ days}} \\
\textit{exact} \text{ time left} &= 77 \text{ days}
\end{aligned}$$

$$\text{Discount} = I = Prt = \$500 \times 12\% \times \frac{77}{360} = \$12.833333$$

$$= \$12.83 \text{ rounded}$$

Net proceeds = \$500 − 12.83 = \$487.17

The following is an interest-bearing note example using bankers' interest.

Example: A \$500, 18% note dated May 18 and due on August 18 is discounted June 2 at 12%. Calculate the maturity value, bank discount, and net proceeds.

Step 1. Calculate the maturity value.

The full time on the note is from May 18 to August 18.

$$\begin{aligned}
\text{May 18 to May 31} = 31 - 18 &= 13 \text{ days} \\
\text{June} &= 30 \text{ days} \\
\text{July} &= 31 \text{ days} \\
\text{to August 18} &= \underline{18 \text{ days}} \\
\text{exact total time} &= 92 \text{ days}
\end{aligned}$$

$$I = Prt = \$500 \times 18\% \times \frac{92}{360} = \$23$$

Maturity value = S = P + I = $500 + 23 = $523

Step 2. Calculate the bank discount.
The time left on the note is from June 2 to August 18.

$$
\begin{array}{ll}
\text{June 2 to June 30} = 30 - 2 = & 28 \text{ days} \\
\text{July} = & 31 \text{ days} \\
\text{to August 18} = & \underline{18 \text{ days}} \\
\text{exact time left} = & 77 \text{ days}
\end{array}
$$

$$
\text{Discount} = I = Srt = \$523 \times 12\% \times \frac{77}{360} = \$13.423666
$$
$$
= \$13.42 \text{ rounded}
$$

Step 3. Net proceeds = maturity value − discount
$$
= \$523 - \$13.42 = \$509.58
$$

For some practice, work the following problems using bankers' interest. The answers are on page 589.

A. Calculate the bank discount and net proceeds for the following non-interest-bearing notes.

	Principal	Date of Note	Due Date	Discount Date	Discount Rate	Bank Discount	Net Proceeds
1.	$900	5/7	7/7	6/5	12%	————	————
2.	750	7/8	10/8	8/3	14%	————	————
3.	800	6/4	9/4	7/12	11%	————	————
4.	1500	8/2	11/2	9/12	11½%	————	————
5.	750	3/8	5/8	3/15	12½%	————	————
6.	1250	4/17	7/17	6/12	15%	————	————
7.	875	8/23	10/23	9/3	16½%	————	————
8.	970	9/17	11/4	9/20	18%	————	————
9.	1230	3/12	5/15	4/5	17½%	————	————
10.	1750	5/26	7/15	6/8	15½%	————	————

B. Calculate the maturing value, bank discount, and net proceeds for the following interest-bearing notes.

	Principal	Rate of Interest	Date of Note	Due Date	Discount Date	Discount Rate	Maturity Value	Bank Discount	Net Proceeds
1.	$900	18%	5/7	7/7	6/5	12%	————	————	————
2.	750	20%	7/8	10/8	8/3	14%	————	————	————
3.	800	21%	6/4	9/4	7/12	11%	————	————	————
4.	1500	18%	8/2	11/2	9/12	11½%	————	————	————
5.	750	21%	3/8	5/8	3/15	12½%	————	————	————
6.	1250	20%	4/17	7/17	6/12	15%	————	————	————
7.	875	18%	8/23	10/23	9/3	16½%	————	————	————
8.	970	20%	9/17	11/4	9/20	18%	————	————	————
9.	1230	21%	3/12	5/15	4/5	17½%	————	————	————
10.	1750	21%	5/26	7/15	6/8	15½%	————	————	————

8 Bank Discount, Compound Interest, and Present Value

PROBLEM SET 1

7   The answers are on page 589.

8.1 Bank Discount

A. Calculate the bank discount and net proceeds for the following non-interest-bearing notes. Use ordinary simple interest at 30-day-month time.

	Principal	Time	Date of Note	Discount Date	Discount Rate	Bank Discount	Net Proceeds
1.	$500	60 days	4/9	4/24	6%	____	____
2.	1200	90 days	5/6	6/6	9%	____	____
3.	750	60 days	3/15	4/15	12%	____	____
4.	200	30 days	8/7	8/17	6%	____	____
5.	250	60 days	9/27	10/12	12%	____	____
6.	575	90 days	6/15	8/15	8%	____	____
7.	800	45 days	2/7	3/7	9%	____	____
8.	1000	75 days	7/21	8/6	6%	____	____
9.	700	40 days	11/6	11/21	10%	____	____
10.	675	120 days	4/8	7/8	9%	____	____
11.	1500	60 days	5/23	6/13	8%	____	____
12.	125	90 days	9/18	10/14	$9\frac{1}{2}$%	____	____
13.	250	60 days	8/27	9/13	$6\frac{1}{4}$%	____	____
14.	900	60 days	10/25	11/16	$11\frac{3}{4}$%	____	____
15.	1100	90 days	3/13	5/5	$12\frac{1}{2}$%	____	____
16.	1500	60 days	5/17	6/12	12%	____	____
17.	1250	90 days	7/3	8/15	13%	____	____
18.	780	90 days	8/4	10/7	$12\frac{1}{2}$%	____	____

Date

Name

Course/Section

B. Calculate the maturity value, bank discount, and net proceeds for the following interest-bearing notes. Use ordinary simple interest at 30-day-month time.

	Principal	Rate of Interest	Time	Date of Note	Discount Date	Discount Rate	Maturity Value	Bank Discount	Net Proceed
1.	$900	12%	60 days	4/7	4/22	10%	——	——	——
2.	250	9%	90 days	6/8	7/8	6%	——	——	——
3.	575	6%	45 days	10/18	11/3	5%	——	——	——
4.	1100	12%	60 days	5/6	6/21	9%	——	——	——
5.	200	9%	45 days	9/18	10/18	8%	——	——	——
6.	1200	6%	90 days	6/20	8/20	9%	——	——	——
7.	125	14%	60 days	10/15	11/30	12%	——	——	——
8.	250	18%	90 days	5/15	5/30	12%	——	——	——
9.	750	6%	60 days	6/21	7/6	6%	——	——	——
10.	1500	12%	120 days	8/19	9/19	10%	——	——	——
11.	500	8%	90 days	4/12	5/15	9%	——	——	——
12.	675	$6\frac{3}{4}$%	60 days	6/29	7/3	6%	——	——	——
13.	1000	$12\frac{1}{2}$%	120 days	5/23	6/14	$11\frac{3}{4}$%	——	——	——
14.	700	$14\frac{1}{2}$%	90 days	7/30	8/3	$12\frac{1}{2}$%	——	——	——
15.	800	$9\frac{1}{4}$%	60 days	2/13	3/8	$8\frac{3}{4}$%	——	——	——
16.	1250	18%	60 days	3/15	4/18	14%	——	——	——
17.	1300	21%	90 days	5/7	7/8	12%	——	——	——
18.	875	20%	90 days	8/3	9/5	$12\frac{1}{2}$%	——	——	——

When you have had the practice you need, either return to the Preview on page 287 or continue in **8** with the study of compound interest.

SECTION 2: COMPOUND INTEREST

8

> **OBJECTIVE**
>
> Calculate compound interest.

Suppose my great-great-great-great-grandfather Jeb "Hurry" Cain, a private in General George Washington's army, had invested $10 at 6% a year in 1776. If he allowed the interest to accumulate over the years, how much would his descendants have had in 1976 to celebrate the bicentennial? The answer is that the original $10 would have grown to $1,151,259.04 and would now be increasing at almost $8 per hour! Unfortunately Jeb spent the money.

Compound interest is merely "interest on interest." An example will help clarify this concept. Let's calculate the interest on $500 at 6% for two years, compounded annually. Compounded annually means we must calculate the interest for each year separately.

First year: $I = Prt = \$500 \times 6\% \times 1 = \$30$
Maturity value $= S = P + I = \$500 + 30 = \$530$

The second year, we use the maturity value to calculate the interest instead of the original principal.

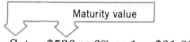
Maturity value

Second year: $I = Srt = \$530 \times 6\% \times 1 = \$31.80$

The total interest is $30 + 31.80 = \$61.80$.

The simple interest on $500 at 6% for two years would be $60.

Simple interest: $I = Prt = \$500 \times 6\% \times 2 = \$60$.

The interest compounded annually is $61.80; the simple interest is $60. The $1.80 difference is the "interest on the interest." The $1.80 is the second year's interest on the first year interest of $30.

That's not too difficult; you try one.

Calculate the maturity value on $200 at 5% compounded annually for three years.

Work the problem carefully. Calculate the interest for each year separately.

Check your work in 9.

---

9   First year: $I = Prt = \$200 \times 5\% \times 1 = \$10$
Maturity value $\rightarrow S = P + I = \$200 + 10 = \$210$

Second year: $I = Srt = \$210 \times 5\% \times 1 = \$10.50$
New maturity value $\rightarrow S = S + I = \$210 + 10.50 = \$220.50$
Previous maturity value

Third year: $I = Srt = \$220.50 \times 5\% \times 1 = \$11.025 = \$11.03$ rounded
New maturity value $\rightarrow S = S + I = \$220.50 + 11.03 = \$231.53$
Previous maturity value

The total interest is $10 + 10.50 + 11.03 = \$31.53$

The maturity value is $231.53.

For each year the new maturity value is found by multiplying the previous total by $1 + r$, where r is the annual interest rate. We can summarize this as follows:

$200 original principal
×1.05 $1 + r$
——————
$210.00 total amount after 1 year
×1.05
——————
$220.50 total amount after 2 years
×1.05
——————
$231.525 or $231.53 total amount after 3 years

This method can be quite long for many compound interest problems. To compute the interest compounded monthly for two years would require an interest calculation for each month—a total of 2×12 or 24 monthly calculations! This method is too time-consuming for such problems. A much easier and quicker method involves using the compound interest table shown on page 299. The following example will explain how to use it.

Example: What is the maturity value of $600 at 6% compounded monthly for two years?

Step 1. Find the number of *compounding periods*. The number of compounding periods, n, is the number of periods per year times the number of years.

The table lists the most common compounding periods.

Compounding Period	Periods per Year
Annual	1
Semiannual	2
Quarterly	4
Monthly	12
Daily	365 or 366 in leap year

Number of monthly periods per year Number of years

$$n = 12 \times 2 = 24$$

Step 2. Find the interest rate per compounding period. The interest rate per compounding period is the annual rate divided by the number of compounding periods per year.

$$\text{Rate} = \frac{6\%}{12} \begin{array}{l} \leftarrow \text{Annual rate} \\ \leftarrow \text{Number of monthly periods per year} \end{array}$$

$$= \frac{1}{2}\%$$

Step 3. Use the compound interest table on page 299. Find the number of time periods (Step 1) and the interest rate per time period (Step 2) to locate the appropriate value in the table. The number of time periods determines the row; the interest rate per time period determines the correct column.

In our example using $n = 24$, and $\frac{1}{2}\%$, the table value is 1.1271598. This is the maturity value of $1 at 6% compounded monthly for two years.

Step 4. Multiply the table value times the principal and round.

$$1.1271598 \times \$600 = \$676.29588$$
$$= \$676.30 \text{ rounded}$$

The maturity value of $600 at 6% compounded monthly for two years is $676.30.

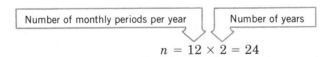

Now, you try one.

What is the maturity value of $250 at 8% compounded quarterly for 5 years?
Check your work in 10.

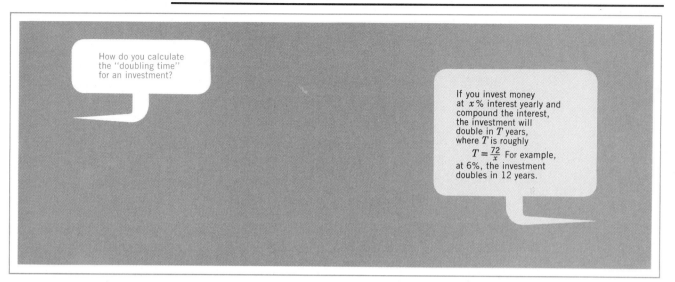

How do you calculate the "doubling time" for an investment?

If you invest money at x % interest yearly and compound the interest, the investment will double in T years, where T is roughly $T = \frac{72}{x}$ For example, at 6%, the investment doubles in 12 years.

COMPOUND INTEREST TABLE
Amount of $1 at Compound Interest

n	$\frac{1}{2}\%$	1%	$1\frac{1}{2}\%$	2%	3%	4%	5%	6%
1	1.0050000	1.0100000	1.0150000	1.0200000	1.0300000	1.0400000	1.0500000	1.0600000
2	1.0100250	1.0201000	1.0302250	1.0404000	1.0609000	1.0816000	1.1025000	1.1236000
3	1.0150751	1.0303010	1.0456784	1.0612080	1.0927270	1.1248640	1.1576250	1.1910160
4	1.0201505	1.0406040	1.0613636	1.0824322	1.1255088	1.1698586	1.2155062	1.2624770
5	1.0252513	1.0510100	1.0772840	1.1040808	1.1592741	1.2166529	1.2762816	1.3382256
6	1.0303775	1.0615202	1.0934433	1.1261624	1.1940523	1.2653190	1.3400956	1.4185191
7	1.0355294	1.0721354	1.1098449	1.1486857	1.2298739	1.3159318	1.4071004	1.5036303
8	1.0407070	1.0828567	1.1264926	1.1716594	1.2667701	1.3685690	1.4774554	1.5938481
9	1.0459106	1.0936853	1.1433900	1.1950926	1.3047732	1.4233118	1.5513282	1.6894790
10	1.0511401	1.1046221	1.1605408	1.2189944	1.3439164	1.4802443	1.6288946	1.7908477
11	1.0563958	1.1156684	1.1779489	1.2433743	1.3842339	1.5394541	1.7103394	1.8982986
12	1.0616778	1.1268250	1.1956182	1.2682418	1.4257609	1.6010322	1.7958563	2.0121965
13	1.0669862	1.1380933	1.2135524	1.2936066	1.4685337	1.6650735	1.8856491	2.1329283
14	1.0723211	1.1494742	1.2317557	1.3194788	1.5125897	1.7316764	1.9799316	2.2609040
15	1.0776827	1.1609690	1.2502321	1.3458683	1.5579674	1.8009435	2.0789282	2.3965582
16	1.0830712	1.1725786	1.2689856	1.3727857	1.6047064	1.8729812	2.1828746	2.5403517
17	1.0834865	1.1843044	1.2880203	1.4002414	1.6528476	1.9479005	2.2920183	2.6927728
18	1.0939289	1.1961475	1.3073406	1.4282462	1.7024331	2.0258165	2.4066192	2.8543392
19	1.0993986	1.2081090	1.3269508	1.4568112	1.7535061	2.1068492	2.5269502	3.0255995
20	1.1048956	1.2201900	1.3468550	1.4859474	1.8061112	2.1911231	2.6532977	3.2071355
21	1.1104201	1.2323919	1.3670578	1.5156663	1.8602946	2.2787681	2.7859626	3.3995636
22	1.1159722	1.2447159	1.3875637	1.5459797	1.9161034	2.3699188	2.9252607	3.6035374
23	1.1215520	1.2571630	1.4083772	1.5768993	1.9735865	2.4647155	3.0715238	3.8197497
24	1.1271598	1.2697346	1.4295028	1.6084372	2.0327941	2.5633042	3.2250999	4.0489346
25	1.1327956	1.2824320	1.4509454	1.6406060	2.0937779	2.6658363	3.3863549	4.2918707
26	1.1384596	1.2952563	1.4727095	1.6734181	2.1565913	2.7724698	3.5556727	4.5493830
27	1.1441519	1.3082089	1.4948002	1.7068865	2.2212890	2.8833686	3.7334563	4.8223459
28	1.1498726	1.3212910	1.5172222	1.7410242	2.2879277	2.9987033	3.9201291	5.1116867
29	1.1556220	1.3345039	1.5399805	1.7758447	2.3565655	3.1186514	4.1161356	5.4183879
30	1.1614001	1.3478489	1.5630802	1.8113616	2.4272625	3.2433975	4.3219424	5.7434912
40	1.2207942	1.4888637	1.8140184	2.2080397	3.2620378	4.8010206	7.0399887	10.2857179
50	1.2832258	1.6446318	2.1052424	2.6915880	4.3839060	7.1066834	11.4673998	18.4201543
60	1.3488502	1.8166967	2.4432198	3.2810308	5.8916031	10.5196274	18.6791859	32.9876908
70	1.4178305	2.0067634	2.8354563	3.9995582	7.9178219	15.5716184	30.4264255	59.0759302
80	1.4903386	2.2167152	3.2906628	4.8754392	10.6408906	23.0497991	49.5614411	105.7959935
90	1.5665547	2.4486327	3.8189485	5.9431331	14.3004671	34.1193333	80.7303650	189.4645112
100	1.6466685	2.7048138	4.4320457	7.2446461	19.2186320	50.5049482	131.5012578	339.3020835

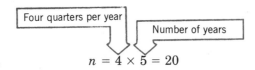

Four quarters per year

Number of years

10 **Step 1.** $n = 4 \times 5 = 20$

Step 2. Rate $= \dfrac{8\%}{4}$ ←—Annual rate

←—Number of quarter periods per year

$= 2\%$

Step 3. The table value is 1.4859474.

Step 4. $1.4859474 \times \$250 = \$371.48685$
$= \$371.49$ rounded

I'm sure you will agree that this method is far easier than the long way. For the above problem, using the long method you would need to calculate the interest for each of the 20 quarters.

Work the following problems using the table.

1. Find the maturity value on \$625 at 6% compounded quarterly for 5 years.
2. What is the maturity value on \$625 at 6% compounded monthly for 5 years?
3. What is the maturity value on \$200 at 8% compounded semiannually for 25 years?

Our work is in **11**.

---

11 1. **Step 1.** $n = 5 \times 4 = 20$

   Step 2. Rate $= \dfrac{6\%}{4} = \dfrac{3}{2}\% = 1\dfrac{1}{2}\%$

   Step 3. Table value $= 1.3468550$

   Step 4. $1.3468550 \times \$625 = \$841.784375$
   $= \$841.78$ rounded

2. **Step 1.** $n = 12 \times 5 = 60$

   Step 2. Rate $= \dfrac{6\%}{12} = \dfrac{1}{2}\%$

   Step 3. Table value $= 1.3488502$

   Step 4. $1.3488502 \times \$625 = \$843.031375$
   $= \$843.03$ rounded

3. **Step 1.** $n = 2 \times 25 = 50$

   Step 2. Rate $= \dfrac{8\%}{2} = 4\%$

   Step 3. Table value $= 7.1066834$

   Step 4. $7.1066834 \times \$200 = \$1421.3366$
   $= \$1421.34$ rounded

It's easy, if you follow the four steps!

The compound interest table included in the text isn't large enough to work all compound interest problems. For example, to compute the maturity value of any principal at $5\frac{3}{4}\%$, compounded monthly for 9 years, the table is too small. The number of compounding periods is $12 \times 9 = 109$ and the interest rate per time period is $\dfrac{5\frac{3}{4}\%}{12} = \dfrac{23}{48}$ %. Neither of these values appear in the table. To work these problems, you will have to obtain a book of interest tables or use another method. One method of working any compound interest problem on a calculator is presented on page 307.

Turn to 12 for a set of practice problems on compound interest.

8 Bank Discount, Compound Interest, and Present Value

PROBLEM SET 2

12

The answers are on page 590.

8.2 Compound Interest

A. Calculate the maturity value for the following compound interest problems using the table.

	Principal	Time	Annual Rate	Compounding Period	Maturity Value
1.	$500	8 years	6%	Annual	_____
2.	625	12 years	5%	Annual	_____
3.	250	9 years	6%	Semiannual	_____
4.	800	10 years	8%	Semiannual	_____
5.	175	6 years	6%	Quarterly	_____
6.	1500	5 years	8%	Quarterly	_____
7.	475	12 years	4%	Semiannual	_____
8.	850	15 years	6%	Semiannual	_____
9.	1750	4 years	8%	Quarterly	_____
10.	400	3 years	6%	Quarterly	_____
11.	225	5 years	6%	Monthly	_____
12.	2000	2 years	12%	Monthly	_____
13.	425	10 years	10%	Semiannual	_____
14.	775	2 years	6%	Monthly	_____
15.	925	5 years	12%	Monthly	_____
16.	1755	12 years	5%	Annual	_____
17.	892	7 years	8%	Semiannual	_____
18.	477	3 years	6%	Monthly	_____

Date _____

Name _____

Course/Section _____

B. Calculate the maturity value for the following compound interest problems using the table.

	Principal	Time	Annual Rate	Compounding Period	Maturity Value
1.	$250	5 years	8%	Quarterly	——————
2.	1500	2 years	6%	Monthly	——————
3.	800	5 years	12%	Monthly	——————
4.	115	6 years	6%	Quarterly	——————
5.	725	3 years	8%	Quarterly	——————
6.	280	25 years	10%	Semiannual	——————
7.	740	25 years	5%	Annual	——————
8.	225	21 years	6%	Annual	——————
9.	289	11 years	6%	Semiannual	——————
10.	143	3 years	12%	Quarterly	——————
11.	878	1 year	12%	Monthly	——————
12.	453	2 years	6%	Monthly	——————
13.	1259	2 years	12%	Quarterly	——————
14.	1567	5 years	6%	Monthly	——————
15.	451	2 years	6%	Monthly	——————
16.	895	5 years	6%	Monthly	——————
17.	623	4 years	6%	Quarterly	——————
18.	749	4 years	6%	Semiannual	——————
19.	1282	4 years	6%	Annual	——————
20.	1575	3 years	8%	Quarterly	——————

When you have had the practice you need, either return to the Preview on page 287 or continue in 13 with the study of present value.

SECTION 3: PRESENT VALUE

13

> **OBJECTIVE**
>
> Calculate present value

How much money must I invest today in order to have $500 ten years from now?

This problem can easily be solved using present value. This is a compound interest problem worked backwards, where the maturity value is known and the amount to be invested (principal) is unknown. This is called a *present value* problem. You solve these problems using the same four steps used in solving compound interest problems.

Let's look at an example.

What is the present value of $500 at 6% compounded monthly for five years? That is, how much must be invested today (present value) at 6% compounded monthly to yield $500 in five years?

Step 1. **Find the number of *compounding periods.*** The number of compounding periods, n, is the number of periods per year times the number of years.

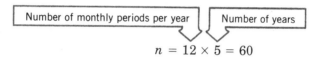

$$n = 12 \times 5 = 60$$

Step 2. **Find the interest rate per compounding period.** The interest rate per compounding period is the annual rate divided by the number of compounding periods per year.

$$\text{Rate} = \frac{6\%}{12} \quad \begin{matrix} \leftarrow \text{Annual rate} \\ \leftarrow \text{Number of monthly periods per year} \end{matrix}$$

$$= \frac{1}{2}\%$$

Step 3. **Use the present value table on page 306.** Find the number of time periods (Step 1) and the interest rate per time period (Step 2) to locate the appropriate value in the present value table. The number of time periods determines the row; the interest rate per time period determines the correct column.

In our example, using $n = 60$, and $\frac{1}{2}\%$, the table value is 0.7413722.

Step 4. **Multiply the table value times the maturity value and round.**

Complete Step 4 and go to **14**.

---

14 **Step 4.** 0.7413722 × $500 = $370.6861
                      = $370.69 rounded

The answer is $370.69. If you invest $370.69 at 6% compounded monthly for five years, you will have a maturity value of $500. Stated differently, the present value of $500 at 6% compounded monthly for 5 years is $370.69.

Check it. Use the compound interest table and calculate the maturity value of $370.69 at 6% compounded monthly for 5 years.

Compare your work with ours in **15**.

PRESENT VALUE TABLE
Present Value of $1

n	$\frac{1}{2}\%$	1%	$1\frac{1}{2}\%$	2%	3%	4%	5%	6%
1	0.9950249	0.9900990	0.9852216	0.9803922	0.9708738	0.9615385	0.9523810	0.9433962
2	0.9900745	0.9802960	0.9706617	0.9611688	0.9425959	0.9245562	0.9070295	0.8899964
3	0.9851488	0.9705902	0.9563170	0.9423223	0.9151417	0.8889964	0.8638376	0.8396193
4	0.9802475	0.9609803	0.9421842	0.9238454	0.8884870	0.8548042	0.8227025	0.7920937
5	0.9753707	0.9514657	0.9282603	0.9057308	0.8626088	0.8219271	0.7835262	0.7472582
6	0.9705181	0.9420452	0.9145422	0.8879714	0.8374843	0.7903145	0.7462154	0.7049605
7	0.9656896	0.9327180	0.9010268	0.8705602	0.8130915	0.7599178	0.7106813	0.6650571
8	0.9608852	0.9234832	0.8877111	0.8534904	0.7894092	0.7306902	0.6768394	0.6274124
9	0.9561047	0.9143398	0.8745922	0.8367553	0.7664167	0.7025867	0.6446089	0.5918985
10	0.9513479	0.9052870	0.8616672	0.8203483	0.7440939	0.6755642	0.6139132	0.5583948
11	0.9466149	0.8963237	0.8489332	0.8042630	0.7224213	0.6495809	0.5846793	0.5267875
12	0.9419053	0.8874492	0.8363874	0.7884932	0.7013799	0.6245970	0.5568374	0.4969694
13	0.9372192	0.8786626	0.8240270	0.7730325	0.6809513	0.6005741	0.5303214	0.4688390
14	0.9325565	0.8699630	0.8118493	0.7578750	0.6611178	0.5774751	0.5050680	0.4423010
15	0.9279169	0.8613495	0.7998515	0.7430147	0.6418619	0.5552645	0.4810171	0.4172651
16	0.9233004	0.8528213	0.7880310	0.7284458	0.6231669	0.5339082	0.4581115	0.3936463
17	0.9187068	0.8443775	0.7763853	0.7141626	0.6050164	0.5133732	0.4362967	0.3713644
18	0.9141362	0.8360173	0.7649116	0.7001594	0.5873946	0.4936281	0.4155206	0.3503438
19	0.9095882	0.8277399	0.7536075	0.6864308	0.5702860	0.4746424	0.3957340	0.3305130
20	0.9050629	0.8195445	0.7424704	0.6729713	0.5536758	0.4563870	0.3768895	0.3118047
21	0.9005601	0.8114302	0.7314980	0.6597758	0.5375493	0.4388336	0.3589424	0.2941554
22	0.8960797	0.8033962	0.7206876	0.6468390	0.5218925	0.4219554	0.3418499	0.2775051
23	0.8916216	0.7954418	0.7100371	0.6341559	0.5066917	0.4057263	0.3255713	0.2617973
24	0.8871857	0.7875661	0.6995439	0.6217215	0.4919337	0.3901215	0.3100679	0.2469786
25	0.8827718	0.7797684	0.6892058	0.6095309	0.4776056	0.3751168	0.2953028	0.2329985
26	0.8783799	0.7720480	0.6790205	0.5975793	0.4636947	0.3606892	0.2812407	0.2198100
27	0.8740099	0.7644039	0.6689857	0.5858620	0.4501891	0.3468166	0.2678483	0.2073680
28	0.8696616	0.7568356	0.6590993	0.5743746	0.4370768	0.3334775	0.2550936	0.1956301
29	0.8653349	0.7493421	0.6493589	0.5631123	0.4243464	0.3206514	0.2429463	0.1845567
30	0.8610297	0.7419229	0.6397624	0.5520709	0.4119868	0.3083187	0.2313774	0.1741101
40	0.8191389	0.6716531	0.5512623	0.4528904	0.3065568	0.2082890	0.1420457	0.0972222
50	0.7792861	0.6080388	0.4750047	0.3715279	0.2281071	0.1407126	0.0872037	0.0542884
60	0.7413722	0.5504496	0.4092960	0.3047823	0.1697331	0.0950604	0.0535355	0.0303143
70	0.7053029	0.4983149	0.3526769	0.2500276	0.1262974	0.0642194	0.0328662	0.0169274
80	0.6709885	0.4511179	0.3038902	0.2051097	0.0939771	0.0433843	0.0201770	0.0094522
90	0.6383435	0.4083912	0.2618522	0.1682614	0.0699278	0.0293089	0.0123869	0.0052786
100	0.6072868	0.3697112	0.2256294	0.1380330	0.0520328	0.0198000	0.0076045	0.0029472

15   **Step 1.** $n = 12 \times 5 = 60$

Step 2.   Rate $= \dfrac{6\%}{12} = \dfrac{1}{2}\%$

Step 3.   Compound Interest Table value = 1.3488502

Step 4.   1.3488502 × $370.69 = $500.00528
                     = $500.01 rounded

The extra penny is due to rounding. It checks, $370.69 invested at 6% compounded monthly for 5 years is about $500.

Calculating present value is easy, just follow the four steps using the present value table.

Try the following problems.

1. What is the present value of $250 at 8% compounded quarterly for five years?
2. How much must be invested today at 8% compounded semiannually to yield $2750 in 25 years?

3. What is the present value of $11,467.40 at 5% compounded annually for 50 years?

Check your work in 16.

COMPOUND INTEREST USING A CALCULATOR

Tables of compound interest are created using the following formula:

$$\text{Maturity value} = P \left(1 + \frac{r}{n}\right)^{Nn}$$

where P = principal
r = annual rate of interest written as a decimal
N = number of years
n = number of compounding periods per year

For example, $100 at 6% compounded quarterly for 5 years gives

$$\text{Maturity value} = \$100 \left(1 + \frac{0.06}{4}\right)^{5 \cdot 4}$$
$$= \$100(1.015)^{20}$$

If your calculator has a $\boxed{y^x}$ key, punch in

$\boxed{1}\boxed{.}\boxed{0}\boxed{1}\boxed{5}\boxed{y^x}\boxed{2}\boxed{0}\boxed{=}\boxed{\times}\boxed{1}\boxed{0}\boxed{0}\boxed{=}$ 134.68550

and the display reads 134.68550

The maturity value is $134.69

If your calculator has no $\boxed{y^x}$ function key, try punching in

$\boxed{1}\boxed{0}\boxed{0}\boxed{\times}\boxed{1}\boxed{.}\boxed{0}\boxed{1}\boxed{5}\boxed{\times}\boxed{1}\boxed{.}\boxed{0}\boxed{1}\boxed{5}\boxed{\times}\boxed{1}\boxed{.}\boxed{0}\boxed{1}\boxed{5}\boxed{\times}$ . . .

repeating the factor 1.015 twenty times.

Present value can be calculated using the same formula.

$$\text{Present value} = \frac{\text{maturity value}}{\left(1 + \frac{r}{n}\right)^{Nn}}$$

For example, the present value of $100 at 5% compounded semiannually for six years given

$$\text{Present value} = \frac{\$100}{\left(1 + \frac{0.05}{2}\right)^{6 \cdot 2}}$$

$$= \frac{\$100}{(1.025)^{12}} = \frac{\$100}{1.3448888} = \$74.36 \text{ rounded}$$

Try a few problems. The answers are in the appendix on page 590.

1. Find the maturity value of $1000 for three years at an annual rate of 6% compounded (a) annually, (b) quarterly, (c) monthly, (d) daily.
2. Find the present value of $1000 for five years at an annual rate of 8% compounded (a) annually, (b) quarterly, (c) monthly, (d) daily.

1. **Step 1.** $n = 4 \times 5 = 20$

   Step 2. Rate $= \dfrac{8\%}{4} = 2\%$

   Step 3. Table value $= 0.6729713$

   Step 4. $0.6729713 \times \$250 = \$168.242825$
   $$= \$168.24 \text{ rounded}$$

2. **Step 1.** $n = 2 \times 25 = 50$

   Step 2. Rate $= \dfrac{8\%}{2} = 4\%$

   Step 3. Table value $= 0.1407126$

   Step 4. $0.1407126 \times \$2,750 = \$386.95965$
   $$= \$386.96 \text{ rounded}$$

3. **Step 1.** $n = 1 \times 50 = 50$

   Step 2. Rate $= \dfrac{5\%}{1} = 5\%$

   Step 3. Table value $= 0.0872037$

   Step 4. $0.0872037 \times \$11,467.40 = \$999.9997094$
   $$= \$1000.00 \text{ rounded}$$

It's easy.

The present value table included in the text isn't large enough to work all present value problems. For other interest rates and compounding periods, you will have to obtain a book on interest and present value tables.

Turn to 17 for a set of practice problems on present value.

---

8 Bank Discount, Compound Interest, and Present Value

PROBLEM SET 3

17 The answers are on page 590.

8.3 Present Value

A. Calculate the present value for the following problems using the table.

	Maturity Value	Time	Annual Rate	Compounding Period	Present Value
1.	$425	8 years	5%	Annual	_____
2.	850	15 years	6%	Annual	_____
3.	300	8 years	8%	Semiannual	_____
4.	1500	11 years	6%	Semiannual	_____
5.	190	3 years	8%	Quarterly	_____
6.	550	5 years	4%	Quarterly	_____
7.	1200	12 years	4%	Semiannual	_____
8.	875	2 years	6%	Monthly	_____
9.	420	5 years	12%	Monthly	_____
10.	575	7 years	8%	Semiannual	_____
11.	925	6 years	6%	Quarterly	_____
12.	320	10 years	8%	Quarterly	_____
13.	1525	3 years	10%	Semiannual	_____
14.	455	2 years	12%	Monthly	_____
15.	880	5 years	6%	Monthly	_____
16.	790	12 years	5%	Annual	_____
17.	845	7 years	8%	Semiannual	_____
18.	655	3 years	6%	Monthly	_____

Date _____

Name _____

Course/Section _____

B. Calculate the present value for the following problems using the table.

	Maturity Value	Time	Annual Rate	Compounding Period	Present Value
1.	$ 500	2 years	6%	Monthly	———
2.	825	6 years	8%	Quarterly	———
3.	435	3 years	6%	Quarterly	———
4.	750	1 year	12%	Monthly	———
5.	1500	7 years	6%	Semiannual	———
6.	250	5 years	6%	Monthly	———
7.	5000	2 years	12%	Monthly	———
8.	1250	7 years	10%	Semiannual	———
9.	700	5 years	6%	Quarterly	———
10.	425	2 years	8%	Quarterly	———
11.	800	2 years	12%	Monthly	———
12.	900	5 years	6%	Monthly	———
13.	1000	12 years	5%	Annual	———
14.	12,000	3 years	12%	Quarterly	———
15.	285	5 years	8%	Quarterly	———
16.	850	5 years	6%	Monthly	———
17.	745	4 years	6%	Quarterly	———
18.	295	4 years	6%	Semiannual	———
19.	655	4 years	6%	Annual	———
20.	870	3 years	8%	Quarterly	———

When you have had the practice you need, turn to 18 for a Self-Test on bank discount, compound interest, and present value.

8 Bank Discount, Compound Interest, and Present Value

18   Self-Test

The answers are on page 590.

You may use the compound interest and present value tables on pages 299 and 306.

1.   A 90-day note for $200, dated October 27, is discounted December 27 at 9%. What are the (a) bank discount and (b) net proceeds?

(a) _____

(b) _____

2.   For a $200, 12%, 90-day note, dated October 5 and discounted December 5 at 9%, calculate the (a) maturity value, (b) bank discount, and (c) net proceeds.

(a) _____

(b) _____

(c) _____

3.   What is the maturity value of $450 invested at 6% compounded annually for three years?

_____

4.   What is the maturity value of $725 invested at 6% compounded monthly for two years?

_____

5.   What is the present value of $300 at 8% compounded semiannually for 9 years?

_____

Date _____

Name _____

Course/Section _____

9 Consumer Math

WHAT ARE YOU DOING?

BALANCING MY CHECKBOOK

Balancing your checkbook as shown above might be a lot easier than actually matching it with your monthly bank statement.

Unfortunately, this kind of balancing would not give you the accurate numbers needed to check your monthly bank summary.

In this unit, you will become acquainted with checking account procedures, installment loans, credit cards, and buying and selling stocks and bonds.

Ready? Good! Let's get "in the know" about borrowing, investing, or spending "dough."

UNIT OBJECTIVE

When you complete this unit successfully, you will be able to complete checking account records including check stubs, a check register and a bank reconciliation form. You will also be able to calculate the total cost, interest and APR for installment loans and add-on loans. You will be able to interchange the APR and monthly rate, and calculate the interest and new balance on credit card or revolving charge accounts. You will be able to calculate stock dividends, the price of a stock or bond transaction, and the current yield of a bond.

PREVIEW 9   This is a sample test for Unit 9. All the answers are placed immediately after the test. Work as many of the problems as you can, then check your answers and look after the answers for further directions.

1. Complete the following check register.

319   1

Check No.	Date	Check Issued to or Deposit	Amount of Check		✔	Amount of Deposit	Balance	
							842	96
156	6/2	Pineview Apts.	257	82				
157	6/4	Electric Company	84	25				
	6/4	Deposit				42	60	
158	6/10	Friendly Finance Co.	124	45				
159	6/15	Sears	26	32				
160	6/20	Department store	42	87				
	6/22					52	87	
161	6/25	Fast Foods	82	91				
162	6/26	Gas Company	41	76				
	6/28	Deposit				25	86	

2. The monthly bank summary for the preceding account shows checks numbered 156, 157, 158, 159, and 162 were processed, the deposits on 6/4 and 6/22 were credited, and there was a service charge of $2.86. The bank balance is $400.87.

Use the check register in Problem 1, and the form on page 315 to complete the bank reconciliation.

324   5

BANK RECONCILIATION FORM

Outstanding Checks		
Number	Amount	
Total		

1. Compare the cancelled checks with your records.
2. List any outstanding checks.
3. Total the outstanding checks.
4. Enter the bank balance here: $ _____

5. Add any deposits not on the summary. +$ _____

 +$ _____

6. Total $ _____
7. Enter outstanding check total and subtract. −$ _____

Corrected Bank Balance $ _____

8. Enter checkbook balance. $ _____
9. Enter any service charge and subtract. −$ _____

Corrected Checkbook Balance $ _____

Corrected checkbook balance and corrected bank balance must be equal.

3. A refrigerator may be purchased for $400 cash for $50 down and 24 monthly payments of $17.50. Calculate the (a) interest and (b) APR.

(a) _____

(b) _____ 337   9

4. A revolving charge statement received in June shows a balance of $258.92. The minimum required payment is $12.95. The APR is 18%. Purchases totaling $52.45 were made in June but not included in the statement. If the minimum payment is made, calculate the (a) unpaid balance, (b) interest on the unpaid balance, and (c) balance for the July statement.

(a) _____

(b) _____

(c) _____ 347   17

Preview 9
Answers

1.

Check No.	Date	Check Issued to or Deposit	Amount of Check		✔	Amount of Deposit		Balance	
								842	96
156	6/2	Pineview Apts.	257	82	✓			585	14
157	6/4	Electric Company	84	25	✓			500	89
	6/4	Deposit			✓	42	50	543	39
158	6/10	Friendly Finance Co.	124	45	✓			418	94
159	6/15	Sears	26	32	✓			392	62
160	6/20	Department Store	42	87				349	75
	6/22	Deposit			✓	52	87	402	62
161	6/25	Fast Foods	82	91				319	71
162	6/26	Gas Company	41	76	✓			277	95
	6/28	Deposit				25	86	303	81

5. The Gibsen Company was unable to give a dividend last year to their stockholders with cumulative preferred 6%, par value $75 stock. This year the company is giving two year's dividends. Calculate the total dividend for 150 shares of stock.

357   26

6. Calculate the current yield for a $1000 New York Telephone Company bond with interest at 8⅜% and currently selling for 99¼. Round your answer to the nearest 0.1%.

362   32

2.

BANK RECONCILIATION FORM

Outstanding Checks		
Number	Amount	
160	42	87
161	82	91
Total	125	78

1. Compare the cancelled checks with your records.
2. List any outstanding checks.
3. Total the outstanding checks.
4. Enter the bank balance here: $ 400.87

5. Add any deposits not on the summary. +$ 25.86

   +$ _____

6. Total $ 426.73
7. Enter outstanding check total and subtract. -$ 125.78

Corrected Bank Balance $ 300.95

8. Enter checkbook balance. $ 303.81
9. Enter any service charge and subtract. -$ 2.86

Corrected Checkbook Balance $ 300.95

Corrected checkbook balance and corrected bank balance must be equal.

3. (a) $70 (b) 19.2%

4. (a) $245.97 (b) $3.69 (c) $302.11

5. $1350

6. 8.7%

★ If you cannot work any of these problems, start with frame 1 on page 319.
★ If you missed some of the problems, turn to the page numbers indicated.
★ If all your answers were correct, turn to page 369 for the unit Self-Test.
★ Super-students—those who want to be certain they learn all of this material—will turn to frame 1 and begin work there.

SECTION 1: CHECKING ACCOUNTS

1 | **OBJECTIVE**

Complete checking account records including check stubs, a check register, and a bank reconciliation form.

The majority of business transactions today are completed with checks. Most businesses and many individuals use checks to pay for their purchases because checks provide a convenient record of transactions, and the cancelled or processed checks provide a receipt of payment.

To open a checking account, the bank requires that you fill out a *signature card* listing your name, address, type of account, and your signature as you will sign your checks. After making an initial deposit, the bank will issue your personalized checks with your name and account number imprinted. You may then make deposits and write checks on your account, always being careful to keep your account balance above $0. If your balance falls below $0, you are *overdrawn* and the bank will not process any more of your checks until you make a further deposit to bring your account balance over $0. Most banks also assess a service charge for being overdrawn.

For its checking service, the bank will usually charge a fee. The amount and type of fee will depend on the type of account selected.

A *check printing charge* is a fee that covers the cost of printing the checks. A *fixed charge per check* is one account option, usually selected by people who write a small number of checks. With this option, a fee is charged for each check.

Another type of account uses a *monthly charge* for the services. Usually there is no limit on the number of checks that may be written each month.

Some accounts are charged a *variable monthly fee*. The amount charged is based on the number of checks written and/or the minimum balance.

Many banks do not have a checking service charge for people who maintain a minimum monthly balance, usually $300 to $500.

Now that we have looked at some of the various types of accounts and service charges, turn to 2 for a discussion on writing checks.

The following is a typical example of a check. On the check, there are six items to be completed. To avoid alteration, all items should be written in nonerasable ink, not pencil.

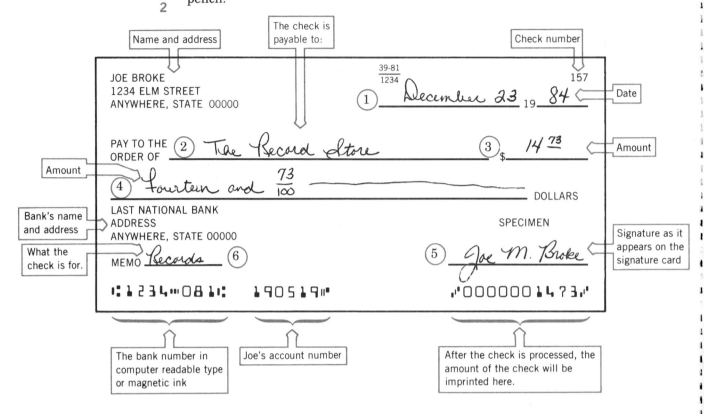

(1) The *date* of the check is entered in the upper right-hand corner.

(2) The person or business to whom the check is issued is entered in the blank denoted *"Pay to the order of."*

(3) The amount of the check, in numerical form, is entered here. To avoid possible alteration, the amount should start immediately following the imprinted $. Do not write an amount as $14.73, for it can be easily altered to $14,730 by changing the period to a comma, and adding a zero to the end. Write the amount as $14.^{73} to avoid this problem.

(4) The amount of the check in word form is entered in (4) The whole number portion of the amount is entered in word form and the decimal portion is entered as a fraction. To avoid alteration, the amount should start at the beginning of the space provided. After the amount, place a wavy line. The amount in (4) must be the same as in (3).

(5) The individual's signature as it appears on the original signature card.

(6) A space to note what the check is for is provided in the line marked "memo" or "for." This space is optional.

Complete the following check for Joe Broke to Quick Groceries for $62.75, dated December 21, 1984.

```
┌─────────────────────────────────────────────────────────────────────┐
│                                        39-81                          │
│  JOE BROKE                             ────                      156  │
│  1234 ELM STREET                       1234                           │
│  ANYWHERE, STATE 00000                 _____ 19_____     │
│                                                                       │
│                                                                       │
│  PAY TO THE                                                           │
│  ORDER OF  _____  $_____   │
│                                                                       │
│                                                                       │
│  _____ DOLLARS   │
│  LAST NATIONAL BANK                                                   │
│  ADDRESS                                          SPECIMEN            │
│  ANYWHERE, STATE 00000                                               │
│                                                                       │
│  MEMO _____                              _____   │
│                                                                       │
│  ⑈1234⑈081⑈   190519⑈                                                │
└─────────────────────────────────────────────────────────────────────┘
```

Turn to 3 to check your work.

3

```
┌─────────────────────────────────────────────────────────────────────┐
│                                        39-81                          │
│  JOE BROKE                             ────                      156  │
│  1234 ELM STREET                       1234                           │
│  ANYWHERE, STATE 00000               December 21  19 84              │
│                                                                       │
│                                                                       │
│  PAY TO THE        Quick Groceries                                   │
│  ORDER OF  _____  $ 62.75        │
│                              75                                       │
│     Sixty - two and ────                                  DOLLARS    │
│  LAST NATIONAL BANK         100                                       │
│  ADDRESS                                          SPECIMEN            │
│  ANYWHERE, STATE 00000                                               │
│                                                                       │
│  MEMO _____                        Joe M. Broke              │
│                                                                       │
│  ⑈1234⑈081⑈   190519⑈                                                │
└─────────────────────────────────────────────────────────────────────┘
```

After the check has been written, the amount of the check must be recorded.

Check Stubs

There are two methods of recording checking account transactions: Check stubs or a check register. If you use *check stubs,* a stub is attached to each check. Both the check and stub should be filled out at the same time. The check stub includes entries for the amount, date, to whom the check is issued, description of purchase, deposits, and the balance.

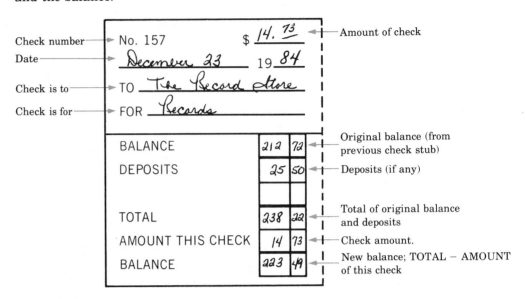

To calculate the new balance, first enter the balance from the bottom of the previous check stub in the top balance position. Enter any deposits, then add the beginning balance and the deposits to obtain the total. Finally, subtract the amount of the check to get the new balance.

Sound confusing? It's really very easy. Just remember to add any deposits and subtract the check amount.

Complete the following check and check stub by calculating the total and new balance.

No. 158	$ *15. 74*		
December 27 19 *84*			
TO *Slow Gas*			
FOR *Gasoline*			
BALANCE		223	49
DEPOSITS		52	95
TOTAL			
AMOUNT THIS CHECK		15	74
BALANCE			

JOE BROKE
1234 ELM STREET
ANYWHERE, STATE 00000

39-81 / 1234

158

_____ 19_____

PAY TO THE
ORDER OF _____ $_____

_____ DOLLARS

LAST NATIONAL BANK
ADDRESS
ANYWHERE, STATE 00000

SPECIMEN

MEMO _____

⑆1234⑈081⑆ 190519⑈

Check your work in 4.

Does my signature on a check need to be exactly the same as the name printed on the check?

No, but you must sign it as you signed the signature card. In fact, a good way to foil check forgers is to make the two names slightly different. If you sign checks "Joan Q. Smith" have your checks printed with "Joan Smith."

4

No. 158	$ *15. 74*		
December 27 19 *84*			
TO *Slow Gas*			
FOR *Gasoline*			
BALANCE	223	49	
DEPOSITS	52	95	
TOTAL	276	44	
AMOUNT THIS CHECK	15	74	
BALANCE	260	70	

JOE BROKE
1234 ELM STREET
ANYWHERE, STATE 00000

$\frac{39-81}{1234}$ 158

December 27 19 *84*

PAY TO THE
ORDER OF _____ *Slow Gas* _____ $ *15.* $\frac{74}{}$

Fifteen and $\frac{74}{100}$ _____ DOLLARS

LAST NATIONAL BANK
ADDRESS
ANYWHERE, STATE 00000   SPECIMEN

MEMO *Gasoline*   *Joe M. Broke*

⑈1234⑈081⑈ 190519⑈

Easy, isn't it? Try a few more problems. Complete the bottom portion of the next two check stubs.

No. 159	$ *47. 52*		
December 28 19 *84*			
TO *Rita's Boutique*			
FOR *Dress*			
BALANCE	260	70	
DEPOSITS			
TOTAL			
AMOUNT THIS CHECK			
BALANCE			

No. 160	$ *14. 75*		
December 30 19 *84*			
TO *Anderson's Vitamin Supply*			
FOR *Vitamin's*			
BALANCE			
DEPOSITS	25	70	
TOTAL			
AMOUNT THIS CHECK			
BALANCE			

Turn to 5 to check your work.

No. 159	$ 47. 59
December 28 19 84	
TO Rita's Boutique	
FOR Dress	

BALANCE	260	70
DEPOSITS		
TOTAL	260	70
AMOUNT THIS CHECK	47	59
BALANCE	213	11

No. 160	$ 14. 75
December 30 19 84	
TO Anderson's Vitamin Supply	
FOR Vitamin's	

BALANCE	213	11
DEPOSITS	25	70
TOTAL	238	81
AMOUNT THIS CHECK	14	75
BALANCE	224	06

Just add the deposit and subtract the check amount. Note that the new balance at the bottom of stub No. 159 is transferred as the beginning balance in stub No. 160.

Check Register

The second method of recording checks is in a *check register*. A check register has entries for the check number, date, to whom the check is issued, amount of check, deposits, and balance.

Check No.	Date	Check Issued to or Deposit	Amount of Check		✔	Amount of Deposit		Balance	
								212	72
	12/23	Deposit				25	50	238	22
157	12/23	The Record Store	14	73				223	49

Deposits or checks are recorded on separate lines. Each deposit is added to the previous balance to obtain the new balance. Each check amount is subtracted.

The ✔ is used to record processed checks in a bank reconciliation. Bank reconciliation is discussed in 6.

Complete the following check register by calculating the balance for each line. Simply add each deposit and subtract each check.

Check No.	Date	Check Issued to or Deposit	Amount of Check		✔	Amount of Deposit	Balance 212	72
	12/23	Deposit				25 50	238	22
157	12/23	The Record Store	14	73			223	49
158	12/27	Slow Gas	15	74				
	12/27	Deposit				52 95		
159	12/28	Rita's Boutique	47	59				
160	12/30	Anderson's Vitamins	14	75				
	12/30	Deposit				25 70		

Turn to 6 to check your calculations.

6

Check No.	Date	Check Issued to or Deposit	Amount of Check		✔	Amount of Deposit	Balance 212	72
	12/23	Deposit				25 50	238	22
157	12/23	The Record Store	14	73			223	49
158	12/27	Slow Gas	15	74			207	75
	12/27	Deposit				52 95	260	70
159	12/28	Rita's Boutique	47	59			213	11
160	12/30	Anderson's Vitamins	14	75			198	36
	12/30	Deposit				25 70	224	06

BANK RECONCILIATION

Once a month, the bank sends a summary of all checking account transactions along with the checks that have been processed, the *cancelled checks*. The account summary lists all checks processed, deposits recorded, and any service fees.

Once the summary is received, either your check stubs or check register should be checked against the summary. This procedure is called *reconciliation*. A reconciliation form is usually included on the back of the summary. The reconciliation procedure is easy, just follow the steps included on the form.

Example: Mr. Broke's bank summary shows that checks numbered 157, 158, and 160 were processed, the deposits on 12/23 and 12/27 were credited to his account, and there was a service charge of $2.50. The bank balance is $243.45.

Use Mr. Broke's check register and the following form to complete his bank reconciliation.

First, check off the items which were processed in the column marked ✔. The items not processed are unchecked; this leaves check number 159 and the deposit on 12/30. Any checks not processed are called *outstanding checks*.

Check No.	Date	Check Issued to or Deposit	Amount of Check		✔	Amount of Deposit		Balance	
								212	72
	12/23	Deposit			✓	25	50	238	22
157	12/23	The Record Store	14	73	✓			223	49
158	12/27	Allow Gas	15	74	✓			207	75
	12/27	Deposit			✓	52	95	260	70
159	12/28	Rita's Boutique	47	59				213	11
160	12/30	Anderson's Vitamins	14	75	✓			198	36
	12/30	Deposit				25	70	224	06

Next, use the above information to complete the bank reconciliation form.

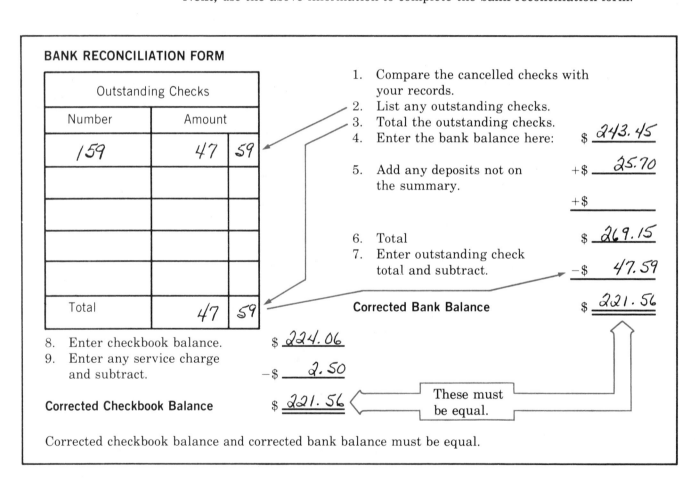

BANK RECONCILIATION FORM

Outstanding Checks		
Number	Amount	
159	47	59
Total	47	59

1. Compare the cancelled checks with your records.
2. List any outstanding checks.
3. Total the outstanding checks.
4. Enter the bank balance here: $ 243.45
5. Add any deposits not on the summary. +$ 25.70

 +$ _____

6. Total $ 269.15
7. Enter outstanding check total and subtract. -$ 47.59

Corrected Bank Balance $ 221.56

8. Enter checkbook balance. $ 224.06
9. Enter any service charge and subtract. -$ 2.50

Corrected Checkbook Balance $ 221.56

These must be equal.

Corrected checkbook balance and corrected bank balance must be equal.

Since both the corrected bank balance and the corrected checkbook balance agree, no errors were made by either the bank or Mr. Broke. If these two amounts do not agree and you can't find any errors, take your check register, bank summary and cancelled checks to your bank's customer service department.

Important

Mr. Broke must be careful not to forget to record the service charge in his check register.

Now, it's your turn.

Ms. Rodriguez's bank summary shows that checks numbered 561, 562, 563, and 566 were processed, that the deposits on 2/5 and 2/25 were credited to her account, and that there was a service charge of $3.75. The bank balance is $513.57.

Complete Ms. Rodriguez's check register and her bank reconciliation form.

Check No.	Date	Check Issued to or Deposit	Amount of Check		✔	Amount of Deposit	Balance 721	86
561	2/3	High - Low Apts.	359	50				
562	2/5	Electric Company	127	83				
	2/5	Deposit				275	00	
563	2/7	Quick Foods	83	42				
564	2/15	Friendly Bank-Auto	156	72				
565	2/20	Gas Company	62	78				
	2/25	Deposit				136	72	
566	2/25	Clarke's Clothes	45	51				
	2/27	Deposit				842	95	

BANK RECONCILIATION FORM

Outstanding Checks		
Number	Amount	
Total		

1. Compare the cancelled checks with your records.
2. List any outstanding checks.
3. Total the outstanding checks.
4. Enter the bank balance here: $ _____

5. Add any deposits not on the summary.   +$ _____

   +$ _____

6. Total $ _____
7. Enter outstanding check total and subtract.   −$ _____

Corrected Bank Balance $ _____

8. Enter checkbook balance. $ _____
9. Enter any service charge and subtract.   −$ _____

Corrected Checkbook Balance $ _____

Corrected checkbook balance and corrected bank balance must be equal.

Turn to **7** to check your work.

Check No.	Date	Check Issued to or Deposit	Amount of Check		✔	Amount of Deposit	Balance	
							721	86
561	2/3	High-Low Apts.	359	50	✓		362	36
562	2/5	Electric Company	127	83	✓		234	53
	2/5	Deposit			✓	275 00	509	53
563	2/7	Quick Foods	83	42	✓		426	11
564	2/15	Friendly Bank Auto	156	72			269	39
565	2/20	Gas Company	62	78			206	61
	2/25	Deposit			✓	136 72	343	33
566	2/25	Clarke's Clothes	45	51	✓		297	82
	2/27	Deposit				842 95	1140	77

BANK RECONCILIATION FORM

Outstanding Checks		
Number	Amount	
564	156	72
565	62	78
Total	219	50

1. Compare the cancelled checks with your records.
2. List any outstanding checks.
3. Total the outstanding checks.
4. Enter the bank balance here: $ _513.57_
5. Add any deposits not on the summary.   +$ _842.95_

       +$ _____

6. Total $ _1356.52_
7. Enter outstanding check total and subtract.   -$ _219.50_

Corrected Bank Balance $ _1137.02_

8. Enter checkbook balance. $ _1140.77_
9. Enter any service charge and subtract.   -$ _3.75_

Corrected Checkbook Balance $ _1137.02_

Corrected checkbook balance and corrected bank balance must be equal.

The corrected bank balance and corrected checkbook balance both are $1137.02. Bank reconciliation is easy: Simply follow the steps on the form.

For a set of practice problems on checking accounts turn to **8**.

9 Consumer Math

PROBLEM SET 1

8 The answers are on page 590.

9.1 Checking Accounts

A. Complete the following checks and check stubs for Joe Broke. Joe's beginning balance is $637.25.

1. Check number 217 to Southland Mortgage Company for $315.89, dated March 2, 1984.

No. 217 $ _____	JOE BROKE	$\frac{39-81}{1234}$	217
_____ 19 ___	1234 ELM STREET		
TO _____	ANYWHERE, STATE 00000	_____ 19 _____	
FOR _____			
	PAY TO THE ORDER OF _____ $ _____		
BALANCE			
DEPOSITS		_____ DOLLARS	
TOTAL	LAST NATIONAL BANK	SPECIMEN	
AMOUNT THIS CHECK	ADDRESS ANYWHERE, STATE 00000		
BALANCE	MEMO _____	_____	

⑈1234⑈081⑈ 190519⑈

2. Check number 218 to South Gas Company for $56.25, dated March 4, 1984.

No. 218 $ _____	JOE BROKE	$\frac{39-81}{1234}$	218
_____ 19 ___	1234 ELM STREET		
TO _____	ANYWHERE, STATE 00000	_____ 19 _____	
FOR _____			
	PAY TO THE ORDER OF _____ $ _____		
BALANCE			
DEPOSITS		_____ DOLLARS	
TOTAL	LAST NATIONAL BANK	SPECIMEN	
AMOUNT THIS CHECK	ADDRESS ANYWHERE, STATE 00000		
BALANCE	MEMO _____	_____	

⑈1234⑈081⑈ 190519⑈

Date _____

Name _____

Course/Section _____

3. Check number 219 to Bell Telephone Company for $15.82, dated March 10, 1984. Also a deposit for $432.50 was made on March 10, 1984.

No. 219 $
————— 19 ———
TO —————————
FOR ————————

BALANCE		
DEPOSITS		
TOTAL		
AMOUNT THIS CHECK		
BALANCE		

JOE BROKE
1234 ELM STREET
ANYWHERE, STATE 00000

39-81
————
1234

219

———————————— 19 ———

PAY TO THE
ORDER OF ————————————————————————— $—————

—————————————————————————————————— DOLLARS

LAST NATIONAL BANK
ADDRESS
ANYWHERE, STATE 00000   SPECIMEN

MEMO —————

⑆ 1234 ⑈ 081 ⑆   190519 ⑈

4. Check number 220 to Slow Foods for $123.36, dated March 19, 1984.

No. 220 $
————— 19 ———
TO —————————
FOR ————————

BALANCE		
DEPOSITS		
TOTAL		
AMOUNT THIS CHECK		
BALANCE		

JOE BROKE
1234 ELM STREET
ANYWHERE, STATE 00000

39-81
————
1234

220

———————————— 19 ———

PAY TO THE
ORDER OF ————————————————————————— $—————

—————————————————————————————————— DOLLARS

LAST NATIONAL BANK
ADDRESS
ANYWHERE, STATE 00000   SPECIMEN

MEMO —————

⑆ 1234 ⑈ 081 ⑆   190519 ⑈

5. Check number 221 to the Money Magician Finance Company for $136.14, dated March 25, 1984.

No. 221 $
————— 19 ———
TO —————————
FOR ————————

BALANCE		
DEPOSITS		
TOTAL		
AMOUNT THIS CHECK		
BALANCE		

JOE BROKE
1234 ELM STREET
ANYWHERE, STATE 00000

39-81
————
1234

221

———————————— 19 ———

PAY TO THE
ORDER OF ————————————————————————— $—————

—————————————————————————————————— DOLLARS

LAST NATIONAL BANK
ADDRESS
ANYWHERE, STATE 00000   SPECIMEN

MEMO —————

⑆ 1234 ⑈ 081 ⑆   190519 ⑈

B. Use the check stubs from part A and the following information to complete the bank reconciliation form.

Mr. Broke's bank summary shows checks numbered 217, 218, and 220 were processed, the deposit on March 10 was credited to his account, and there was a service charge of $4.15. The bank balance was $570.10.

BANK RECONCILIATION FORM

Outstanding Checks		
Number	Amount	
Total		

1. Compare the cancelled checks with your records.
2. List any outstanding checks.
3. Total the outstanding checks.
4. Enter the bank balance here: $ _____

5. Add any deposits not on the summary.   +$ _____

   +$ _____

6. Total $ _____
7. Enter outstanding check total and subtract.   −$ _____

Corrected Bank Balance $ _____

8. Enter checkbook balance. $ _____
9. Enter any service charge and subtract.   −$ _____

Corrected Checkbook Balance $ _____

Corrected checkbook balance and corrected bank balance must be equal.

C. Complete the following check register.

Check No.	Date	Check Issued to or Deposit	Amount of Check		✔	Amount of Deposit	Balance 641	87
517	4/2	Faront Apts.	317	55				
518	4/2	Electric Company	45	17				
519	4/5	Gas Company	52	51				
	4/7	Deposit				275	58	
520	4/10	Quick Foods	117	48				
521	4/12	Money Magic Co.	219	86				
522	4/15	Rita's Boutique	48	92				
523	4/20	Wilson Company	89	14				
524	4/21	Lon's Surplus	17	95				
	4/25	Deposit				215	16	

D. Use the check register from part C and the following information to complete the bank reconciliation.

The bank summary shows checks numbered 517, 518, 519, 520, 522, and 524 were processed, the deposit on April 7 was credited to the account, and there was a service charge of $5.42. The bank balance was $312.45.

BANK RECONCILIATION FORM

Outstanding Checks		
Number	Amount	
Total		

1. Compare the cancelled checks with your records.
2. List any outstanding checks.
3. Total the outstanding checks.
4. Enter the bank balance here: $ _____

5. Add any deposits not on the summary.   + $ _____

   + $ _____

6. Total $ _____
7. Enter outstanding check total and subtract.   − $ _____

Corrected Bank Balance $ _____

8. Enter checkbook balance. $ _____
9. Enter any service charge and subtract.   − $ _____

Corrected Checkbook Balance $ _____

Corrected checkbook balance and corrected bank balance must be equal.

When you have had the practice you need, either return to the Preview on page 314 or continue in 9 with the study of installment loans.

SECTION 2: INSTALLMENT LOANS

9

OBJECTIVE

Calculate the total cost, interest and Annual Percentage Rate for installment loans. Calculate the interest and APR for an add-on loan.

It may sometimes be inconvenient, if not impossible, to pay cash for all your purchases. You may want to "charge" an item on your credit card and pay for it the following month. Or, you might wish to purchase an item, such as a car, that would require more cash than you have on hand. In this case, borrowing money from a lending institution can help solve your "short of cash" problem.

Whatever your reasons for borrowing money, you should know how much to borrow, what interest you will pay (for the "privilege" of borrowing), what the amount and length of payments are, and the rate of interest for the loan.

The Federal Reserve Board, in *Regulation Z,* also known as the *Truth-In-Lending Act,* requires all lenders to state the *effective interest rate.* This rate, the *Annual Percentage Rate* (APR), is the true annual percentage rate charged to borrowers.

Interest rates are set by individual states and may vary widely depending on the type of lending institution (e.g., bank, finance company, and pawn shop) and the state.

INSTALLMENT LOANS

Most "buy now, pay later" loans are installment loans. The *installment loan* is a loan to be repaid in fixed regular payments or installments for a certain length of time. This type of loan is frequently used for a variety of purchases ranging from household appliances to vehicles.

The *total amount of installment payments* is the product of the number of payments times the amount of each payment.

Total amount of
installment payments = number of payments × amount of each payment

---

Many installment loans require an initial payment called the *down payment.* The *total installment cost* is the sum of the down payment and the total installment payments.

Total installment cost = down payment + total installment payments

---

Example: A new washing machine may be purchased for $305 cash or $35 down and 12 monthly payments of $25. Calculate the total installment cost.

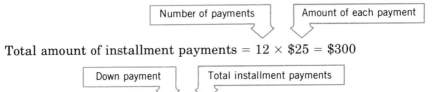

Total amount of installment payments = 12 × $25 = $300

Total installment cost = $35 + 300 = $335

The total installment cost is $335.

The next question is how much interest is being paid on the loan. The interest is the difference between the total installment cost and the cost if paid by cash, called the *cash cost*. The interest is the amount you pay for the privilege of buying on an installment plan.

> Interest = total installment cost − cash cost

Calculate the interest in our washing machine example, then turn to 10.

10

Interest = $335 − 305 = $30

The interest is $30. Is that a good deal, a rip-off, or OK? The best way to determine if the interest is too high is to calculate the APR (interest rate).

First, work another problem.

An automobile sold for $5765.86 cash or $700 down and 36 monthly payments of $168. Calculate the total installment cost and the interest.

Turn to 11 to check your work.

11

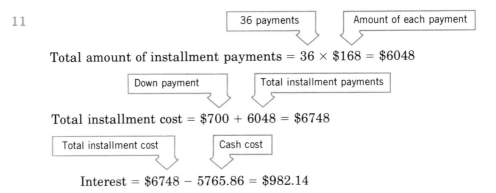

Total amount of installment payments = 36 × $168 = $6048

Total installment cost = $700 + 6048 = $6748

Interest = $6748 − 5765.86 = $982.14

The total installment cost is $6748 and the interest is $982.14. Again the question arises: Is $982.14 a reasonable amount of interest? To answer this question, you must look at the annual percentage rate or APR. Remember, on *any* loan, the APR must be stated.

Use the following formulas to approximate the annual percentage rate or APR. The formula only gives an approximation. The exact APR can be found by using a set of tables or a special business calculator.

$$\text{Loan amount} = \text{cash cost} - \text{down payment}$$

$$\text{APR} = \frac{2 \times (\text{number of payments per year}) \times (\text{interest})}{(\text{loan amount}) \times (\text{total number of payments} + 1)}$$

The loan amount is simply the total amount borrowed. The number of payments per year is 12 if paid monthly or 52 if paid weekly.

Example: A new washing machine may be purchased for $305 cash or $35 down and 12 monthly payments of $25. Calculate the APR and round to the nearest 0.1%.

In frames **9** and **10**, we calculate the interest to be $30.

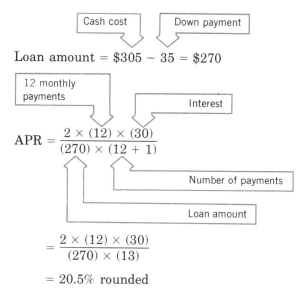

Loan amount = $305 − 35 = $270

$$\text{APR} = \frac{2 \times (12) \times (30)}{(270) \times (12 + 1)}$$

$$= \frac{2 \times (12) \times (30)}{(270) \times (13)}$$

$$= 20.5\% \text{ rounded}$$

The APR is 20.5%. This must be stated in the loan agreement. Does 20.5% sound a little high? Shop around and see if a better interest rate is available.

Now, it's your turn.

Calculate the APR for our automobile problem of frame **10**. The car's cost is $5765.86 or $700 down and 36 monthly payments of $168.

You calculated the interest to be $982.14. What is the APR rounded to the nearest 0.1%?

Turn to **12** to check your work.

12    Loan amount = $5765.86 − 700 = $5065.86

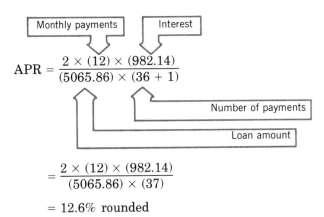

$$\text{APR} = \frac{2 \times (12) \times (982.14)}{(5065.86) \times (36 + 1)}$$

$$= \frac{2 \times (12) \times (982.14)}{(5065.86) \times (37)}$$

$$= 12.6\% \text{ rounded}$$

The APR is 12.6%.

Add-on Interest   For automobile loans and many other installment loans, some lenders may quote an *add-on interest rate*. The add-on interest rate is used to calculate the interest using the simple interest formula.

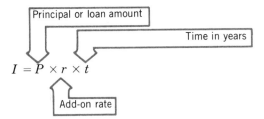

$$I = P \times r \times t$$

Careful! The add-on interest rate is *not* the true interest rate (APR). In fact, the add-on rate is always much lower than the APR.

The simple interest formula may be used to calculate the interest, but when using the add-on rate, this is not a simple interest loan. Recall, with a simple interest loan, only one payment is made at the end of the loan period. With add-on interest, the APR must be calculated to obtain the true interest rate. the add-on rate is only convenient to calculate the interest.

Example: A used car is offered at a cash price of $2900 or with a down payment of $500 and the balance in 24 monthly payments at an add-on interest rate of 9%.

Calculate the loan amount and interest.

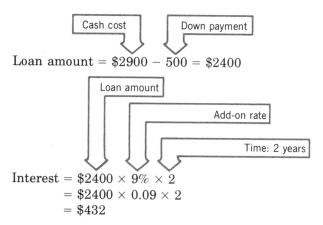

Loan amount = $2900 − 500 = $2400

Interest = $2400 × 9% × 2
= $2400 × 0.09 × 2
= $432

The interest is $432.

Next, estimate the true interest rate using our APR formula. Round to the nearest 0.1%.

Check your work in **13**.

What is a loan shark?

A loan shark lends money at an excessive — and illegal — interest rate. And because your body is the collateral, he has a payment plan you can't refuse.

13

$$APR = \frac{2 \times (12) \times (432)}{(2400) \times (24 + 1)}$$

$$= \frac{2 \times (12) \times (432)}{(2400) \times (25)}$$

$$= 17.3\%$$

The APR of 17.3% is much higher than the add-on rate of 9%. Always ask for the APR, the effective or true interest rate.

Need some additional practice? Work the following problem.

A used car is offered at a cash price of $6498.75 or $1000 down and the balance in 36 monthly payments at an add-on interest rate of 8%.

Calculate the loan amount, interest, and APR. Round the APR to the nearest 0.1%.

Turn to **14** to check your work.

14   Loan amount = $6498.75 − 1000 = $5498.75

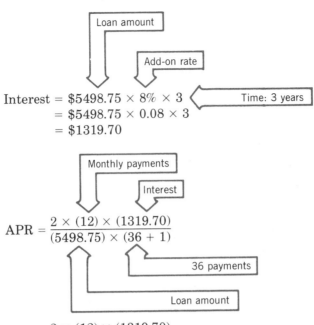

Interest = $5498.75 × 8% × 3
        = $5498.75 × 0.08 × 3
        = $1319.70

$$APR = \frac{2 \times (12) \times (1319.70)}{(5498.75) \times (36 + 1)}$$

$$= \frac{2 \times (12) \times (1319.70)}{(5498.75) \times (37)}$$

$$= 15.6\% \text{ rounded}$$

The interest is $1319.70 and the APR is 15.6%.

How high do interest rates go? In some cases, the sky seems to be the limit. A friend of ours, Lou, borrowed $25 for one month from a pawn shop. He made one payment of $30 the next month.

Calculate the APR, then turn to **15**.

---

15   Loan amount = $25

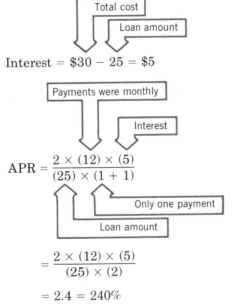

Interest = $30 − 25 = $5

$$APR = \frac{2 \times (12) \times (5)}{(25) \times (1 + 1)}$$

$$= \frac{2 \times (12) \times (5)}{(25) \times (2)}$$

$$= 2.4 = 240\%$$

Does the APR of 240% seem a little high? You bet! This is *not* a fictitious case; and the APR of 240% was on the back of Lou's pawn stub.

For a set of practice problems on installment loans, turn to **16**.

Consumer Math

PROBLEM SET 2

16   The answers are on page 595.

9.2   Installment Loans   **A. Calculate the total installment cost and interest.**

	Cash Cost	Down Payment	Number of Monthly Payments	Amount of Each Payment	Total Installment Cost	Interest
1.	$500	$100	12	$36	——	——
2.	450	50	18	25	——	——
3.	75	10	6	12	——	——
4.	150	25	12	12	——	——
5.	250	50	24	10	——	——
6.	400	200	24	10	——	——
7.	350	50	18	20	——	——
8.	418	50	24	18	——	——
9.	126	25	18	7	——	——
10.	79	10	6	13	——	——
11.	895	100	6	140	——	——
12.	950	350	12	55	——	——
13.	1200	250	24	48	——	——
14.	750	150	12	56	——	——
15.	750	150	18	40	——	——
16.	525	120	9	49	——	——
17.	2350	250	24	113	——	——
18.	5778	290	36	212	——	——
19.	845	125	18	48	——	——
20.	570	50	12	51	——	——

Date

Name

Course/Section

B. Calculate the total installment cost, interest, and estimate the APR.

	Cash Cost	Down Payment	Number of Payments	Frequency of Payments	Amount of Each Payment	Total Installment Cost	Interest	APR
1.	$450	$150	12	Monthly	$27.50	——	——	——
2.	200	50	12	Weekly	12.75	——	——	——
3.	1250	400	18	Monthly	56.00	——	——	——
4.	600	200	12	Monthly	36.75	——	——	——
5.	175	50	6	Weekly	21.00	——	——	——
6.	800	200	20	Weekly	31.25	——	——	——
7.	925	250	24	Monthly	35.00	——	——	——
8.	772	200	24	Monthly	28.55	——	——	——
9.	985	250	6	Monthly	130.00	——	——	——
10.	156	50	18	Monthly	6.75	——	——	——
11.	237	25	6	Weekly	35.50	——	——	——
12.	189	10	12	Weekly	15.25	——	——	——
13.	1350	250	24	Monthly	56.85	——	——	——
14.	250	50	6	Monthly	35.75	——	——	——
15.	525	75	20	Weekly	22.60	——	——	——
16.	750	125	24	Weekly	27.35	——	——	——
17.	2575	225	18	Monthly	156.75	——	——	——
18.	3750	375	24	Monthly	178.25	——	——	——
19.	820	125	12	Monthly	65.50	——	——	——
20.	370	75	6	Monthly	52.35	——	——	——

C. Calculate the loan amount, interest, and estimate the APR.

	Price	Down Payment	Number of Monthly Installments	Add-on Rate	Loan Amount	Interest	APR
1.	$850	$150	12	6%	——	——	——
2.	1200	250	24	7%	——	——	——
3.	125	10	12	7%	——	——	——
4.	550	50	12	$6\frac{1}{2}\%$	——	——	——
5.	5650	1250	24	8%	——	——	——
6.	625	125	12	$8\frac{1}{2}\%$	——	——	——
7.	450	50	18	$7\frac{1}{2}\%$	——	——	——
8.	6550	500	36	7%	——	——	——
9.	7225	2000	24	8%	——	——	——
10.	4282	935	36	$8\frac{1}{2}\%$	——	——	——
11.	296	50	12	8%	——	——	——
12.	553	125	12	$7\frac{1}{2}\%$	——	——	——
13.	5650	500	24	9%	——	——	——
14.	9375	1250	36	$8\frac{1}{2}\%$	——	——	——
15.	7625	550	30	$9\frac{1}{2}\%$	——	——	——
16.	7500	1200	30	$9\frac{1}{4}\%$	——	——	——
17.	3250	450	18	$7\frac{1}{2}\%$	——	——	——
18.	6720	570	24	$8\frac{1}{4}\%$	——	——	——
19.	8725	975	42	$9\frac{1}{2}\%$	——	——	——
20.	6270	575	36	$8\frac{3}{4}\%$	——	——	——

When you have had the practice you need, either return to the Preview on page 314, or continue in 17 with the study of credit cards.

17

OBJECTIVE

Interchange the APR and the monthly interest rate, and calculate the interest and new balance on revolving charge or credit card accounts.

Credit card accounts or *revolving charge plans* are available from a variety of business, including VISA, Mastercard and many national and local department stores.

With your credit card, purchases may be "charged" or added to your account. Any purchases made during a month appear on the next month's statement. If the balance is paid within 30 days from the statement date, no interest or *finance charge* is added. If the entire balance isn't paid, interest is charged on the remaining or *unpaid balance*. Since the maximum interest rates are set by the states, the APR may vary from a low of 12% to a high of 24%. The most common rate is 21%. Several states base their rates on the balance of the account. In these instances, two different rates may be applicable. In addition to the interest, several of the major credit cards charge an annual fee.

The interest rate may be stated as an APR (annual percentage rate) or a monthly rate. **To convert the APR to a monthly rate, simply divide by 12, the number of months in a year.**

Example: Convert an APR of 21% to a monthly rate.

$$\text{Monthly rate} = \frac{21\%}{12} = \frac{7}{4}\% = 1\frac{3}{4}\%$$

(APR / 12 months)

Your turn.

Convert 10% APR to a monthly rate.

Check your answer in 18.

---

18 $$\text{Monthly rate} = \frac{10\%}{12} = \frac{5}{6}\%$$

To convert from monthly rate to the APR, simply multiply by 12.

Example: Convert the monthly rate of $1\frac{3}{4}\%$ to an APR.

$$\text{APR} = 12 \times 1\frac{3}{4}\% = \frac{\overset{3}{\cancel{12}}}{1} \times \frac{7}{\underset{1}{\cancel{4}}}\% = 21\%$$

A monthly rate of $1\frac{3}{4}\%$ is equivalent to an APR of 21%.

Convert the monthly rate of $1\frac{1}{4}\%$ to an APR.

Turn to 19 to check your answer.

---

19 $$APR = 12 \times 1\frac{1}{4}\% = \frac{\overset{3}{\cancel{12}}}{1} \times \frac{5}{\underset{1}{\cancel{4}}}\% = 15\%$$

A monthly rate of $1\frac{1}{4}\%$ is the same as 15% APR.

Once you have the monthly rate, the finance charge (interest) is found as the product of the unpaid balance and the monthly rate.

Interest = unpaid balance × monthly rate

Example: The unpaid balance on a credit card account is $182.35. The APR is 15%. Calculate the interest.

First, calculate the monthly rate.

Monthly rate $= \dfrac{15\%}{12} = \dfrac{5}{4}\% = 1\dfrac{1}{4}\%$

Next, calculate the interest.

Interest $= \$182.35 \times 1\frac{1}{4}\%$ ← Unpaid balance / Monthly rate

$= \$182.35 \times 1.25\%$
$= \$182.35 \times 0.0125$
$= \$2.28$ rounded

The interest is $2.28.

If you need a review of the process of changing fractional percents to decimal numbers, return to page 126.

Otherwise work the following problem.

The unpaid balance on a revolving charge account is $156.73. The APR is 18%. Calculate the interest.

Turn to **20** to check your work.

---

20   Monthly rate $= \dfrac{18\%}{12} = \dfrac{3}{2}\% = 1\dfrac{1}{2}\%$

Interest $= \$156.73 \times 1\frac{1}{2}\%$ ← Unpaid balance / Monthly rate
$= \$156.73 \times 1.5\%$
$= \$156.73 \times 0.015$
$= \$2.35$ rounded

The finance charge (interest) is $2.35.

When you receive the monthly statement, you have three choices for payment.

1. You may pay the entire balance. In this case, there is no finance charge.
2. You pay the minimum required amount. A minimum payment is usually required, varying from 5 to 10% of the balance, but never less than $10. In this case, interest is charged on the remaining or unpaid balance on the next month's statement.

3. You may pay an amount larger than the minimum but less than the balance. This reduces the balance more than the minimum amount and results in less interest.

Example: A revolving charge statement is received with a balance of $241.79. The minimum required payment is $12, and the APR is 18%. If the minimum payment is made, calculate (a) the unpaid balance, and (b) the interest on the unpaid balance for next month's statement.

(a) Unpaid balance = $241.79 − 12.00 = $229.79

   Monthly rate = $\dfrac{18\%}{12} = \dfrac{3}{2}\% = 1\dfrac{1}{2}\%$

(b) Interest = $\$229.79 \times 1\dfrac{1}{2}\%$ Monthly rate

   = $229.79 × 1.5%
   = $229.79 × 0.015
   = $3.45 rounded

The new balance is $229.79 and the interest charged on the next month's statement is $3.45.

Work the following problem.

A revolving charge statement was received with a balance of $463.81. The minimum required payment is $23.19, and the APR is 15%. If a payment of $25.50 is made, calculate (a) the unpaid balance, and (b) the interest on the unpaid balance.

Check your calculations in 21.

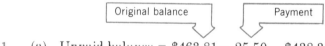

21   (a) Unpaid balance = $463.81 − 25.50 − $438.31

   Monthly rate = $\dfrac{15\%}{12} = \dfrac{5}{4}\% = 1\dfrac{1}{4}\%$

(b) Interest = $\$438.31 \times 1\dfrac{1}{4}\%$ Monthly rate

   = $438.31 × 1.25%
   = $438.31 × 0.0125
   = $5.48 rounded

The unpaid balance is $438.31 and the interest is $5.48.

In many states the interest rate charged depends on the account balance. For example, in several states the interest rate is

APR	Balance
18%	$0 to $500
12%	Over $500

The interest rate is 18% on the first $500, and 12% on any amount over $500.

Work the following problem.

A charge card statement was received with a balance of $762.15. The minimum required payment is $38.11. The APR is determined as follows:

APR	Balance
18%	$0 to $500
12%	Over $500

If the minimum payment is made, calculate (a) the unpaid balance and (b) the interest.

Check your answer in 22.

22   (a)   Unpaid balance = $762.15 − 38.11 = $724.04

Monthly rate: (First $500) $= \dfrac{18\%}{12} = \dfrac{3}{2}\% = 1\dfrac{1}{2}\%$

Monthly rate: (Over $500) $= \dfrac{12\%}{12} = 1\%$

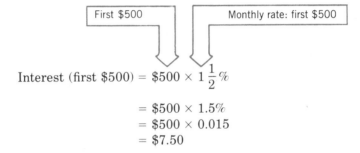

Interest (first $500) $= \$500 \times 1\dfrac{1}{2}\%$

$$= \$500 \times 1.5\%$$
$$= \$500 \times 0.015$$
$$= \$7.50$$

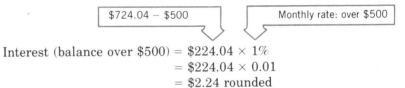

Interest (balance over $500) $= \$224.04 \times 1\%$
$$= \$224.04 \times 0.01$$
$$= \$2.24 \text{ rounded}$$

(b)   Total interest = $7.50 + 2.24 = $9.74

The unpaid balance is $724.04. The interest is $9.74; $7.50 on the first $500 and $2.24 on the amount over $500.

Once the unpaid balance and interest are calculated, the next month's balance is the sum of the unpaid balance and the interest.

Calculate the next month's balance for the above problem.

Turn to 23 to check your answer.

23   Next month's balance = $724.04 + 9.74 = $733.78

The new balance is $733.78.

If any new purchases were made, these amounts must also be added to the next month's balance.

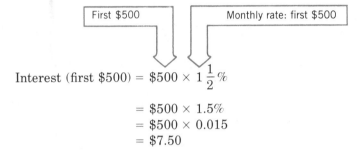

> Next month's balance = unpaid balance + interest + new purchases

Example: A revolving charge account statement received in March shows a balance of \$435.17. The minimum required payment is \$43.52. The APR is 15%. Purchases totaling \$22.95 were made in March but not included in the statement. If the minimum payment is made, calculate the (a) unpaid balance, (b) interest on the unpaid balance, and (c) balance for the April statement.

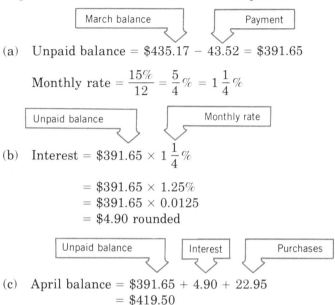

(a) Unpaid balance = \$435.17 − 43.52 = \$391.65

$$\text{Monthly rate} = \frac{15\%}{12} = \frac{5}{4}\% = 1\frac{1}{4}\%$$

(b) Interest = \$391.65 × $1\frac{1}{4}\%$

$\qquad$ = \$391.65 × 1.25%
$\qquad$ = \$391.65 × 0.0125
$\qquad$ = \$4.90 rounded

(c) April balance = \$391.65 + 4.90 + 22.95
$\qquad\qquad\qquad$ = \$419.50

Now it's your turn. Work the following problem.

A revolving charge account statement received in April from the Faster-Charge Company shows a balance of \$419.50. The minimum payment is \$41.95. The APR is 15%. Purchases for \$72.86 were made in April but not included in the statement. If the minimum payment is made, calculate the (a) unpaid balance, (b) interest on the unpaid balance, and (c) balance for the May statement.

Check your calculations in **24**.

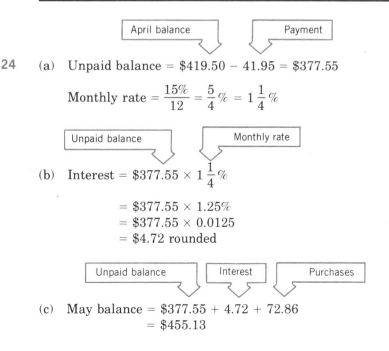

24 (a) Unpaid balance = \$419.50 − 41.95 = \$377.55

$$\text{Monthly rate} = \frac{15\%}{12} = \frac{5}{4}\% = 1\frac{1}{4}\%$$

(b) Interest = \$377.55 × $1\frac{1}{4}\%$

$\qquad$ = \$377.55 × 1.25%
$\qquad$ = \$377.55 × 0.0125
$\qquad$ = \$4.72 rounded

(c) May balance = \$377.55 + 4.72 + 72.86
$\qquad\qquad\qquad$ = \$455.13

Most major credit cards do not calculate the interest on the unpaid balance at the end of the month. The balance on the last day of the month may not reflect the actual amount of money on loan during the rest of the month. It is more reasonable to calculate the interest as a percent of the *average daily balance*. The average daily balance is the average of the account's balance throughout the month.

For a set of practice problems on credit card accounts, turn to 25.

Consumer Math

PROBLEM SET 3

25

9.3   Credit Cards

The answers are on page 595.

A.   Interchange the following APRs and monthly interest rates.

	APR	Monthly Rate			APR	Monthly Rate
1.	6%	——		11.	——	$1\frac{5}{6}$%
2.	8%	——		12.	——	$1\frac{1}{3}$%
3.	——	1%		13.	18%	——
4.	——	2%		14.	21%	——
5.	——	$1\frac{1}{2}$%		15.	——	$1\frac{1}{6}$%
6.	9%	——		16.	20%	——
7.	——	$1\frac{2}{3}$%		17.	——	$2\frac{1}{3}$%
8.	——	$2\frac{1}{2}$%		18.	——	$2\frac{1}{6}$%
9.	10%	——		19.	22%	——
10.	——	3%		20.	19%	——

B.   For the following charge accounts, calculate the unpaid balance and the interest on the unpaid balance.

	Balance	Payment	APR	Unpaid Balance	Interest
1.	$370.51	$19.00	18%	———————	———————
2.	125.32	25.00	18%	———————	———————
3.	894.50	45.00	21%	———————	———————
4.	29.72	29.72	18%	———————	———————
5.	152.89	50.00	15%	———————	———————
6.	216.70	25.00	18%	———————	———————
7.	342.91	18.00	18%	———————	———————
8.	192.18	25.00	15%	———————	———————
9.	395.27	32.50	12%	———————	———————
10.	502.16	62.00	18%	———————	———————

Date

Name

Course/Section

	Balance	Payment	APR	Unpaid Balance	Interest
11.	526.78	57.00	21%	————	————
12.	436.25	125.00	22%	————	————
13.	892.14	90.00	18%	————	————
14.	126.82	42.50	18%	————	————
15.	278.42	45.00	21%	————	————
16.	57.23	10.00	20%	————	————
17.	192.25	25.00	21%	————	————
18.	976.13	46.00	20%	————	————
19.	843.79	42.00	18%	————	————
20.	625.32	75.00	20%	————	————

C. For the following problems the APR is determined as follows:

APR	Balance
18%	$0 to $500
15%	Over $500

Calculate the unpaid balance and the interest on the unpaid balance.

	Balance	Payment	Unpaid Balance	Interest
1.	$785	$150	————	————
2.	630	35	————	————
3.	825	100	————	————
4.	950	48	————	————
5.	1275	125	————	————
6.	895	90	————	————
7.	530	125	————	————
8.	650	60	————	————
9.	932	85	————	————
10.	827	125	————	————

	Balance	Payment	Unpaid Balance	Interest
11.	755	125	—————	—————
12.	525	35	—————	—————
13.	620	150	—————	—————
14.	780	85	—————	—————
15.	2350	220	—————	—————
16.	420	24	—————	—————
17.	870	75	—————	—————
18.	926	85	—————	—————
19.	732	36	—————	—————
20.	678	33	—————	—————

D. For the following charge accounts, calculate the unpaid balance, interest on the unpaid balance, and the new balance.

	Balance	Payment	APR	New Purchases	Unpaid Balance	Interest	New Balance
1.	$535	$27	18%	$55.72	———	———	———
2.	126	10	18%	26.50	———	———	———
3.	895	150	21%	132.87	———	———	———
4.	450	50	15%	59.65	———	———	———
5.	375	45	18%	278.25	———	———	———
6.	155	25	15%	12.98	———	———	———
7.	280	115	21%	92.75	———	———	———
8.	972	125	18%	89.12	———	———	———
9.	152	50	18%	14.06	———	———	———
10.	207	25	21%	281.57	———	———	———
11.	275	14	21%	42.95	———	———	———
12.	765	35	22%	23.50	———	———	———

	Balance	Payment	APR	New Purchase	Unpaid Balance	Interest	New Balance
13.	472	150	18%	16.95	———	———	———
14.	836	75	20%	32.76	———	———	———
15.	972	250	21%	126.18	———	———	———
16.	378	125	21%	35.26	———	———	———
17.	422	23	22%	26.29	———	———	———
18.	528	27	20%	18.19	———	———	———
19.	735	225	18%	136.72	———	———	———
20.	125	50	18%	75.19	———	———	———

When you have had the practice you need, either return to the Preview on page 314, or continue in **26** with the study of stocks and bonds.

SECTION 4: STOCKS AND BONDS

<table>
<tr><td>26</td><td>

OBJECTIVE

Calculate stock dividends, the price of a stock or bond transaction, and the current yield of a bond.

</td></tr>
</table>

Stock   A corporation issues *stock* in order to raise money. The stock is issued in shares, where each share represents a portion of the company ownership. For example, if the Widget Company has issued a total of 1000 shares of stock and you own 20 shares, then you own $\frac{20}{1000}$ or $\frac{1}{50}$ of the company. You share proportionally in company growth or decline, and in making corporate decisions.

Stock ownership offers other benefits. When the company makes a profit, it may declare a *dividend* to the stockholders, passing the profit on. This will entitle the stockholders to a dividend for each share owned.

A company may issue as little as a few hundred shares of stock or as many as several hundred thousand shares. When the stock is issued, the company assigns a value called the *par value*. The par value is used only for accounting purposes. The actual value of the stock depends not on the par value but on the supply and demand of the stock on the "stock market."

There are two basic types of stock, common and preferred. Most stock issued is *common stock*. Each share of common stock entitles the owner to one vote during company elections. If you own 20 shares of a company's stock, then you are entitled to 20 votes. Stock ownership also entitles you to dividends, when declared by the company.

Preferred stock has precedence over common stock in the payment of dividends. In case of company liquidation (going out of business), preferred stockholders will be given preference over common share stockholders when "cashing in" their stock. Preferred stockholders usually do not have voting privileges.

The dividend rate of preferred stock is often fixed and stated as a percent of the par value. Dividends generally do not exceed the stated rate.

Example: The Fisher Corporation gives a 6% dividend on their preferred stock, based on a $50 par value. Calculate the total dividend on 10 shares of stock.

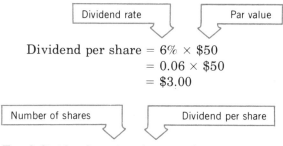

$$\text{Dividend per share} = 6\% \times \$50$$
$$= 0.06 \times \$50$$
$$= \$3.00$$

Total dividend = $10 \times \$3.00 = \$30.00$

Your turn. Work the following problem.

The Taylor Company gives a $6\frac{1}{2}\%$ dividend on their preferred stock, based on a $70 par value. Calculate the total dividend on 325 shares of stock.

Check your work in **27**.

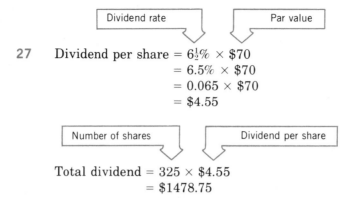

27 Dividend per share = $6\frac{1}{2}\% \times \$70$
= $6.5\% \times \$70$
= $0.065 \times \$70$
= $\$4.55$

Total dividend = $325 \times \$4.55$
= $\$1478.75$

The total dividend for 325 shares is $1478.75.

When a company has little or no profit, it may be unable to declare a dividend. But with *cumulative preferred stock,* unpaid dividends are paid at a later date, at the first opportunity. If one year, no dividend is given, then the next year the company will pay two years' dividends. When there is sufficient profit, cumulative preferred stockholders are paid first, then the common share stockholders receive their dividends, that is, if profits permit.

The Owen Company was unable to give a dividend last year to their stockholders with cumulative preferred 5%, par value $30. This year the company is giving two years' dividends. Calculate the total dividend for 250 shares of stock.

Turn to **28** to check your work.

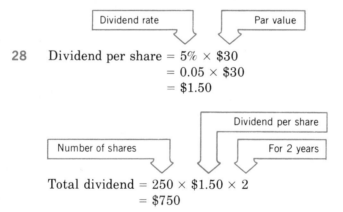

28 Dividend per share = $5\% \times \$30$
= $0.05 \times \$30$
= $\$1.50$

Total dividend = $250 \times \$1.50 \times 2$
= $\$750$

Stock dividends for common stock are not based on the par value or the market value. The dividend is simply a certain amount per share.

The Dorn Company has declared a common stock dividend of 12¢ per share. If you own 200 shares, what is your total dividend?

After you complete the calculation, turn to **29**.

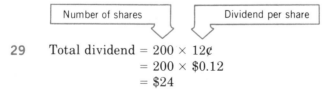

29 Total dividend = $200 \times 12¢$
= $200 \times \$0.12$
= $\$24$

Some companies may decide to retain their dividends to provide funds for company expansion. In this case, they may declare a *stock dividend,* where each stockholder receives a dividend of additional stock rather than cash.

Example: The Maples Company has declared a 10% stock dividend. Judy Bailor has 270 shares of stock. Calculate the number of additional shares of stock she is issued.

Dividend rate | Number of shares

$$\text{Stock dividend} = 10\% \times 270$$
$$= 0.10 \times 270$$
$$= 27$$

Ms. Bailor is issued 27 additional shares of stock as a dividend.

Work the following problem.

The Arnold Corporation has declared a 5% stock dividend. If you have 380 shares of stock, how many additional shares will you receive?

Check your work in 30.

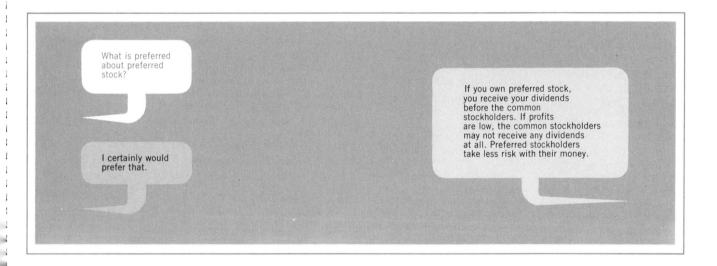

What is preferred about preferred stock?

I certainly would prefer that.

If you own preferred stock, you receive your dividends before the common stockholders. If profits are low, the common stockholders may not receive any dividends at all. Preferred stockholders take less risk with their money.

30   $$\text{Stock dividend} = 5\% \times 380$$
$$= 0.05 \times 380$$
$$= 19 \text{ shares}$$

Your stock dividend is 19 shares.

If the number of shares of stock to be issued isn't a whole number, the shares are rounded down. For example, if the number of shares is calculated to be 19.8, only 19 shares are issued. A cash amount is usually issued in lieu of the additional 0.8 shares.

Stock is usually purchased and sold by a broker at a stock exchange, such as the New York Stock Exchange. For his services, the broker charges a commission based on the number of shares of stock "traded" (bought or sold) and the total price.

Stocks are traded in either round lots or odd lots. A *round lot* is a multiple of a 100, such as 100, 200, 300, etc. An *odd lot* is any number of shares less than 100.

Traditionally, stock prices are usually stated in eighths of a dollar, such as $42\frac{3}{8}$. The current value of a certain stock may be obtained from your broker or the current stock listing. The listing of all stocks on the New York Stock Exchange appears daily in most major newspapers. One of the most comprehensive listings of stock and bond transactions is *The Wall Street Journal*.

The total cost of a stock purchase (excluding commission and taxes) is the product of the number of shares and the cost per share.

Total cost = number of shares × cost per share

Calculate the cost of 300 shares of McDonald's at $45\frac{1}{8}$ per share.

Check your work in **31**.

___

31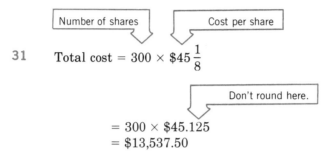

Total cost = $300 \times \$45\frac{1}{8}$

= $300 \times \$45.125$

= $\$13,537.50$

The cost of 300 shares is $13,537.50.

In addition to the current price, the stock listings on page 361 contain several other items of information.

TYPICAL STOCK LISTING

Yearly High	Low	Stock	Dividends	P–E	Sales in 100s	High	Low	Close	Change
(1)		(2)	(3)	(4)	(5)	(6)	(7)	(8)	(9)
$40\frac{7}{8}$	$35\frac{1}{8}$	Coca Cola	1.74	14	493	$37\frac{3}{8}$	$36\frac{3}{4}$	$37\frac{1}{4}$	$+\frac{1}{4}$
$72\frac{3}{8}$	$57\frac{1}{8}$	General Motors	6.95	5	782	$58\frac{7}{8}$	$58\frac{3}{8}$	$58\frac{1}{2}$	$+\frac{1}{8}$
$285\frac{5}{8}$	238	IBM	11.52	13	1712	$247\frac{3}{8}$	$243\frac{3}{4}$	$244\frac{1}{4}$	$-1\frac{7}{8}$
$52\frac{7}{8}$	$37\frac{3}{4}$	McDonald's	.20	13	589	$45\frac{1}{2}$	$44\frac{3}{4}$	$45\frac{1}{8}$	$-\frac{1}{8}$

(1) The yearly high and low.

(2) This column contains the company's name, often abbreviated.

(3) The annual dividend, if any. Coca Cola's annual dividend for this year is $1.74.

(4) P–E is the *price-earnings ratio*. It is the closing price of the stock divided by the earnings per share.

$$\text{P–E ratio} = \frac{\text{closing price}}{\text{earnings per share}}$$

The earnings per share is not the dividend per share. The P–E ratio for Coca Cola is 14, that is, the current selling price is 14 times their annual earnings per share.

(5) The day's sales in 100 lot shares. Today's sales of Coca Cola stock totaled 493×100 or 49,300 shares.

(6) The high price of the day.

(7) The low price of the day.

(8) The closing price of the day.

(9) The change from the previous day's closing price. Coca Cola closed at $37\frac{1}{4}$, up $\frac{1}{4}$ from the previous day.

If you had 50 shares of IBM stock, what would be their value using the sample stock listing given here?

Turn to **32** after completing your calculation.

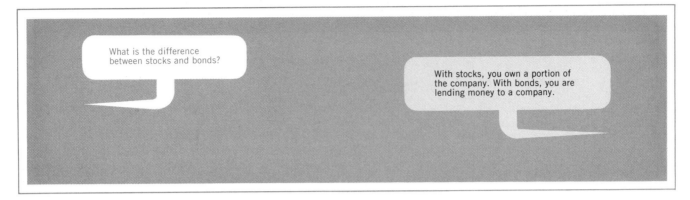

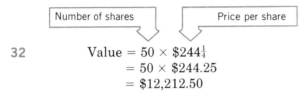

32
$$\text{Value} = 50 \times \$244\tfrac{1}{4}$$
$$= 50 \times \$244.25$$
$$= \$12,212.50$$

The value of 50 shares of IBM stock is \$12,212.50.

BONDS

In addition to selling stock, a company or other corporate entity can raise money by selling bonds. A *bond* is a long-term debt with a stated interest rate. Most bonds have a par value of \$1,000. As with stock, the value of a bond can fluctuate. The price of a bond is usually stated as a percent of the par value.

Municipal bonds are often issued at a lower interest rate than corporate bonds, but are exempt from federal income tax and certain local taxes. These tax exempt bonds can be a great savings, especially if you are in a high income tax bracket.

Example: A certain corporate bond has a par value of \$1000 and is currently selling for 82 (that's 82% of the par value, *not* \$82). Calculate the current price of the bond.

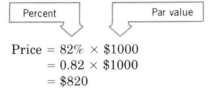

$$\text{Price} = 82\% \times \$1000$$
$$= 0.82 \times \$1000$$
$$= \$820$$

Calculate the current price of an Evansville City municipal bond with a par value \$1000 and selling at $102\tfrac{1}{2}$. Is this bond selling at a discount or a premium?

Turn to 33 to check your work.

Percent   Par value

33   Price $= 102\frac{1}{2}\% \times \$1000$
           $= 102.5\% \times \$1000$
           $= 1.025 \times \$1000$
           $= \$1025$

The bond is currently selling at a premium for $1025.

Bonds are a loan to the company and simple interest is paid based on the par value. The interest rate is fixed and specified on the bond document.

Example: Calculate the annual interest on a DuPont $1000 bond at 8%.

$$I = \text{principal} \times \text{rate} \times \text{time}$$
$$= \$1000 \times 8\% \times 1$$
$$= \$1000 \times 0.08 \times 1$$
$$= \$80$$

The annual interest is $80.

For a review of calculating simple interest, return to page 259.

Interest payments on bonds are usually made twice a year or semiannually. There are two methods of collecting the interest. First, with a *registered bond,* your name appears on the bond and is registered with the company issuing the bond. In this case, the interest is paid directly to you by the company.

The second type of bond is a *coupon bond* or *bearer bond.* With this type of bond, you "clip" a coupon attached to the bond and collect your interest due.

In the case of the DuPont bond, the semiannual payment is

Annual interest

Semiannual interest $= \dfrac{\$80}{2} = \$40$

Calculate the annual interest and semiannual interest payment on a Goodyear $1000 bond at 8.6%.

Check your work in 34.

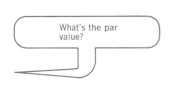

What's the par value?

The par value is the bookkeeping value that is assigned to a stock or bond by the company. It is used to calculate dividends or interest, but it is not the actual selling price of the security.

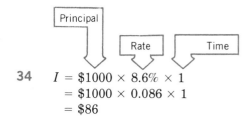

Principal

Rate   Time

34
$$I = \$1000 \times 8.6\% \times 1$$
$$= \$1000 \times 0.086 \times 1$$
$$= \$86$$

$$\text{Semiannual payment} = \frac{\$86}{2} = \$43$$

The annual interest is $86 and the semiannual interest payment is $43.

Much like stocks, bonds are sold through a bond market. The listing of bonds appears in many newspapers and in *The Wall Street Journal*.

Below is a sample bond listing.

TYPICAL BOND LISTING

① Bond	② Sales in $1000	③ High	④ Low	⑤ Close	⑥ Change
DuPont 8s91	15	$101\frac{3}{8}$	$100\frac{3}{4}$	$100\frac{3}{4}$	$-\frac{5}{8}$
Goodyear 8.60s95	4	$98\frac{5}{8}$	$98\frac{3}{8}$	$98\frac{1}{2}$	$+\frac{1}{8}$
Texaco $8\frac{7}{8}$ s05	19	102	101	102	-1

① The first column contains the company's name, bond interest rate, and the maturity date of the bond.

DuPont's bond yields 8% interest and the bond matures or is paid off by the company in 1991. The Texaco bonds pay $8\frac{7}{8}\%$ interest and mature in 2005.

② The daily sales in units of one thousand dollars. Goodyear sold $4000 in bonds.

③ The high price of the day expressed as a percent.

④ The low price of the day expressed as a percent.

⑤ The closing price of the day expressed as a percent.

⑥ The change from the previous day's closing price expressed as a percent. Texaco bonds dropped 1% from yesterday's closing.

Using the closing price from our sample bond listing, calculate the price of a Texaco bond.

Turn to 35 to check your answer.

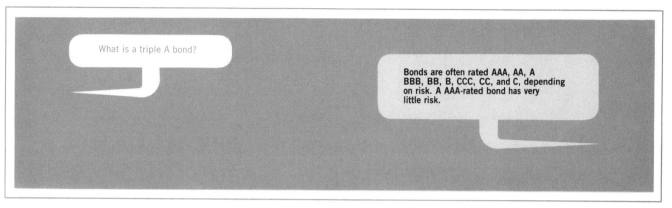

What is a triple A bond?

Bonds are often rated AAA, AA, A BBB, BB, B, CCC, CC, and C, depending on risk. A AAA-rated bond has very little risk.

35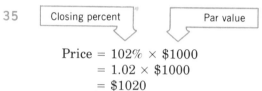

Closing percent   Par value

$$\text{Price} = 102\% \times \$1000$$
$$= 1.02 \times \$1000$$
$$= \$1020$$

The current price of a Texaco bond is $1020.

Because the interest earned by the bond is based on the par value rather than the current price, we need a way to determine the actual earning capacity of the bond.

One measure of the rate of return of a bond is its *current yield.* The current yield is the annual interest divided by the current price of the bond.

$$\text{Current yield} = \frac{\text{annual interest}}{\text{current price of bond}}$$

Example: Calculate the current yield of the DuPont $1000 bond from our typical bond listing in 34. Round to the nearest 0.01%.

First, calculate the interest.   **Second,** calculate the current price.

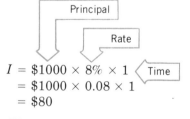

Principal

Rate

$$I = \$1000 \times 8\% \times 1 \quad \boxed{\text{Time}}$$
$$= \$1000 \times 0.08 \times 1$$
$$= \$80$$

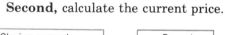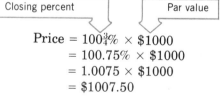

Closing percent   Par value

$$\text{Price} = 100\tfrac{3}{4}\% \times \$1000$$
$$= 100.75\% \times \$1000$$
$$= 1.0075 \times \$1000$$
$$= \$1007.50$$

Third, calculate the current yield.

Interest

$$\text{Current yield} = \frac{\$80}{\$1007.50} \quad \boxed{\text{Current price}}$$

$$= 0.0794 \text{ rounded}$$
$$= 7.94\%$$

The current yield of the bond is 7.9%, less than the annual interest rate of 8%. Calculate the current yield of the Goodyear $1000 bond from our typical bond listing in **34**. Round your answer to the nearest 0.01%.

Turn to **36** to check your solution.

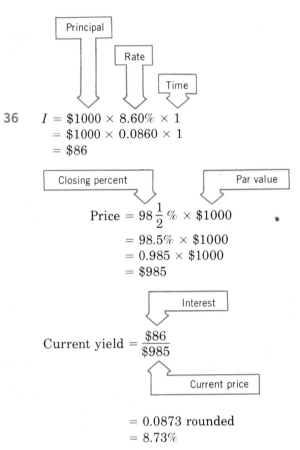

36

$$I = \$1000 \times 8.60\% \times 1$$
$$= \$1000 \times 0.0860 \times 1$$
$$= \$86$$

$$\text{Price} = 98\frac{1}{2}\% \times \$1000$$
$$= 98.5\% \times \$1000$$
$$= 0.985 \times \$1000$$
$$= \$985$$

$$\text{Current yield} = \frac{\$86}{\$985}$$

$$= 0.0873 \text{ rounded}$$
$$= 8.73\%$$

The current yield of 8.73% is higher than the bond's interest rate of 7.9%, because the bond is selling for less than 100%.

If the bond was selling for more than 100%, the current yield would be less than the bond's interest rate.

For a set of practice problems on stocks and bonds, turn to **37**.

---

9 Consumer Math

9.4 Stocks and Bonds

The answers are on page 596.

A. Calculate the dividend for the following preferred stocks.

	Par Value	Dividend Rate	Dividend per Share	Number of Shares	Total Dividends
1.	$50	5%	——————	150	——————
2.	75	7%	——————	200	——————
3.	40	12%	——————	50	——————
4.	15	6%	——————	25	——————
5.	170	10%	——————	70	——————
6.	80	12%	——————	85	——————
7.	250	9%	——————	300	——————
8.	175	5%	——————	40	——————
9.	85	7%	——————	175	——————
10.	120	8%	——————	250	——————

B. Calculate the cost of the following stocks.

	Number of Shares	Cost per Share	Total Cost
1.	200	$235\frac{1}{2}$	——————
2.	150	$15\frac{7}{8}$	——————
3.	75	$124\frac{1}{2}$	——————
4.	80	$89\frac{3}{4}$	——————
5.	300	$75\frac{1}{8}$	——————
6.	500	$42\frac{5}{8}$	——————
7.	50	$18\frac{1}{2}$	——————
8.	700	$27\frac{3}{8}$	——————
9.	70	$45\frac{1}{8}$	——————
10.	200	$37\frac{1}{2}$	——————

Date

Name

Course/Section

C. **Calculate the semiannual interest payment for the following bonds.**

	Par Value	Annual Rate	Annual Payment	Semiannual Payment
1.	$1000	8%	_____	_____
2.	1000	8.7%	_____	_____
3.	1000	8.6%	_____	_____
4.	1000	9.2%	_____	_____
5.	1000	7.6%	_____	_____
6.	5000	$8\frac{7}{8}$%	_____	_____
7.	5000	$8\frac{1}{2}$%	_____	_____
8.	1000	$9\frac{1}{4}$%	_____	_____
9.	1000	9.3%	_____	_____
10.	1000	9.4%	_____	_____

D. **For the following bonds, calculate the annual interest, closing price, and the current yield.**

	Par Value	Annual Rate	Closing Percent	Interest	Closing Price	Current Yield
1.	$1000	8.2%	$102\frac{1}{4}$	_____	_____	_____
2.	1000	8.6%	$98\frac{7}{8}$	_____	_____	_____
3.	1000	9.2%	$99\frac{1}{2}$	_____	_____	_____
4.	1000	7.3%	101	_____	_____	_____
5.	1000	$8\frac{7}{8}$%	$100\frac{1}{2}$	_____	_____	_____
6.	5000	$8\frac{1}{4}$%	$102\frac{1}{2}$	_____	_____	_____
7.	1000	$9\frac{1}{4}$%	$99\frac{3}{4}$	_____	_____	_____
8.	5000	9.2%	$98\frac{1}{4}$	_____	_____	_____
9.	1000	9.5%	$97\frac{1}{8}$	_____	_____	_____
10.	1000	$8\frac{3}{4}$%	$99\frac{3}{8}$	_____	_____	_____

When you have had the practice you need, turn to 38 for the Self-Test on this unit.

Consumer Math

The answers are on page 596.

1. Complete the following check register.

Check No.	Date	Check Issued to or Deposit	Amount of Check		✔	Amount of Deposit		Balance 655	29
942	8/1	Badview Apts.	325	50					
943	8/4	Electric Co.	58	92					
944	8/7	Acme Finance Co.	182	75					
	8/10	Deposit				255	86		
945	8/12	Rita's Boutique	53	18					
946	8/20	Wards	32	95					
947	8/20	Fast Foods	82	17					
948	8/26	Has Company	45	89					
	8/27	Deposit				114	08		
949	8/29	Lou's Surplus	29	86					

2. The monthly bank summary for the above account shows that checks numbered 942, 943, 944, 946, and 948 were processed, the deposit on 8/10 was credited, and there was a service charge of $3.17. The bank balance is $261.97. Use the check register in Problem 1 and the form below to complete the bank reconciliation.

Date _____

Name _____

Course/Section _____

BANK RECONCILIATION FORM

Outstanding Checks		
Number	Amount	

Total

1. Compare the cancelled checks with your records.
2. List any outstanding checks.
3. Total the outstanding checks.
4. Enter the bank balance here: $ _____

5. Add any deposits not on the summary.   + $ _____

    + $ _____

6. Total $ _____
7. Enter outstanding check total and subtract.   – $ _____

Corrected Bank Balance $ ══════

8. Enter checkbook balance. $ _____
9. Enter any service charge and subtract.   – $ _____

Corrected Checkbook Balance $ ══════

Corrected checkbook balance and corrected bank balance must be equal.

3. A new stereo may be purchased for $250 cash or $50 down and 26 weekly payments of $8.10. Calculate the (a) interest and (b) estimate the APR.

    (a) _____

    (b) _____

4. Kathy purchased a car for $5000. She made a down payment of $750 and financed the remaining amount for 30 months at $6\frac{1}{2}\%$ add-on. Calculate the (a) interest and (b) estimate the APR.

    (a) _____

    (b) _____

5. Convert the monthly rate of $1\frac{2}{3}\%$ to an APR.

    _____

6. A revolving charge statement in August shows a balance of $375.82. The minimum required payment is $18.79. The APR is 18%. Purchases totaling $82.17 were made in August but not included in the statement. If the minimum payment is made, calculate the (a) unpaid balance, (b) interest on the unpaid balance, and (c) balance for the September statement.

    (a) _____

    (b) _____

    (c) _____

7. Virgil's Welding Supply Company was unable to give a dividend last year to their stockholders who owned cumulative preferred 6%, par value $55 stock. This year the company is giving two years' dividends. Calculate the total dividend for 25 shares of stock.

   _____

8. Calculate the cost of 120 shares of stock selling for $37⅛.

   _____

9. Calculate the current yield for a $1000 bond with interest at $8\frac{3}{4}\%$ and currently selling for $98\frac{1}{2}$. Round your answer to the nearest 0.01%.

   _____

Real Estate Mathematics

Can you imagine buying a house like the one above? At least the taxes and the mortgage should not be a worry. Buying a house is one of the biggest financial decisions of your life. It involves more than simply finding one you like. Without careful consideration of all real estate aspects involved, you could make a very expensive mistake. Sufficient knowledge of real estate transactions is an asset to potential home buyers as well as to those who are thinking of selling their homes.

This chapter will help familiarize you with some of the business aspects of real estate: size, cost, principal and interest payments, prorations, and property taxes.

Don't panic! This is a fun chapter. Its "real" value will be apparent when and if you're ready to purchase your "estate."

UNIT OBJECTIVE

After you complete this unit successfully, you will be able to calculate area and volume, convert units of measurement, and solve problems involving price per unit of area. You will also be able to calculate property tax, principal and interest payments, points, mortgage discount, and interest, taxes, and insurance prorations.

PREVIEW 10 This is a sample test for Unit 10. All the answers are placed immediately after the test. Work as many of the problems as you can, then check your answers and look after the answers for further directions.

In Problems 4, 5, 6, and 7 use 30-day-month time and ordinary interest.

1. Calculate the area of a rectangular lot 92 feet by 145 feet.   ——————————   375   1

2. What is the price per square foot of a house with 1570 square feet priced at $50,240?   ——————————   382   9

3. How many cubic yards of concrete are needed to construct a sidewalk 4 feet wide, $175\frac{1}{2}$ feet long and 6 inches deep?   ——————————   384   12

4. For a mortgage with principal balance $79,872 at $12\frac{1}{2}\%$ and monthly *P&I* payment $854, calculate the (a) interest payment, (b) principal payment, and (c) new principal.

    (a) ——————————

    (b) ——————————

    (c) ——————————   391   16

5. The Trawick family is negotiating an FHA insured loan for $52,700. The current market rate is 13% and the maximum federal rate is $12\frac{3}{4}\%$. Calculate the (a) points and the (b) discount amount.

    (a) ——————————

    (b) ——————————   393   19

6. A buyer is purchasing a house with the closing date August 26. The yearly taxes of $576.82 were prepaid on January 1. Prorate the taxes.   ——————————   397   23

7. The market value of a house is $49,700. If the assessment rate is 55% and the tax rate is $3.12 per $100 of assessed value, calculate the tax.   ——————————   405   31

* If you cannot work any of these problems, start with frame 1 on page 375.
* If you missed some of the problems, turn to the page numbers indicated.
* If all of your answers were correct, you are probably ready to proceed to Unit 11. (If you would like more practice on Unit 10 before turning to Unit 11, try the self-test on page 411.)
* Super-students—those who want to be certain they learn all of this material—will turn to frame 1 and begin work there.

Answers

1. 13,340 square feet

2. $32 per square foot

3. 13 cubic yards

4. (a) $832.00   (b) $22.00   (c) $79,850.00

5. (a) 2 points   (b) $1,054

6. $200.28   7. $852.85

Preview 10

SECTION 1: BASIC CALCULATIONS

1

> **OBJECTIVE**
>
> Calculate the area and volume of simple geometric figures and convert units of measurement. Solve problems involving price per unit of area.

Some of the basic calculations used in real estate mathematics include finding area, volume, and cost per unit of area.

Area is commonly used to measure the extent of a house, building, or land. Once the area has been found, the cost per square foot can be determined. Volume is frequently used to calculate the capacity of a storage building or the amount of concrete necessary for a sidewalk or driveway.

The dimensions of a room, building, or lot may be given in many different units: feet, inches, yards, or even meters if you use the metric system. It is important that when you do any calculations involving units you use the *same* units for each dimension.

Example: Two adjacent lots have frontages of 84 feet and 30 yards. Calculate the total frontage of the two lots.

To find the total frontage distance we need to add the distances 84 ft and 30 yd. In order to add these numbers, they must be converted to the same units—either the feet to yards or yards to feet. Remember, *always* use the same units of measurement.

To convert yards to feet use the following table of conversion:

> 1 ft = 12 in.
> 1 yd = 36 in.
> 1 yd = 3 ft
> 1 mile = 5280 ft
> 1 mile = 1760 yd

$$30 \text{ yards} = 30 \text{ yd} \times \frac{3 \text{ ft}}{1 \text{ yd}} = 30 \times 3 \text{ ft} = 90 \text{ ft}$$

Notice since 1 yd = 3 ft, then $\frac{3 \text{ ft}}{1 \text{ yd}} = 1$. When you multiply a number by 1, its value is unchanged. There are, however, two possible fractions: $\frac{3 \text{ ft}}{1 \text{ yd}}$ and $\frac{1 \text{ yd}}{3 \text{ ft}}$.

We used $\frac{3 \text{ ft}}{1 \text{ yd}}$ in order to cancel the yd units, leaving ft units. To cancel the yd units, place the yd units of our fraction in the denominator.

Now that both distances are given in feet, we can find the total frontage distance.

Complete the calculation, then continue in 2.

2 Total = 84 feet + 90 feet

= 174 feet

Notice that it is easy to convert from one unit of measurement to another by multiplying by the appropriate fraction. We have included no metric units in the table in 1. You will find a discussion of metric system conversions in Appendix B on page 563.

Use the table to work the following problem.

The width of a room is 13 yards. Convert this measurement to (a) feet, (b) inches.

Check your work in 3.

---

3 (a) $13 \text{ yards} = 13 \text{ yd} \times \dfrac{3 \text{ ft}}{1 \text{ yd}} = 13 \times 3 \text{ ft} = 39 \text{ ft}$

(b) $13 \text{ yards} = 13 \text{ yd} \times \dfrac{36 \text{ in.}}{1 \text{ yd}} = 13 \times 36 \text{ in.} = 468 \text{ in.}$

The conversions are easy when you use the right fraction. In our previous examples we converted to smaller units. In conversions to larger units use the same method.

Example: Use the table in 1 to convert 168 inches to feet.

$$168 \text{ inches} = \overset{14}{168} \text{ in.} \times \dfrac{1 \text{ ft}}{\cancel{12} \text{ in.}} = 14 \times 1 \text{ ft} = 14 \text{ ft}$$

Try a few problems involving unit conversion. Change to the units indicated.

1. 3834 feet = _____ yards

2. 7040 yards = _____ miles

3. 180 feet = _____ inches

4. 540 inches = _____ feet

5. 540 inches = _____ yards

6. 2 miles = _____ feet

Our solutions are in 4.

I write 6 ft as 6' and 4 inches as 4". Isn't that OK?

That's an old abbreviation still much used in practical work and in the trades, but it can get confusing. In this book we'll use the abbreviations in. and ft for inches and feet.

4 1. $3834 \text{ ft} = \overset{1278}{3834} \text{ ft} \times \dfrac{1 \text{ yd}}{\cancel{3} \text{ ft}} = 1278 \times 1 \text{ yd} = 1278 \text{ yd}$

2. $7040 \text{ yd} = \overset{4}{7040} \text{ yd} \times \dfrac{1 \text{ mile}}{\cancel{1760} \text{ yd}} = 4 \times 1 \text{ mile} = 4 \text{ miles}$

3. $180 \text{ ft} = 180 \text{ ft} \times \dfrac{12 \text{ in.}}{1 \text{ ft}} = 180 \times 12 \text{ in.} = 2160 \text{ in.}$

4. $540 \text{ in.} = \overset{45}{540} \text{ in.} \times \dfrac{1 \text{ ft}}{\cancel{12} \text{ in.}} = 45 \times 1 \text{ ft} = 45 \text{ ft}$

5. $540 \text{ in.} = \overset{15}{\cancel{540}} \text{ in.} \times \dfrac{1 \text{ yd}}{\cancel{36} \text{ in.}} = 15 \times 1 \text{ yd} = 15 \text{ yd}$

6. $2 \text{ miles} = 2 \text{ miles} \times \dfrac{5280 \text{ ft}}{1 \text{ mile}} = 2 \times 5280 \text{ ft} = 10{,}560 \text{ ft}$

It's easy, just use the right conversion fraction.

AREA

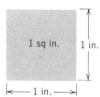

1 sq in.   1 in.

1 in.

The size of a plot of land, a lot, or a room can be measured in terms of area. Area is always expressed in square units, such as square inches (abbreviated sq in.), square feet (sq ft), square yards (sq yd), square miles (sq mi), or in special units such as acres, which are defined in square units. One square unit is an area one unit by one unit. For example, one square inch is the area of a square region one inch on each side.

A *rectangle* is a four-sided geometric figure whose sides meet at right angles.

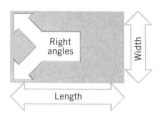

Right angles

Width

Length

The area of a rectangle is the product of its length and width.

> Area of rectangle = length × width

The area of this rectangle garden:

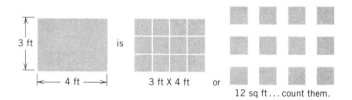

3 ft

4 ft   is   3 ft X 4 ft   or   12 sq ft... count them.

Example: Calculate the area of a rectangular lot 85 ft by 140 ft.

Area = length × width

Area = 140 ft × 85 ft

Area = 11,900 sq ft

Work the following problem.

Calculate the area of a rectangular lot 75 feet by 132 feet.

Turn to **5** to check your work.

---

5   Area = length × width

= 75 feet × 132 feet

= 9900 square feet

> **Careful!** When calculating area, both measurements must be expressed in the same units. When you multiply feet by feet, the area is in square feet. When you multiply yards by yards, the result is in square yards. But you *cannot* multiply feet × yards.

Example: Calculate the area of a rectangular lot 96 feet by 59 yards.

First, change yards to feet:

$$59 \text{ yd} = 59 \text{ yd} \times \frac{3 \text{ ft}}{1 \text{ yd}} = 59 \times 3 \text{ ft} = 177 \text{ ft}$$

Second, multiply to find the area:

Area = length × width
     = 177 feet × 96 feet
     = 16,992 square feet

Calculate the area of the following rectangular lots.

1. 93 feet by 46 yards. Express the area in square feet.
2. 37 yards by 183 feet. Express the area in square yards.

Check your work in **6**.

---

6   1. $46 \text{ yd} = 46 \text{ yd} \times \dfrac{3 \text{ ft}}{1 \text{ yd}} = 46 \times 3 \text{ ft} = 138 \text{ ft}$

        Area = length × width
             = 138 feet × 93 feet
             = 12,834 square feet

     2. $183 \text{ ft} = \overset{61}{\cancel{183}} \text{ ft} \times \dfrac{1 \text{ yd}}{\underset{1}{\cancel{3}} \text{ ft}} = 61 \times 1 \text{ yd} = 61 \text{ yd}$

        Area = length × width
            ≐ 61 yards × 37 yards
            = 2257 square yards

In Problem 2, once the area is calculated in square yards it can easily be converted to square feet or to acres. Use the following table to convert area units.

> 1 acre = 43,560 sq ft
> 1 acre = 4840 sq yd
> 1 sq yd = 9 sq ft
> 1 sq ft = 144 sq in.

Example: Convert the area 2257 sq yd in Problem 2 to (a) sq ft, and (b) acres.

(a) $2257 \text{ sq yd} = 2257 \text{ sq yd} \times \dfrac{9 \text{ sq ft}}{1 \text{ sq yd}}$

                   = $2257 \times 9 \text{ sq ft} = 20{,}313 \text{ sq ft}$

(b) $2257 \text{ sq yd} = 2257 \text{ sq yd} \times \dfrac{1 \text{ acre}}{4840 \text{ sq yd}}$

                   = $\dfrac{2257}{4840} \times 1 \text{ acre}$

                   = 0.466 acre   rounded

In the following problems, change to the units indicated.

1. 19,485 square feet = _____ square yards.

2. 108,900 square feet = _____ acres.

3. 236 square yards = _____ square feet.

4. 4 acres = _____ square yards = _____ square feet.

Turn to **7** to check your answers.

---

7 1. $19,485 \text{ sq ft} = \overset{2165}{\cancel{19,485}} \text{ sq ft} \times \dfrac{1 \text{ sq yd}}{\underset{1}{\cancel{9 \text{ sq ft}}}}$

$= 2165 \times 1 \text{ sq yd} = 2165 \text{ sq yd}$

2. $108,900 \text{ sq ft} = 108,900 \cancel{\text{ sq ft}} \times \dfrac{1 \text{ acre}}{43,560 \cancel{\text{ sq ft}}}$

$= \dfrac{108,900}{43,560} \times 1 \text{ acre}$

$= 2.5 \text{ acres}$

3. $236 \text{ sq yd} = 236 \cancel{\text{ sq yd}} \times \dfrac{9 \text{ sq ft}}{1 \cancel{\text{ sq yd}}}$

$= 236 \times 9 \text{ sq ft} = 2124 \text{ sq ft}$

4. $4 \text{ acres} = 4 \cancel{\text{ acres}} \times \dfrac{4840 \text{ sq yd}}{1 \cancel{\text{ acre}}}$

$= 4 \times 4840 \text{ sq yd} = 19,360 \text{ sq yd}$

$4 \text{ acres} = 4 \cancel{\text{ acres}} \times \dfrac{43,560 \text{ sq ft}}{1 \cancel{\text{ acre}}}$

$= 4 \times 43,560 \text{ sq ft} = 174,240 \text{ sq ft}$

Another common geometric figure used in practical situations is the *triangle*. A triangle is a three-sided figure.

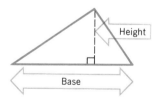

The *height* of a triangle is the perpendicular distance from one vertex or corner to the opposite side.

The area of any triangle is given by the formula:

Area of triangle = $\frac{1}{2}$ × base × height

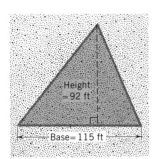

Example: Calculate the area of a lot in the shape of a triangle with base 115 ft and height 92 ft.

$$\text{Area} = \frac{1}{2} \times \text{base} \times \text{height}$$

$$\text{Area} = \frac{1}{2} \times 115 \text{ ft} \times 92 \text{ ft}$$

$$= 5290 \text{ sq ft}$$

Calculate the area of the following triangles.

  1. Base = 175 feet; height = 52 feet.   2. Base = 85 feet; height = 128 feet.

Check your calculations in **8**.

---

8   1. $\text{Area} = \frac{1}{2} \times \text{base} \times \text{height}$

$$= \frac{1}{2} \times 175 \text{ feet} \times 52 \text{ feet}$$

$$= 4550 \text{ square feet}$$

    2. $\text{Area} = \frac{1}{2} \times \text{base} \times \text{height}$

$$= \frac{1}{2} \times 85 \text{ feet} \times 128 \text{ feet}$$

$$= 5440 \text{ square feet}$$

Unfortunately, all figures are not simple rectangles or triangles. Lots and buildings are often more complex figures. The areas of these figures can be calculated by breaking them down into rectangles and triangles.

Example: Calculate the area of the following figure.

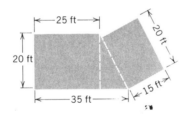

This figure can be divided into three smaller figures.

The area of the first rectangle is

$$\text{Area} = \text{length} \times \text{width}$$
$$= 20 \text{ feet} \times 25 \text{ feet} = 500 \text{ square feet}$$

The area of the triangle is

$$\text{Area} = \frac{1}{2} \times \text{base} \times \text{height}$$

$$= \frac{1}{2} \times 10 \text{ feet} \times 20 \text{ feet}$$

$$= 100 \text{ square feet}$$

The area of the second rectangle is

Area = length × width
   = 15 feet × 20 feet
   = 300 square feet

The total area is the sum of the area of the three parts.

Total area = 500 + 100 + 300 square feet
       = 900 square feet

Calculate the area of the following figures.

1.

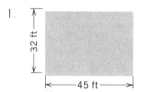

2.

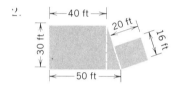

Check your work in 9.

9   1.

The area of the first rectangle is

Area = length × width = 32 feet × 45 feet = 1440 square feet

The area of the second rectangle is

Area = length × width = 20 feet × 15 feet = 300 square feet

Total area = 1440 square feet + 300 square feet = 1740 square feet

2.

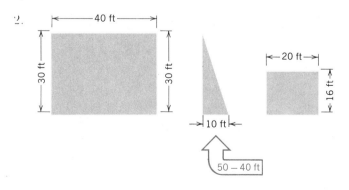

The area of the first rectangle is

Area = length × width = 40 feet × 30 feet = 1200 square feet

The area of the triangle is

Area = $\frac{1}{2}$ × base × height = $\frac{1}{2}$ × 10 feet × 30 feet = 150 square feet

The area of the second rectangle is

Area = length × width = 20 feet × 16 feet = 320 square feet

Total area = 1200 square feet + 150 square feet + 320 square feet
 = 1670 square feet

A common problem in real estate transactions is to determine the cost using the area and price per unit of area. For example, we can calculate the cost of a house using the area and price per square foot. The basic formula is

> Cost = area × price per unit area

Example: What is the total cost of a house with 1850 square feet of area at $51 per square foot?

Cost = area × price per unit area
 = 1850 × $51 = $94,350

Try a few problems.

1. What is the cost of a house with 1560 square feet of area at $49.75 per square foot?

2. Calculate the cost of a $5\frac{1}{2}$ acre lot at $4750 per acre.

Check your work in **10**.

---

10   1. Cost = area × price per unit
 = 1560 × $49.75 = $77,610

2. Cost = area × price per unit
 = $5\frac{1}{2}$ × $4750 = 5.5 × $4750 = $26,125

A related problem involves calculating the price per unit area when the cost and area are known. This procedure can be used to calculate the price per square foot of a house if the cost and floor area are known. The per unit cost is extremely useful when comparing similar houses.

The formula for calculating price per unit is a different form of the basic formula in **9**. The price per unit can be found either by solving the formula algebraically or with the circle diagrams used previously. (If you want a review of percent circle diagrams, turn to page 141.)

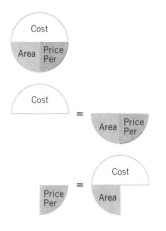

$$\text{Cost} = \text{area} \times \text{price per unit}$$

$$\boxed{\text{Price per unit area} = \frac{\text{cost}}{\text{area}}}$$

Example: What is the cost per square foot of a house that sells for $76,725 and has 1650 square feet of floor area?

$$\text{Price per square foot} = \frac{\text{cost}}{\text{area}} = \frac{\$76,725}{1650} = \$46.50$$

Work the following problems.

1. Calculate the price per square foot of a building that costs $149,500 and has 5200 square feet of floor area.
2. What is the price per acre of a $12\frac{1}{2}$ acre lot priced at $36,250?

Check your calculations in **11**.

11

1. $\text{Price per square foot} = \dfrac{\text{cost}}{\text{area}} = \dfrac{\$149,500}{5200} = \$28.75$

2. $\text{Price per acre} = \dfrac{\text{cost}}{\text{area}} = \dfrac{\$36,250}{12\frac{1}{2}} = \dfrac{\$36,250}{12.5} = \$2900$

A similar problem involves calculating the area when the cost and price per unit are known. Another form of the cost formula in **9** is used.

$$\boxed{\text{Area} = \frac{\text{cost}}{\text{price per unit area}}}$$

Example: Houses in a certain subdivision are selling for $42 per square foot. What is the area of a house selling for $77,240?

$$\text{Area} = \frac{\text{cost}}{\text{price per sq. ft}} = \frac{\$77,240}{\$42} = 1720 \text{ square feet}$$

Work the following problems.

1. What is the area of a lot priced at $14,875 and selling for $1.25 per square foot?
2. What is the area of a building selling for $95,625 and priced at $22.50 per square foot?

Turn to **12** to check your work.

---

12 1. $\text{Area} = \dfrac{\text{cost}}{\text{price per sq ft}} = \dfrac{\$14{,}875}{\$1.25} = 11{,}900$ square feet

2. $\text{Area} = \dfrac{\text{cost}}{\text{price per sq ft}} = \dfrac{\$95{,}625}{\$22.50} = 4250$ square feet

VOLUME

The capacity or size of a building or container is measured by its *volume*. Volume is expressed in cubic units such as cubic inches (abbreviated cu in.), cubic feet (cu ft), or cubic yards (cu yd).

One cubic inch is the volume of a cube one inch on each side.

The volume of a rectangular box is the product of its length, width, and height.

Volume of a box = length × width × height

For example, the volume of this box:

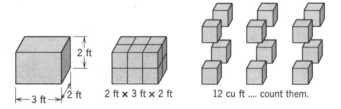

2 ft × 3 ft × 2 ft   12 cu ft .... count them.

Example: Calculate the volume of a storage building 10 ft by 12 ft by 6 ft.

Volume = length × width × height
Volume = 12 ft × 10 ft × 6 ft
Volume = 720 cu ft

Work the following problems.

1. What is the volume of a room 12 feet wide, 16 feet long, and 8 feet high?
2. How many cubic feet of cement are needed to construct a sidewalk 4 feet wide, 108 feet long, and 6 inches deep? (**Hint:** Remember to use the same units for all measurements.)

Check your calculations in **13**.

---

13 1. Volume = length × width × height
        = 12 feet × 16 feet × 8 feet = 1536 cubic feet

2. First, convert 6 inches to feet:

$$6 \,\cancel{\text{in.}} = 6 \text{ in.} \times \dfrac{1 \text{ ft}}{12 \,\cancel{\text{in.}}} = \dfrac{6}{12} \text{ ft} = 0.5 \text{ ft}$$

Volume = length × width × height
        = 4 feet × 108 ft × 0.5 feet = 216 cubic feet

In Problem 2 above, feet are the most convenient units of measurement to use in the volume calculation. But because concrete is usually ordered in cubic yards rather than cubic feet, we should convert the answer to cu yd. The most commonly used volume conversions are:

$$1 \text{ cu yd} = 27 \text{ cu ft} \qquad (3 \text{ ft} \times 3 \text{ ft} \times 3 \text{ ft} = 27 \text{ cu ft})$$
$$1 \text{ cu ft} = 1728 \text{ cu in.} \qquad (12 \text{ in.} \times 12 \text{ in.} \times 12 \text{ in.} = 1728 \text{ cu in.})$$

Example: Convert the 216 cu ft from Problem 2 above to cubic yards.

$$216 \text{ cu ft} = \overset{8}{\cancel{216 \text{ cu ft}}} \times \frac{1 \text{ cu yd}}{\underset{1}{\cancel{27 \text{ cu ft}}}}$$

$$= 8 \times 1 \text{ cu yd} = 8 \text{ cu yd}$$

In the following problems, change to the units indicated.

1. $2\frac{1}{2}$ cubic feet = _____ cubic inches

2. 81 cubic feet = _____ cubic yards

3. 5616 cubic inches = _____ cubic feet

4. 3 cubic yards = _____ cubic feet

Check your work in **14**.

14

1. $2\frac{1}{2}$ cu ft $= 2.5 \, \cancel{\text{cu ft}} \times \dfrac{1728 \text{ cu in.}}{1 \, \cancel{\text{cu ft}}}$

    $= 2.5 \times 1728 \text{ cu in.} = 4320 \text{ cu in.}$

2. 81 cu ft $= \overset{3}{\cancel{81 \text{ cu ft}}} \times \dfrac{1 \text{ cu yd}}{\underset{1}{\cancel{27 \text{ cu ft}}}}$

    $= 3 \times 1 \text{ cu yd} = 3 \text{ cu yd}$

3. 5616 cu in. $= 5616 \, \cancel{\text{cu in.}} \times \dfrac{1 \text{ cu ft}}{1728 \, \cancel{\text{cu in.}}}$

    $= \dfrac{5616}{1728} \text{ cu ft} = 3.25 \text{ cu ft}$

4. 3 cu yd $= 3 \, \cancel{\text{cu yd}} \times \dfrac{27 \text{ cu ft}}{1 \, \cancel{\text{cu yd}}}$

    $= 3 \times 27 \text{ cu ft} = 81 \text{ cu ft}$

For a set of practice problems on area and volume in real estate calculations, turn to **15**.

---

10 Real Estate Mathematics

15

10.1 Basic Calculations

The answers are on page 597.

A. Change to the units indicated. Use the tables in the text.

1. 222 feet = _____ yards
2. 11,880 feet = _____ miles
3. 2640 yards = _____ miles
4. 5.5 yards = _____ feet
5. 9 feet = _____ inches
6. 4 yards = _____ inches
7. 3 miles = _____ feet
8. 2 miles = _____ yards
9. 180 inches = _____ yards
10. 204 inches = _____ feet
11. 108 square feet = _____ square yards
12. 65,340 square feet = _____ acres
13. 14,520 square yards = _____ acres
14. 14 square yards = _____ square feet
15. 2 acres = _____ square feet
16. $3\frac{1}{2}$ acres = _____ square yards
17. 17 cubic yards = _____ cubic feet
18. 6048 cubic inches = _____ cubic feet
19. 3 cubic feet = _____ cubic inches
20. 135 cubic feet = _____ cubic yards

B. Find the area of the following rectangular lots.

1. 82 feet by 140 feet _____
2. 105 feet by 178 feet _____
3. 95 feet by 145 feet _____
4. 28 yards by 47 yards _____
5. 65 feet by 130 feet _____
6. 32 yards by 53 yards _____
7. 205 feet by 95 yards _____
8. 55 yards by 264 feet _____
9. 63 feet by 60 yards _____
10. 29 yards by 180 feet _____
11. 85 feet by 125 feet _____
12. 37 yards by 48 yards _____
13. 29 yards by 129 feet _____
14. 87 yards by 252 feet _____
15. 97 feet by 47 yards _____
16. 126 feet by 42 yards _____

C. Calculate the area of the following triangles.

1. Base = 12 feet; height = 15 feet _____
2. Base = 9 feet; height = 11 feet _____

Date

Name

Course/Section

3. Base = 10 feet; height = 20 feet _____

4. Base = 114 feet; height = 150 feet _____

5. Base = 5 yards; height = 6 yards _____

6. Base = 52 feet; height = 35 feet _____

7. Base = 13 feet; height = 23 feet _____

8. Base = 42 yards; height = 78 yards _____

9. Base = 127 feet; height = 32 yards _____

10. Base = 18 feet; height = 4 yards _____

11. Base = 127 feet; height = 92 feet _____

12. Base = 18 feet; height = 4 yards _____

13. Base = 25 yards; height = 12 feet _____

14. Base = 253 feet; height = 175 feet _____

15. Base = 32 yards; height = 15 feet _____

D. Find the area of the following figures.

1.
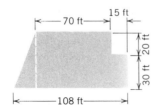

2.
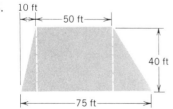

3.
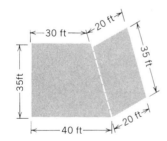

4.
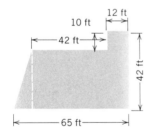

5.
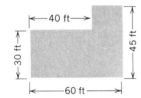

6.
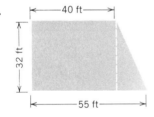

E. Work the following problems.

1. Calculate the volume of a storage building 10 feet by 8 feet by 6 feet.

   _____

2. How many cubic yards of cement are necessary to construct a sidewalk 4 feet wide, 81 feet long and 6 inches deep?

   _____

3. Calculate the volume of a room 11 feet by 15 feet by 7 feet.

   _____

4. What is the cost of a house with 1575 square feet of floor area at $38 per square foot?

   _____

5. Calculate the price per square foot of a house that costs $51,975 and has 1650 square feet of floor area.

   _____

6. What is the area of a home selling for $40,710 and priced $29.50 per square foot?

   _____

7. What is the price per square foot of a building that costs $95,850 and has 3550 square feet of floor area?

   _____

8. What is the cost of a house with 1970 square feet of floor area at $33.75 per square foot?

   _____

9. Calculate the area of a building selling for $75,400 and priced at $26 per square foot.

   _____

10. Calculate the cost of a building with 3750 square feet of floor area priced at $28.20 per square foot.

    _____

11. What is the cost of a house with 1250 square feet of floor area at $52 per square foot?

    _____

12. What is the price per square foot of a house that costs $70,290 and has 1420 square feet of floor area?

    _____

13. Calculate the area of a house selling for $54,375 and priced at $37.50 per square foot.

_____

14. Calculate the volume of a storage building 12 feet by 9 feet by $7\frac{1}{2}$ feet.

_____

15. How many cubic yards of cement are necessary to construct a sidewalk 3 feet wide, 54 feet long, and 4 inches deep?

_____

When you have had the practice you need, either return to the Preview on page 374, or continue in frame **16** with the study of mortgages and points.

SECTION 2: MORTGAGES AND POINTS

16

> **OBJECTIVE**
>
> Calculate the principal and interest payments on mortgages and the number of points and amount of discount on a mortgage.

With the high cost of housing, few individuals can pay the full price of a home. Most people must borrow some portion of the price of a home. Any loan secured by real estate is called a *mortgage.*

P&I Payments   Monthly payments for principal and interest (*P&I*) can be easily determined using mortgage payment tables. The monthly payments for *P&I* are fixed for the entire length of the loan. In addition to principal and interest, taxes and insurance charges may be added to the monthly charge. Taxes and insurance may change from year to year.

At the beginning of the mortgage, most of the payment is for the interest and a relatively small amount goes to reduce the principal.

For example, on a mortgage of $42,000 at 14% for 30 years, the P&I payment is $497.65. The interest on the loan for the first month may be calculated using ordinary simple interest at 30-day-month time. If you need a review of simple interest, return to page 259.

$$I = P \times r \times t = \$42,000 \times 14\% \times \frac{1}{12}$$

$$= \$42,000 \times 0.14 \times \frac{1}{12}$$

$$= \$490.00$$

of the $497.65 payment, $490.00 in the *interest payment*. The remainder, called the *principal payment,* reduces the principal.

> Principal payment = *P&I* payment − interest payment

The new principal is the previous principal less the principal payment.

> New principal = previous principal − principal payment

Calculate the principal payment and the new principal for our example.

Turn to 17 to check your work.

17   Principal payment = P&I payment − interest payment

           = \$497.65 − 490.00 = \$7.65

New principal = previous principal − principal payment

           = \$42,000 − 7.65 = \$41,992.35

Of the \$497.65 payment, only \$7.65 is applied to the principal!

Calculate the second month's interest and principal payments. Use the new principal of \$41,992.35.

Check your calculations in **18**.

How do I know how expensive a house I can afford to buy?

Hmm. Now all I need is the down payment.

Most people can afford to buy a house costing about 2 times their gross annual income. If your income is \$22,000, you can afford about 2 times \$22,000 = \$44,000.

New principal	Rate	One month

18 $I = P \times r \times t = \$41,992.35 \times 14\% \times \dfrac{1}{12}$

           $= \$41,992.35 \times 0.14 \times \dfrac{1}{12}$

           $= \$489.910 \ldots = \$489.91$ rounded

Principal payment = P&I payment − interest payment

           = \$497.65 − 489.91 = \$7.74

New principal = previous principal − principal payment

           = \$41,992.35 − 7.74 = \$41,984.61

After two months, the principal has been reduced from \$42,000 to \$41,984.61. It is only after many years that the principal is reduced to the point where most of the payment is applied to the principal. In fact, for this loan with principal \$42,000, interest rate 14%, the total interest is \$137,154!

A complete listing of all P&I payments is called an *amortization schedule*. To calculate a complete amortization schedule by our method would require 360 monthly calculations. This would test anyone's patience! At the closing time of a real estate transaction, the buyer is usually given (at a fee) a complete amortization schedule for the property.

Try one more calculation by hand (or hand calculator).

On a new loan of \$55,200 at 12¾% for 25 years, the monthly payments are \$612.20. Calculate the principal and interest payments for the first two months.

Check your calculations in **19**.

---

19   *First Month:* $I = P \times r \times t = \$55,200 \times 12\dfrac{3}{4}\% \times \dfrac{1}{12}$

           $= \$55,200 \times 0.1275 \times \dfrac{1}{12}$

           $= \$586.50$

Principal payment = P&I payment − interest payment
$$= \$612.20 - 586.50 = \$25.70$$

New principal = previous principal − principal payment
$$= \$55,200 - 25.70 = \$55,174.30$$

Second Month: $I = P \times r \times t = \$55,174.30 \times 12\frac{3}{4}\% \times \frac{1}{12}$

$$= \$55,174.30 \times 0.1275 \times \frac{1}{12}$$

$$= \$586.226 \ldots = \$586.23 \text{ rounded}$$

Principal payment = P&I payment − interest payment
$$= \$612.20 - 586.23 = \$25.97$$

New principal = previous principal − principal payment
$$= \$55,174.30 - 25.97 = \$55,148.33$$

POINTS

A large number of mortgages are insured by the FHA (Federal Housing Administration) and the VA (Veterans Administration). The FHA and VA do not lend money but only guarantee the loans to the lender.

When the loan rate determined by the market is higher than the maximum rate permitted by the FHA or VA, *points* are charged by the lending institution. Each point is one percent of the mortgage principal to be paid at settlement. Since federal regulations limit the amount of charges to the borrower, the points are paid by the seller. The points are paid by the seller to the buyer's mortgage company.

One point is charged for each $\frac{1}{8}\%$ the market rate exceeds the maximum federal rate.

Example: If the current market rate is $12\frac{3}{4}\%$ and the maximum federal rate is $12\frac{1}{2}\%$, calculate the points.

Step 1. **Calculate the difference in interest rates.**

$$12\frac{3}{4}\% - 12\frac{1}{2}\% = \frac{1}{4}\% \qquad \text{(For a review of subtracting fractions, turn to page 69).}$$

Step 2. **Convert the difference to eighths.**

$$\frac{1}{4}\% = \frac{2}{8}\% \qquad \text{(For a review of renaming fractions, turn to page 45.)}$$

Step 3. **Determine the number of points. Each $\frac{1}{8}\%$ is equivalent to one point.**

$$\frac{2}{8}\% = 2 \text{ points.}$$

Work the following problem.

With the market rate at 12% and the maximum federal rate at $11\frac{1}{4}\%$, calculate the points.

Check your work in **20**.

---

20   **Step 1.** $12\% - 11\frac{1}{4}\% = \frac{3}{4}\%$

Step 2. $\frac{3}{4}\% = \frac{6}{8}\%$

Step 3. $\frac{6}{8}\% = 6 \text{ points}$

After the points have been calculated, the discount amount is found. Each point is equivalent to a one percent discount rate. The amount of discount charged to the seller is the product of the discount rate and the mortgage principal.

Example: The Carlile family is negotiating an FHA insured mortgage for $59,500. The current market rate is $12\frac{3}{4}\%$ and the maximum federal rate is $12\frac{1}{4}\%$. Calculate the points and the discount amount.

Step 1. $12\frac{3}{4}\% - 12\frac{1}{4}\% = \frac{1}{2}\%$

Step 2. $\frac{1}{2}\% = \frac{4}{8}\%$

Step 3. $\frac{4}{8}\% = 4$ points

Step 4. Find the discount rate. Each point is equivalent to one percent.

4 points = 4%

Step 5. Calculate the discount.

Discount = discount rate × mortgage principal
= 4% × $59,500 = 0.04 × $59,500
= $2380

The discount of $2380 is paid by the seller to the Carlile's mortgage company.

Try the following problem.

The Campbell family is negotiating a VA insured loan for $67,250. The current market rate is $12\frac{3}{4}\%$ and the maximum federal rate is 12%. Calculate the points and discount amount.

Turn to **21** to check your calculations.

I'm a veteran. Can I borrow money for a house from the VA?

No. The VA and FHA only insure loans. They do not lend money.

21   **Step 1.** $12\frac{3}{4}\% - 12\% = \frac{3}{4}\%$

Step 2. $\frac{3}{4}\% = \frac{6}{8}\%$

Step 3. $\frac{6}{8}\% = 6$ points

Step 4. 6 points = 6%

Step 5. Discount = discount rate × mortgage principal
= 6% × $67,250 = 0.06 × $67,250
= $4035

The discount amount is $4035. This is paid by the seller to the Campbell's mortgage company.

For a set of practice problems on mortgages and points, turn to **22**.

10 Real Estate Mathematics

The answers are on page 598.

A. Calculate the monthly interest payment, principal payment and new principal. Use ordinary simple interest at 30-day-month time.

	Current Principal	Interest Rate	P&I Payment	Interest Payment	Principal Payment	New Principal
1.	$30,000.00	$10\frac{1}{2}\%$	$274.42	——	——	——
2.	58,200.00	$10\frac{3}{4}\%$	543.29	——	——	——
3.	48,900.00	$10\frac{1}{2}\%$	447.31	——	——	——
4.	50,000.00	10%	438.79	——	——	——
5.	32,516.14	9%	281.62	——	——	——
6.	76,189.92	$9\frac{1}{2}\%$	672.68	——	——	——
7.	11,832.50	$9\frac{3}{4}\%$	214.79	——	——	——
8.	7,412.17	$8\frac{1}{4}\%$	203.86	——	——	——
9.	23,872.27	9%	224.79	——	——	——
10.	45,631.42	$8\frac{1}{2}\%$	456.26	——	——	——
11.	36,126.34	9%	382.73	——	——	——
12.	22,842.95	$10\frac{3}{4}\%$	252.76	——	——	——
13.	59,500.00	$12\frac{1}{2}\%$	635.02	——	——	——
14.	62,250.00	13%	688.61	——	——	——
15.	75,900.00	$14\frac{1}{2}\%$	929.44	——	——	——
16.	82,320.00	$15\frac{3}{4}\%$	1090.42	——	——	——
17.	61,842.97	$13\frac{1}{4}\%$	697.98	——	——	——
18.	51,946.82	$12\frac{3}{4}\%$	552.50	——	——	——

Date

Name

Course/Section

B. For the following loans calculate the points, discount rate, and discount.

	Mortgage Principal	Current Market Rate	Maximum Federal Rate	Points	Discount Rate	Discount
1.	$57,500	$10\frac{3}{4}\%$	$10\frac{1}{2}\%$	——	——	——
2.	35,000	10%	$9\frac{1}{2}\%$	——	——	——
3.	52,250	$10\frac{1}{4}\%$	$9\frac{3}{4}\%$	——	——	——
4.	42,000	$10\frac{3}{4}\%$	$10\frac{1}{4}\%$	——	——	——
5.	63,000	$10\frac{3}{4}\%$	10%	——	——	——
6.	39,500	$10\frac{1}{2}\%$	$9\frac{3}{4}\%$	——	——	——
7.	78,000	$10\frac{1}{2}\%$	10%	——	——	——
8.	38,500	11%	$10\frac{3}{4}\%$	——	——	——
9.	45,000	11%	$10\frac{1}{4}\%$	——	——	——
10.	56,000	$10\frac{1}{4}\%$	$9\frac{1}{2}\%$	——	——	——
11.	49,550	$10\frac{1}{2}\%$	$9\frac{1}{2}\%$	——	——	——
12.	39,400	$10\frac{3}{4}\%$	$10\frac{1}{4}\%$	——	——	——
13.	62,800	$12\frac{3}{4}\%$	$12\frac{1}{4}\%$	——	——	——
14.	75,200	$12\frac{1}{2}\%$	$11\frac{3}{4}\%$	——	——	——
15.	52,800	$11\frac{3}{4}\%$	$11\frac{1}{2}\%$	——	——	——
16.	48,900	13%	$12\frac{3}{4}\%$	——	——	——
17.	63,850	$12\frac{1}{4}\%$	12%	——	——	——
18.	55,100	$11\frac{1}{2}\%$	$10\frac{3}{4}\%$	——	——	——

When you have had the practice you need, either return to the Preview on page 374, or continue in frame **23** with the study of prorations.

SECTION 3: PRORATIONS

23

OBJECTIVE

Prorate interest, taxes, and insurance.

At the time of closing or completion of a real estate transaction, certain continuing expenses must be *prorated* or divided between the buyer and seller. Items prorated include interest, insurance, and taxes.

Most proration calculations use 30-day-month time, where each month is assumed to have 30 days and each year, 360 days. For a review of 30-day-month time, turn to page 262.

Prorating interest

Interest is always paid *in arrears*, that is, for the previous month. A mortgage payment due on February 1 pays the interest for January. (In this section, we will assume all payments are due the first of the month and will include interest through that day. We will also assume all interest calculations use ordinary simple interest at 30-day-month time.)

If the real estate transaction is closed on the first of the month, no proration is necessary. For example, if a mortgage is assumed by the buyer on February 1, the seller must make the February 1 payment. But, if the closing date is later than the first of February, the seller will owe the buyer the unpaid interest. The interest will be calculated from February 1 until the closing date. Only the interest is prorated, not the principal.

Example: A buyer assumes a $87,000 mortgage at $12\frac{1}{2}\%$ on February 17. The seller made the last *P&I* payment on February 1 (interest in arrears). Calculate the accrued interest due the buyer.

The time from February 1 to February 17 is 16 days. (Remember—don't count the first day.)

$$I = P \times r \times t = \$87,000 \times 12\frac{1}{2}\% \times \frac{16}{360}$$

$$= \$87,000 \times 12.5\% \times \frac{16}{360}$$

$$= \$87,000 \times 0.125 \times \frac{16}{360} = \$483.33 \text{ rounded}$$

The seller owes the buyer $483.33 for the unpaid interest from February 1 to February 17.

Your turn. Work the following problem.

A buyer assumes a $57,258 mortgage at $11\frac{1}{2}\%$ on June 23. The seller made the last *P&I* payment on June 1. Calculate the accrued interest due the buyer.

Check your calculations in 24.

What is a "conventional" mortgage?

A conventional mortgage is a non-FHA or non-VA mortgage. It's a bank loan not insured by either FHA or VA.

24 The time from June 1 to June 23 is 22 days.

$$I = P \times r \times t = \$57{,}258 \times 11\frac{1}{2}\% \times \frac{22}{360}$$

$$= \$57{,}258 \times 11.5\% \times \frac{22}{360}$$

$$= \$57{,}258 \times 0.115 \times \frac{22}{360}$$

$$= \$402.3965 \ldots = \$402.40 \text{ rounded}$$

Be careful when counting the days; don't count the first day. Also, use ordinary simple interest at 30-day-month time.

Need some more practice? Work the following problem.

Andrew and Teresa Turner assumed a \$42,856.72 mortgage at $12\frac{1}{4}\%$ interest on December 18. The seller made the last *P&I* payment on December 1. Calculate the accrued interest due the Turners.

Turn to **25** to check your work.

---

25 The time from December 1 to December 18 is 17 days.

$$I = P \times r \times t = \$42{,}856.72 \times 12\frac{1}{4}\% \times \frac{17}{360}$$

$$= \$42{,}856.72 \times 12.25\% \times \frac{17}{360}$$

$$= \$42{,}856.72 \times 0.1225 \times \frac{17}{360}$$

$$= \$247.914 \ldots = \$247.91 \text{ rounded}$$

PRORATING TAXES

Taxes may be paid either in arrears, where the payment is for the previous year's taxes, or *pre-paid,* where the taxes are paid in advance. Although many regions of the country have different due dates, in this section we will assume all taxes are paid on the calendar year and due January 1. Use 30-day-month time on all time calculations.

When taxes are paid in arrears, the January 1 payment is for the previous year. If the closing date is after the first of the year, the taxes must be prorated. In this case, since the taxes are unpaid for the period from January 1 until the closing date, the prorated amount is due the buyer.

Example: Pat Pruitt is purchasing a house with the closing date April 1, 1984. The yearly taxes of $704.04 were paid in arrears on January 1, 1984. Prorate the taxes.

Since the tax amount for 1984 was unknown at the time Pat bought the house, the 1983 taxes were used as an estimate.

The time from January 1 to April 1 is 3 months; therefore, the prorated taxes were for 3 of 12 months for 1984 or $\frac{3}{12}$ of the total.

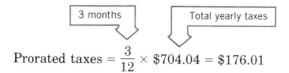

$$\text{Prorated taxes} = \frac{3}{12} \times \$704.04 = \$176.01$$

The amount due Ms. Pruitt for the unpaid taxes in 1984 was $176.01.

Your turn. Work the following problem.

A buyer is purchasing a house with the closing date August 1. The yearly taxes of $476.59 were paid in arrears on January 1. Prorate the taxes.

Check your calculations in **26**.

---

26   The time from January 1 to August 1 is 7 months.

$$\text{Prorated taxes} = \frac{7}{12} \times \$476.59 = \$278.01 \text{ rounded}$$

The amount due the buyer is $278.01 for the unpaid taxes.

If the closing date is not the first of the month, the number of days from the first of the year must be counted. Remember to use 30-day-month time. (If you need a review of 30-day-month time, return to page 262.)

Work the following problem.

Charles Blaine is purchasing a house with a closing date of May 17. The yearly taxes of $786.45 were paid in arrears on January 1. Prorate the taxes.

Turn to **27** to check your work.

---

27   First, calculate the number of days from January 1 to May 17 using 30-day-month time.

January 1 to May   1 = 4 months = 4 × 30 days = 120 days
       May 1 to May 17 = 17 − 1 days   =   16 days
                                           136 days

$$\text{Prorated taxes} = \frac{136}{360} \times \$786.45$$

$$= \$297.103 \ldots = \$297.10 \text{ rounded}$$

The amount due the buyer is $297.10.

If taxes are pre-paid, the January 1 payment is for the next year. If the closing date is after the first of the year, the taxes must again be prorated. In this case, since the taxes are paid from the closing date until January 1 of the next year, the prorated amount is due the seller.

Example: A buyer is purchasing a house with a closing date of September 5. The yearly taxes of $784.26 were prepaid on January 1. Prorate the taxes.

Since the taxes are prepaid, the number of days from the closing date until January 1 must be found.

September 5 to January 5 = 4 months = 4 × 30 days = 120 days
January 1 to January 5 =   −4 days
                                                            116 days

Prorated taxes = $\frac{116}{360}$ × $784.26

= $252.706 = $252.71 rounded

The amount due the seller for the prepaid taxes is $252.71.

Your turn again. Work the following problem.

A buyer is purchasing a house with a closing date of April 17. The yearly taxes of $356.75 were prepaid on January 1. Prorate the taxes.

Check your work in 28.

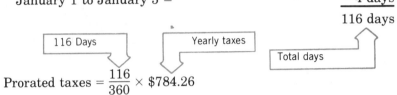

Should I put a large down payment on a new home if I can afford it?

The larger the down payment the less your interest costs will be, the greater your equity, and you will have smaller monthly payments. But maybe you need that money for other expenses.

28   April 17 to January 17 = 9 months = 9 × 30 days = 270 days
      January 1 to January 17 =   −16 days
                                                                 254 days

Prorated taxes = $\frac{254}{360}$ × $356.75

= $251.706 . . . = $251.71 rounded

The amount due the seller is $251.71.

Remember ⟩ If the taxes are prepaid, the prorated amount is due the seller. If the taxes are in arrears, the prorated amount is due the buyer.

PRORATING
INSURANCE

Property hazard insurance is required by most all lending institutions. Insurance is usually written for one or three years with the premium paid in advance. If the mortgage is assumed by the buyer, the insurance is usually transferred to the new owner.

Since the insurance is prepaid, the prorated amount is due to the seller from the buyer.

Example: On July 17, 1983, Vincent Hughes paid $598 for a three-year insurance policy on his house. He sold his house August 5, 1984. Calculate the insurance proration.

First, calculate the number of days remaining on the policy using 30-day-month time. The policy expires three years from July 17, 1983 or July 17, 1986. The time remaining on the policy is from August 5, 1984 to July 1986.

August 5, 1984 to August 5, 1985 = 1 year =   360 days
August 5, 1985 to July 5, 1986   = 11 months = 11 × 30 days = 330 days
July 5   to July 17   =   12 days

                                                Total   702 days

The total length of the policy is for 3 years or $3 \times 360 = 1080$ days. The time remaining on the policy is 702 of the 1080 days or $\frac{702}{1080}$ of the policy.

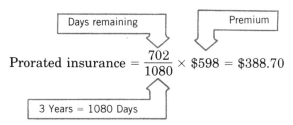

$$\text{Prorated insurance} = \frac{702}{1080} \times \$598 = \$388.70$$

3 Years = 1080 Days

The amount due the seller is $388.70.

Work the following problem.

Mr. Terry Huber paid $637 for a three-year insurance policy on his house on September 25, 1982. He sold his house on May 7, 1984. Calculate the insurance proration.

Check your calculations in **29**.

---

29   The policy expires September 25, 1985. The time remaining on the policy is from May 7, 1984 to September 25, 1985.

May 7, 1984 to May 7, 1985 = 1 year   = 360 days
May 7, 1985 to September 7, 1985 = 4 months = 4 × 30 days = 120 days
September 7 to September 25 =   18 days

                                                   498 days

                                                   Total

$$\text{Prorated insurance} = \frac{498}{1080} \times \$637 = \$293.73 \text{ rounded}$$

Length of policy
3 years = 1080 days

The amount due the seller is $293.73.

For a set of practice problems on proration, turn to **30**.

---

10 Real Estate Mathematics

PROBLEM SET 3

30   The answers are on page 598.

10.3   Prorations   **A.   In the following problems, interest is in arrears. Calculate the accrued interest due the buyer. Use ordinary simple interest at 30-day-month time.**

	Principal	Interest Rate	Last P&I Payment	Closing Date	Prorated Time	Prorated Interest
1.	$47,000.00	$10\frac{3}{4}\%$	August 1	August 16		
2.	42,500.00	$10\frac{1}{2}\%$	April 1	April 3		
3.	56,000.00	10%	January 1	January 27		
4.	54,862.00	$10\frac{3}{4}\%$	July 1	July 5		
5.	48,953.00	10%	February 1	February 1		
6.	36,124.05	$10\frac{1}{2}\%$	October 1	October 2		
7.	29,562.78	$9\frac{3}{4}\%$	March 1	March 20		
8.	59,415.17	$8\frac{3}{4}\%$	December 1	December 18		
9.	24,856.29	$8\frac{1}{2}\%$	September 1	September 10		
10.	31,004.89	9%	May 1	May 29		
11.	56,892.45	$12\frac{1}{2}\%$	October 1	October 13		
12.	62,172.14	14%	January 1	January 3		
13.	45,237.92	$13\frac{1}{4}\%$	November 1	November 11		
14.	36,926.35	$12\frac{3}{4}\%$	April 1	April 28		
15.	52,162.78	$11\frac{3}{4}\%$	August 1	August 21		
16.	46,825.49	$13\frac{1}{2}\%$	December 1	December 19		
17.	72,816.71	$11\frac{1}{2}\%$	March 1	March 5		
18.	59,982.38	$12\frac{1}{2}\%$	June 1	June 29		

Date _____

Name _____

Course/Section _____

B. In the following problems, the yearly tax payment was made January 1. Calculate the prorated taxes and determine whether this amount is due the seller or buyer.

	Yearly Taxes	Last Payment	Closing Date	Prorated Time	Prorated Taxes	Due To
1.	$327.00	In arrears	February 1	————	————	————
2.	456.25	Pre-paid	March 12	————	————	————
3.	813.94	Pre-paid	March 5	————	————	————
4.	1256.17	In arrears	September 19	————	————	————
5.	212.04	In arrears	December 2	————	————	————
6.	937.50	Pre-paid	February 12	————	————	————
7.	1598.42	Pre-paid	July 11	————	————	————
8.	742.86	In arrears	October 13	————	————	————
9.	345.76	Pre-paid	June 23	————	————	————
10.	435.19	In arrears	April 15	————	————	————

C. In the following problems, the property hazard insurance is prepaid. Calculate the prorated amount due the seller. Use 30-day-month time.

	Insurance Premium	Number of Years	Beginning Date	Settlement Date	Prorated Time	Proration
1.	$376	1	January 14, 1979	April 13, 1979	——	——
2.	255	1	April 3, 1979	October 20, 1979	——	——
3.	560	3	September 19, 1977	December 11, 1978	——	——
4.	742	3	June 21, 1978	March 12, 1979	——	——
5.	450	1	November 8, 1979	January 15, 1980	——	——
6.	675	3	February 17, 1979	June 9, 1979	——	——
7.	349	1	October 26, 1979	July 12, 1980	——	——
8.	762	3	May 12, 1979	February 21, 1980	——	——
9.	301	1	July 24, 1979	August 15, 1979	——	——
10.	947	3	March 15, 1978	May 3, 1980	——	——

When you have had the practice you need, either return to the Preview on page 374, or continue in frame 31 with the study of real estate taxes.

SECTION 4: TAXES

31

> **OBJECTIVE**
>
> Calculate assessed value and real property tax.

Real property taxes are levied by state and local governments. Two steps are required in calculating the property tax.

Step 1. Calculate the *assessed value*. We start with the market value, because the assessed value is a given percentage of the market value.

> Assessed value = assessment rate × market value

For example, if the market value is $47,000 and the assessment rate is 45%, the assessment value is 45% of $47,000.

Assessed value = assessment rate × market value
$$= 45\% \times \$47,000 = 0.45 \times \$47,000$$
$$= \$21,150$$

The assessed value is used in Step 2 to complete the tax calculation.

Step 2. Calculate the tax using the assessed value.

> Tax = tax rate × assessed value

The tax rate is usually stated per $100 of the assessed value. To complete our example, assume the tax rate is $3.28 per $100 of the assessed value.

Since the tax rate is $3.28 per $100, the rate per $1 is $\frac{\$3.28}{100} = \$0.0328$

Tax = tax rate × assessed value

$$= \frac{\$3.28}{100} \times \$21,150 \qquad \text{From Step 1}$$

$$= 0.0328 \times \$21,150$$

$$= \$693.72$$

Your turn. Work the following problem.

The market value of a home is $72,500. If the assessment rate is 60% and the tax rate is $2.56 per $100, calculate the tax.

Check your calculation in 32.

32   **Step 1.** Assessed value = assessment rate × market value
$$= 60\% \times \$72{,}500 = 0.60 \times \$72{,}500$$
$$= \$43{,}500$$

Step 2. Tax = tax rate × assessed value

$$= \frac{2.56}{100} \times \$43{,}500 = 0.0256 \times \$43{,}500$$

$$= 1113.60$$

Occasionally, the tax rate is stated per $1000. For example, in a tax rate of $5.62 per $1000, the rate per $1 will be

$$\frac{\$5.62}{1000} \times \$0.00562 \text{ per } \$1$$

Work the following problem.

The market value of a home is $62,000. If the assessment rate is 65% and the tax rate is $7.58 per $1000, calculate the tax.

Turn to **33** to check your work.

---

33   **Step 1.** Assessed value = assessment rate × market value
$$= 65\% \times \$62{,}000 = 0.65 \times \$62{,}000$$
$$= \$40{,}300$$
Step 2. Tax = tax rate × assessed value

$$= \frac{7.58}{1000} \times \$40{,}300 = 0.00758 \times \$40{,}300$$

$$= \$305.474 = \$305.47 \text{ rounded}$$

Some areas express the tax rate in "mills." One mill is one thousandth of a dollar, or a tenth of a cent.

$$1 \text{ mill} = \$0.001 = 0.1¢$$
$$10 \text{ mills} = \$0.010 = 1¢$$

To change mills to a decimal number, move the decimal point three places to the left. This is equivalent to dividing by 1000.

128 mills = 0.128

39 mills = 039 mills = 0.039

$42\frac{1}{2}$ mills = 42.5 mills = 042.5 mills = 0.0425

Change the following mills to decimal numbers.

1. 98 mills = _____   2. 279 mills = _____

3. $14\frac{3}{4}$ mills = _____   4. $42\frac{1}{2}$ mills = _____

Check your work in **34**.

---

34   1. 98 mills = 098 mills = 0.098

2. 279 mills = 0.279

3. $14\frac{3}{4}$ mills = 14.75 mills = 014.75 mills = 0.01475

4. $42\frac{1}{2}$ mills = 42.5 mills = 042.5 mills = 0.0425

Now that you can change mills to a decimal number, work the following property tax problem.

The market value of a house is $57,000. If the assessment rate is 70%, and the tax rate is $20\frac{1}{2}$ mills, calculate the tax.

Check your work in **35**.

---

35   **Step 1.**   Assessed value = assessment rate × market value
$$= 70\% \times \$57,000 = 0.70 \times \$57,000$$
$$= \$39,900$$

Step 2.   Tax = tax rate × assessed value
$$= 20\frac{1}{2} \text{ mills} \times \$39,900 = 20.5 \text{ mills} \times \$39,900$$
$$= 0.0205 \times \$39,900 = \$817.95$$

Remember *mills* means *thousandths*.

Turn to **36** for a set of practice problems on property tax.

---

10 Real Estate Mathematics

PROBLEM SET 4

36

10.4 Taxes

The answers are on page 598.

A. Calculate the assessed value and property tax.

	Market Value	Assessment Rate	Assessed Value	Tax Rate	Property Tax
1.	$56,000	60%	——	$2.37 per $100	——
2.	29,500	50%	——	2.03 per $100	——
3.	36,000	30%	——	3.42 per $100	——
4.	22,700	45%	——	3.86 per $100	——
5.	45,000	65%	——	1.83 per $100	——
6.	78,000	35%	——	3.49 per $100	——
7.	123,000	60%	——	9.36 per $1000	——
8.	47,500	70%	——	4.14 per $100	——
9.	39,200	45%	——	3.18 per $100	——
10.	83,000	65%	——	9.87 per $1000	——
11.	56,300	32%	——	5.37 per $100	——
12.	73,000	55%	——	3.86 per $100	——
13.	72,000	30%	——	9.42 per $1000	——
14.	56,500	65%	——	5.65 per $100	——
15.	82,300	25%	——	8.65 per $100	——
16.	95,800	40%	——	2.96 per $100	——
17.	36,500	45%	——	8.52 per $1000	——
18.	45,000	55%	——	9.17 per $1000	——

Date

Name

Course/Section

B. Calculate the assessed value and property tax.

	Market Value	Assessment Rate	Assessed Value	Tax Rate	Property Tax
1.	$42,800	35%	———	29 mills	———
2.	56,300	65%	———	21 mills	———
3.	22,000	50%	———	19 mills	———
4.	37,000	60%	———	34 mills	———
5.	82,000	40%	———	$38\frac{1}{2}$ mills	———
6.	26,900	70%	———	$23\frac{1}{2}$ mills	———
7.	65,000	55%	———	18 mills	———
8.	114,000	50%	———	23 mills	———
9.	35,000	45%	———	$24\frac{3}{4}$ mills	———
10.	93,000	55%	———	$29\frac{1}{4}$ mills	———
11.	45,000	30%	———	$32\frac{1}{2}$ mills	———
12.	85,500	60%	———	$21\frac{1}{2}$ mills	———
13.	52,500	35%	———	27 mills	———
14.	45,200	40%	———	19 mills	———
15.	52,900	25%	———	$18\frac{1}{2}$ mills	———
16.	63,500	65%	———	$31\frac{3}{4}$ mills	———
17.	62,900	60%	———	$18\frac{1}{4}$ mills	———
18.	73,500	55%	———	$29\frac{1}{2}$ mills	———

When you have had the practice you need, turn to **37** for the Self-Test on this unit.

10 Real Estate Mathematics

37   Self-Test

The answers are on page 599.

In the following problems use 30-day-month time and ordinary simple interest.

1. Find the area of a rectangular lot 85 feet by 145 feet.

   _____

2. Find the area of the following figure.

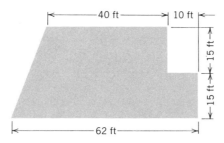

   _____

3. (a) What is the area of a lot 300 feet by 363 feet?
   (b) Convert the area to acres.

   (a) _____

   (b) _____

4. Calculate the cost of a house with 1750 square feet of floor area at $32.50 per square foot.

   _____

5. What is the price per acre of a 12-acre lot priced at $75,300?

   _____

6. How many cubic yards of concrete are needed to construct a sidewalk 4 feet wide, 135 feet long, and 6 inches deep?

   _____

Date

Name

Course/Section

7. For a mortgage with principal balance $26,536 at $8\frac{1}{2}\%$ and monthly *P&I* payment of $250, calculate the (a) interest payment, (b) principal payment, and (c) new principal.

    (a) _____

    (b) _____

    (c) _____

8. The Rodriquez family is negotiating a VA insured loan for $48,500. The current market rate is $10\frac{1}{2}\%$ and the maximum federal rate is $9\frac{3}{4}\%$. Calculate the (a) points, and (b) discount amount.

    (a) _____

    (b) _____

9. A buyer assumes a $32,862 mortgage at $8\frac{3}{4}\%$ on August 17. The seller made the last *P&I* payment (in arrears) on August 1. Calculate the accrued interest due the buyer.

    _____

10. A buyer is purchasing a house with the closing date September 8. The yearly taxes of $836.95 were prepaid on January 1. Prorate the taxes.

    _____

11. Ms. McDown paid $637 on March 17, 1984 for a three-year property insurance policy on her house. She sold the house on September 8, 1985. Calculate the insurance proration.

    _____

12. The market value of a house is $57,000. If the assessment rate is 60% and the tax rate is $2.89 per $100, calculate the tax.

    _____

This is a sample test for Unit 11. All the answers are placed immediately after the test. Work as many of the problems as you can, then check your answers and look after the answers for further directions. Use the necessary tables from this unit.

	Where to Go for Help	
	Page	Frame

1. A building is insured for $45,000. The annual fire insurance rate is $0.37 per $100. Calculate the (a) annual premium and (b) three-year premium.

(a) _____

(b) _____   417   1

2. Ms. Riddle insured her house for one year starting March 25 at a cost of $358. She cancelled the policy on August 10. Calculate the amount of the premium (a) retained by the insurance company, and (b) the refund to Ms. Riddle.

(a) _____

(b) _____   419   4

3. The Cody Corporation carries three fire insurance policies: Company A $10,000, Company B $25,000, and Company C $15,000. The corporation sustained a $27,000 fire loss. Calculate the amount paid by each company.

_____   424   7

4. Steve Pollard, age 19, drives his car to work, 7 miles one way. Calculate his automobile insurance premium for 20/40/10 liability, $2000 medical payments, and $100 deductible collision and comprehensive. His car is price group F, age group 1.

_____   433   13

5. Calculate the premium for a $15,000 20-payment life insurance policy for David Holmes, age 24.

_____   443   24

6. Pam Kolman is discontinuing her $12,000 ordinary life insurance policy issued at age 23. The policy has been in force 10 years. Calculate the (a) cash surrender value, (b) paid-up insurance, and (c) extended term insurance.

(a) _____

(b) _____

(c) _____   445   26

Insurance

It isn't always easy to find someone to predict your future, much less to [...] impending danger. It is easy, however, to find insurance agencies who wi[...] you and your property against loss.

Insurance can't prevent a loss, but it can provide financial protection whe[...] occurs. For instance, fire insurance will not guarantee that your house won't [...] it should burn, though, insurance will provide the finances to rebuild or repl[...] house.

This chapter is designed to familiarize you with all aspects of insurance: fi[...] and life. Once you have completed the unit, you will be able to read insuranc[...] and to calculate insurance rates.

Ready? Good! Put on your thinking cap and get ready to "absorb" this infor[...] packed chapter. It's just good "insurance" for the future!

UNIT OBJECTIVE

After you complete this unit successfully, you will be able to calcul[...] premiums for fire, automobile, and life insurance policies. You will also [...] able to calculate prorations due to cancellation and the amount of loss pa[...] by the insurers for multiple carriers and for coinsurance policies. You w[...] be able to determine what portion of damages are paid by the auto insuran[...] company in the cases of bodily injury and auto damage.

This is a sample test for Unit 11. All the answers are placed immediately after the test. Work as many of the problems as you can, then check your answers and look after the answers for further directions. Use the necessary tables from this unit.

Where to Go for Help

	Page	Frame

1. A building is insured for $45,000. The annual fire insurance rate is $0.37 per $100. Calculate the (a) annual premium and (b) three-year premium.

(a) _____

(b) _____   417   1

2. Ms. Riddle insured her house for one year starting March 25 at a cost of $358. She cancelled the policy on August 10. Calculate the amount of the premium (a) retained by the insurance company, and (b) the refund to Ms. Riddle.

(a) _____

(b) _____   419   4

3. The Cody Corporation carries three fire insurance policies: Company A $10,000, Company B $25,000, and Company C $15,000. The corporation sustained a $27,000 fire loss. Calculate the amount paid by each company.

_____   424   7

4. Steve Pollard, age 19, drives his car to work, 7 miles one way. Calculate his automobile insurance premium for 20/40/10 liability, $2000 medical payments, and $100 deductible collision and comprehensive. His car is price group F, age group 1.

_____   433   13

5. Calculate the premium for a $15,000 20-payment life insurance policy for David Holmes, age 24.

_____   443   24

6. Pam Kolman is discontinuing her $12,000 ordinary life insurance policy issued at age 23. The policy has been in force 10 years. Calculate the (a) cash surrender value, (b) paid-up insurance, and (c) extended term insurance.

(a) _____

(b) _____

(c) _____   445   26

Insurance

It isn't always easy to find someone to predict your future, much less to warn of impending danger. It is easy, however, to find insurance agencies who will insure you and your property against loss.

Insurance can't prevent a loss, but it can provide financial protection when a loss occurs. For instance, fire insurance will not guarantee that your house won't burn. If it should burn, though, insurance will provide the finances to rebuild or replace your house.

This chapter is designed to familiarize you with all aspects of insurance: fire, auto, and life. Once you have completed the unit, you will be able to read insurance tables and to calculate insurance rates.

Ready? Good! Put on your thinking cap and get ready to "absorb" this information-packed chapter. It's just good "insurance" for the future!

UNIT OBJECTIVE

After you complete this unit successfully, you will be able to calculate premiums for fire, automobile, and life insurance policies. You will also be able to calculate prorations due to cancellation and the amount of loss paid by the insurers for multiple carriers and for coinsurance policies. You will be able to determine what portion of damages are paid by the auto insurance company in the cases of bodily injury and auto damage.

---

1. (a) $166.50   (b) $449.55

2. (a) $171.84   (b) $186.16

3. Company A: $5400; Company B: $13,500; Company C: $8100

4. Bodily injury: $237; Property damage: $163; medical payments: $18; collision: $280; comprehensive: $59; annual premium: $757.

5. $374.10

6. (a) $984,   (b) $2376,   (c) 12 years, 341 days

★ If you cannot work any of these problems, start with frame 1 on page 417.

★ If you missed some of the problems, turn to the page numbers indicated.

★ If all your answers were correct, you are probably ready to proceed to Unit 12. (If you would like more practice on Unit 11 before turning to Unit 12, try the Self-Test on page 449.)

★ Super-students—those who want to be certain they learn all of this material—will turn to frame 1 and begin work there.

SECTION 1: FIRE INSURANCE

1

> **OBJECTIVE**
>
> Calculate fire insurance premiums and prorations due to cancellation by the insurer or insured. Calculate the amount of loss paid by insurers for multiple carriers and for coinsurance policies.

Most individuals or businesses need some type of financial protection for personal or property losses. Insurance provides this financial protection against loss or damage by some contingency such as death, accident, fire, etc.

The *insurance policy* is a written contract where the *insurer* or company providing the insurance agrees to compensate the *insured* for specific losses or damages. The insured, the person or business obtaining the insurance, holds the insurance policy.

The *face value* of a policy is the maximum amount payable by the insurer for a loss to the insured.

Insurance does not guarantee prevention of a loss or damage. It only provides financial reimbursement in the case of a loss. For example, automobile insurance cannot prevent your auto from being stolen. If it is stolen, however, the insurance provides compensation for the financial loss.

Fire insurance provides protection against fire loss and related damage such as water damage caused when the fire was distinguished.

Fire insurance rates, which vary widely, are determined by the type of structure, location, maintenance, occupancy (owner or tenant), and the proximity of fire hydrants.

CALCULATING
PREMIUMS

Fire Insurance

Most fire insurance policies are written for either one or three years. A few policies are written for two years, and under certain conditions a four- or five-year policy may be obtained. The insurance cost is called the *premium*.

Premium rates are usually expressed as some given amount per $100 of insurance coverage. For a one-year policy the premium is

> Annual premium = rate $\times$ amount of insurance

Example: A building is insured for $45,000. If the annual fire insurance is $0.35 per $100, calculate the annual premium.

The annual rate is $0.35 per $100. The rate per $1 is $\dfrac{0.35}{100} = 0.0035$.

Annual premium = rate $\times$ amount of insurance

$$= \frac{0.35}{100} \times \$45,000 = 0.0035 \times \$45,000$$

$$= \$157.50 \qquad \text{This is the amount paid yearly}$$
$$\text{for insuring the building.}$$

It's easy! Your turn.

A house is insured for $60,000. The fire insurance rate is $0.27 per $100. Calculate the annual premium.

Turn to 2 to check your calculation.

---

2   Annual premium = rate × amount of insurance

$$= \frac{0.27}{100} \times \$60,000 = 0.0027 \times \$60,000$$

$$= \$162$$

When fire insurance is written for more than one year, most companies offer reduced premiums.

Number of Years	Premium
2	1.85 × annual premium
3	2.7 × annual premium

For a two-year policy, rather than 2 × annual premium, the reduced rate is 1.85 × annual premium.

Example: A building is insured for $72,000. The annual fire insurance rate is $0.42 per $100. Calculate (a) the annual premium, (b) two-year premium, and (c) three-year premium.

(a)   Annual premium = rate × amount of insurance

$$= \frac{0.42}{100} \times \$72,000 = 0.0042 \times \$72,000$$

$$= \$302.40$$

(b)   Two-year premium = 1.85 × annual premium
$$= 1.85 \times \$302.40 = \$559.44$$

(c)   Three-year premium = 2.7 × annual premium
$$= 2.7 \times \$302.40 = \$816.48$$

Work the following problems.

1.   A house is insured for $57,000. If the annual fire insurance rate is $0.31 per $100, calculate the two-year premium.
2.   A building is insured for $73,000. The annual fire insurance rate is $0.37 per $100. Calculate the three-year premium.

Check your work in 3.

---

3   1.   Annual premium = rate × amount of insurance

$$= \frac{0.31}{100} \times 57,000 = 0.0031 \times \$57,000 = \$176.70$$

Two-year premium = 1.85 × annual premium
$$= 1.85 \times \$176.70 = \$326.895 = \$326.90 \text{ rounded}$$

2. Annual premium = rate × amount of insurance

$$= \frac{0.37}{100} \times \$73{,}000 = 0.0037 \times \$73{,}000 = \$270.10$$

Three-year premium = 2.7 × annual premium
= 2.7 × $270.10 = $729.27

Occasionally it is desirable to obtain an insurance policy for less than one year. These policies, called *short-term* policies, are written at a higher rate than an annual policy. The standard short-rate table is given on page 421.

Example: A building is insured for $55,000. The annual insurance rate is $0.45 per $100. Calculate the premium on a 90 day policy.

Step 1. First, calculate the annual premium.
Annual premium = rate × amount of insurance

$$= \frac{0.45}{100} \times \$55{,}000 = 0.0045 \times \$55{,}000 = \$247.50$$

Step 2. Find the number of days and the rate in the short-rate table. For 90 days, the rate is 35%.

Step 3. The short-term premium is the product of the short-rate found in **Step 2** and the annual premium.

Short-term premium = short-rate × annual premium
= 35% × $247.50 = 0.35 × $247.50
= $86.625 = $86.63 rounded

The insurance would cost $247.50 for one year or $86.63 for 90 days.

Work the following problem.

A building is insured for $73,000. The annual insurance rate is $0.34 per $100. Calculate the premium on a 182-day policy.

Check your calculations in **4**.

---

4   **Step 1.** Annual premium = rate × amount of insurance

$$= \frac{0.34}{100} \times \$73{,}000 = 0.0034 \times \$73{,}000 = \$248.20$$

Step 2. For 182 days, the short-rate is 60%.

Step 3. Short-term premium = short-rate × annual premium
= 60% × $248.20 = 0.60 × $248.20
= $148.92

CANCELLATIONS

Insurance may be cancelled either by the insured or the insurer. In either case, the premium already paid must be *prorated* or divided between the two parties. A different proration method is used in each case.

When the *insured* cancels a policy, the cost of the insurance for the reduced time period is calculated using the short-rate table. This results in the insured paying for the policy at a rate higher than the annual rate. The exact number of days is used to calculate the time the policy was in force. For a review of exact time, return to page 251.

Example: Mr. Holloway insured his house for one year starting March 17 at a cost of $378. He cancelled his policy on August 3. Calculate the amount of premium (a) retained by the insurance company, and (b) the refund to the insured.

Step 1. Calculate the exact number of days from March 17 to August 3.

$$\begin{array}{rl}
\text{March 17 to March 31:} & 14 \text{ days} \longleftarrow 31 - 17 = 14 \text{ days} \\
\text{April:} & 30 \\
\text{May:} & 31 \\
\text{June:} & 30 \\
\text{July:} & 31 \\
\text{to August 3:} & \underline{3} \\
& 139 \text{ days} \quad \text{Total days policy in force}
\end{array}$$

Step 2. Find the number of days and the corresponding rate in the short-rate table. For 139 days the rate is 49%.

Step 3. The amount of premium retained by the insurance company is the product of the short-rate found in **Step 2** and the annual premium.

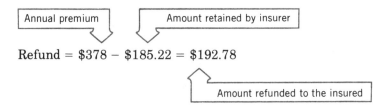

Premium retained by insurer = 49% × $378 = 0.49 × $378
= $185.22

Step 4. The refund to the insured is the difference between the annual premium and the amount retained by the insurer.

Refund = $378 − $185.22 = $192.78

Amount refunded to the insured

Your turn. Work the following problem.

Mr. Baggins insured his house for one year starting May 23 at an annual cost of $432. He cancelled his policy on September 15. Calculate the amount of premium (a) retained by the insurer, and (b) the refund to Mr. Baggins.

Check your calculations in **5**.

Why do short-term policies have a higher rate than annual policies?

Although the term is shorter, the company's expense for doing the paperwork is the same.

SHORT-RATE TABLE

Days Policy in Force	Percent	Days Policy in Force	Percent	Days Policy in Force	Percent
1	5	95–98	37	219–223	69
2	6	99–102	38	224–228	70
3–4	7	103–105	39	229–232	71
5–6	8	106–109	40	233–237	72
7–8	9	110–113	41	238–241	73
9–10	10	114–116	42	242–246	74
11–12	11	117–120	43	247–250	75
13–14	12	121–124	44	251–255	76
15–16	13	125–127	45	256–260	77
17–18	14	128–131	46	261–264	78
19–20	15	132–135	47	265–269	79
21–22	16	136–138	48	270–273	80
23–25	17	139–142	49	274–278	81
26–29	18	143–146	50	279–282	82
30–32	19	147–149	51	283–287	83
33–36	20	150–153	52	288–291	84
37–40	21	154–156	53	292–296	85
41–43	22	157–160	54	297–301	86
44–47	23	161–164	55	302–305	87
48–51	24	165–167	56	306–310	88
52–54	25	168–171	57	311–314	89
55–58	26	172–175	58	315–319	90
59–62	27	176–178	59	320–323	91
63–65	28	179–182	60	324–328	92
66–69	29	183–187	61	329–332	93
70–73	30	188–191	62	333–337	94
74–76	31	192–196	63	338–342	95
77–80	32	197–200	64	343–346	96
81–83	33	201–205	65	347–351	97
84–87	34	206–209	66	352–355	98
88–91	35	210–214	67	356–360	99
92–94	36	215–218	68	361–365	100

5   **Step 1.**   Calculate the time.

$$
\begin{array}{rl}
\text{May 23 to May 31:} & 8 \text{ days} \longleftarrow 31 - 23 = 8 \text{ days} \\
\text{June:} & 30 \\
\text{July:} & 31 \\
\text{August:} & 31 \\
\text{to September 15:} & \underline{15} \\
& 115 \text{ days} \longleftarrow \text{Total}
\end{array}
$$

Step 2.   From the short-rate table, for 115 days the short-rate is 42%.

Short-rate   Annual premium

Step 3.   Premium retained by insurer = 42% × \$432 = 0.42 × \$432
= \$181.44

Step 4.

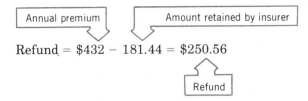

Refund = $432 − 181.44 = $250.56

It's easy if you follow the four steps. If you had any trouble calculating the number of days, be sure to review this carefully (starting on page 251) before continuing.

When a policy is cancelled by the insurance company, most states require a given number of days' notice. This allows time for the insured to obtain new insurance.

In this case, the amount of premium retained by the insurer is prorated using the exact number of days the policy is in force. The short-rate table is *not* used when the policy is cancelled by the insurer.

Example: Mr. Mohler insured his house for one year starting March 18 at an annual cost of $397. The policy was cancelled by the insurance company on June 30. Calculate the amount of the premium (a) retained by the insurer and (b) the refund to Mr. Mohler.

Step 1. Calculate the number of days from March 18 to June 30.

$$\begin{array}{ll} \text{March 18 to March 31}: & 13 \text{ days} \quad \longleftarrow 31 - 18 = 13 \text{ days} \\ \text{April}: & 30 \\ \text{May}: & 31 \\ \text{to June 30}: & \underline{30} \\ & 104 \text{ days} \quad \longleftarrow \text{Total} \end{array}$$

Step 2. Calculate the amount of premium retained by the insurance company. The company provided insurance for 104 of 365 days or $\frac{104}{365}$ of the year. The amount due the insurer will be the product of $\frac{104}{365}$ and the annual premium.

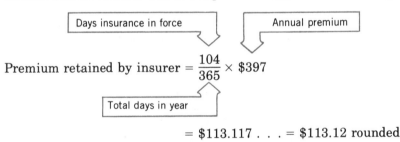

$$\text{Premium retained by insurer} = \frac{104}{365} \times \$397$$

$$= \$113.117 \ldots = \$113.12 \text{ rounded}$$

Step 3. The refund to the insured is the difference between the annual premium and the amount retained by the insurer.

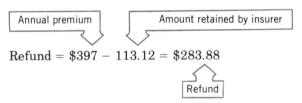

Refund = $397 − 113.12 = $283.88

Your turn to work one.

Ms. Esparza insured her house for one year starting May 13 at a cost of $455. The policy was cancelled by the insurance company on September 15. Calculate the amount of premium (a) retained by the insurer and (b) the refund to Ms. Esparza.

Check your work in 6.

Step 1. Calculate the time.

$$
\begin{aligned}
\text{May 13 to May 31}: \quad & 18 \text{ days} \longleftarrow 31 - 13 = 18 \text{ days}\\
\text{June}: \quad & 30\\
\text{July}: \quad & 31\\
\text{August}: \quad & 31\\
\text{to September 15}: \quad & \underline{15}\\
& 125 \text{ days} \longleftarrow \text{Total}
\end{aligned}
$$

Step 2.

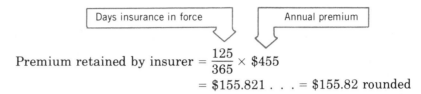

$$
\text{Premium retained by insurer} = \frac{125}{365} \times \$455
$$
$$
= \$155.821 \ldots = \$155.82 \text{ rounded}
$$

Step 3.

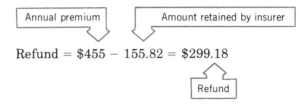

$$
\text{Refund} = \$455 - 155.82 = \$299.18
$$

Refund

Important When the insurance is cancelled by the insured, always use the short-rate table. When insurance is cancelled by the insurer, use the exact number of days for the proration.

Need some additional practice? Try two more.

1. Ms. Duvall insured her house for one year starting August 27 at a cost of $370. Ms. Duvall cancelled the policy on November 7. Calculate the amount of premium (a) retained by the insurance company and (b) the refund to Ms. Duvall.

2. Mr. Brown insured his business for one year starting March 20 at a cost of $552. The policy was cancelled by the insurance company on October 10. Calculate the amount of premium (a) retained by the insurer and (b) the refund to Mr. Brown.

Turn to **7** to check your calculations.

---

7 1. **Step 1.** Calculate the time.

$$
\begin{aligned}
\text{August 27 to August 31}: \quad & 4 \text{ days} \longleftarrow 31 - 27 = 4 \text{ days}\\
\text{September}: \quad & 30\\
\text{October}: \quad & 31\\
\text{to November 7}: \quad & \underline{7}\\
& 72 \text{ days} \longleftarrow \text{Total}
\end{aligned}
$$

Step 2. Since the policy was cancelled by the insured, the short-rate table is used. For 72 days, the rate is 30%.

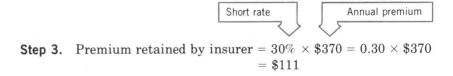

Step 3. Premium retained by insurer = $30\% \times \$370 = 0.30 \times \$370$
$$
= \$111
$$

Step 4.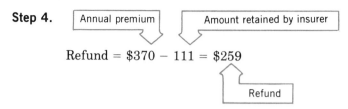

Refund = $370 − 111 = $259

2. **Step 1.** Calculate the time.

March 20 to March 31 : 11 days ⟵ 31 − 20 = 11 days
April : 30
May : 31
June : 30
July : 31
August : 31
September : 30
to October 10 : 10
204 days ⟵ Total

Step 2. Since the policy was cancelled by the insurance company, the exact number of days is used in the proration calculation.

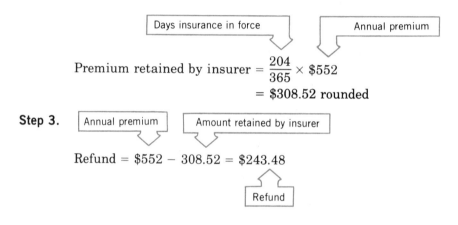

$$\text{Premium retained by insurer} = \frac{204}{365} \times \$552$$

$$= \$308.52 \text{ rounded}$$

Step 3.

Refund = $552 − 308.52 = $243.48

MULTIPLE INSURERS

Insurance on a piece of property may be divided among two or more insurance companies. This is done for several reasons. First, the insured may want to divide his business between several agents. Second, the value of the property may be too large for one company to assume all the risk.

In the case of multiple insurers, each insurance company pays its prorata share of the total insurance for a claim. But in no case will any insurance company pay more than the face value of the policy.

$$\text{Each insurer's share} = \frac{\text{face value of insurer's policy}}{\text{total insurance}} \times \text{loss}$$

Example: Price's Auto Supply carries two fire insurance policies: Company A, $20,000, and Company B, $30,000. Price's sustained a $36,000 fire loss. Calculate the amount of loss paid by each company.

Total insurance = $20,000 + 30,000 = $50,000

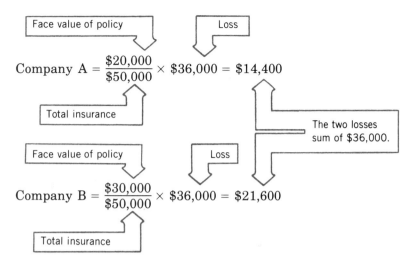

Company A $= \dfrac{\$20,000}{\$50,000} \times \$36,000 = \$14,400$

Company B $= \dfrac{\$30,000}{\$50,000} \times \$36,000 = \$21,600$

The two losses sum of $36,000.

Work the following problem.

The Shallhorn Corporation carries two fire insurance policies: Company A, $25,000, Company B, $15,000. The Shallhorn Corporation sustained a $19,000 fire loss. Calculate the amount paid by each company.

Check your work in **8.**

---

8   Total insurance = $25,000 + 15,000 = $40,000

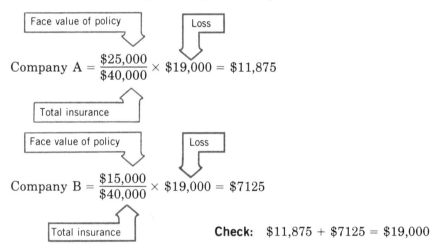

Company A $= \dfrac{\$25,000}{\$40,000} \times \$19,000 = \$11,875$

Company B $= \dfrac{\$15,000}{\$40,000} \times \$19,000 = \$7125$

Check: $11,875 + $7125 = $19,000

No insurance company will pay more than the face value of its policy. If the loss is larger than the total insurance carried, each insurance company will only pay the face value. (See also coinsurance in **9.**)

Example: Lou's Portrait Studio carries two fire insurance policies: Company A, $15,000 and Company B, $10,000. The studio sustained a $27,000 fire loss. Calculate the amount paid by each company.

Total insurance = $15,000 + 10,000 = $25,000

Since the total loss of $27,000 is larger than the total insurance, each company will pay the face value of their policy: Company A, $15,000 and Company B, $10,000.

Work the following problems.

1. The Pratter Company carries three fire insurance policies: Company A, $20,000, Company B, $35,000, and Company C, $25,000. The Pratter Company had a $57,000 fire loss. Calculate the amount paid by each company.
2. The Donihoe Corporation carries two fire insurance policies: Company X, $25,000, and Company Y, $15,000. The Corporation had a $43,000 fire loss. Calculate the amount paid by each company.

Check your work in **9**.

---

9   1.   Total insurance = $20,000 + 35,000 + 25,000 = $80,000

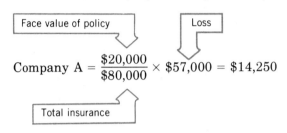

$$\text{Company A} = \frac{\$20,000}{\$80,000} \times \$57,000 = \$14,250$$

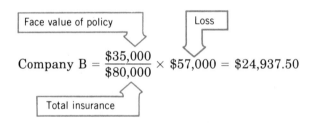

$$\text{Company B} = \frac{\$35,000}{\$80,000} \times \$57,000 = \$24,937.50$$

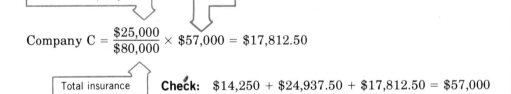

$$\text{Company C} = \frac{\$25,000}{\$80,000} \times \$57,000 = \$17,812.50$$

Check: $14,250 + $24,937.50 + $17,812.50 = $57,000

2.   Total insurance = $25,000 + 15,000 = $40,000
Since the loss is larger than the total insurance, each insurance company will pay the face value of its policy.

Company X = $25,000

Company Y = $15,000

COINSURANCE

Since few fires result in a total loss, many businesses prefer to insure their property for only a portion of the total value. In order to compensate for underinsuring, most insurance companies require a *coinsurance* clause where the insured assumes some of the risk. The most common coinsurance rate is 80%. If the property is insured for at least 80% of its value, the insurer will pay for any loss up to the policy's face value. But if the business is insured for less than 80%, the insurance company will only pay a portion of the loss and the business must pay the rest. The amount the insurance company will pay is

$$\text{Insurer pays} = \frac{\text{face value of policy}}{80\% \times \text{property value}} \times \text{loss}$$

Remember The insurer will never pay more than the face value of the policy or the loss.

Example: The Babett Company carries $25,000 in fire insurance with an 80% coinsurance clause. The value of the property is $40,000. The company had a $12,000 fire loss. Calculate the amount paid by the insurer.

Face value of policy   Loss

$$\text{Insurer pays} = \frac{\$25,000}{80\% \times \$40,000} \times \$12,000 = \frac{\$25,000}{\$32,000} \times \$12,000$$

Property value

$$= \$9375$$

Your turn again.

The Frost Corporation carries $30,000 in fire insurance with an 80% coinsurance clause. The value of the property is $45,000. The corporation had a $20,000 fire loss. Calculate the amount paid by the insurer.

Turn to **10** to check your work.

Face value of policy   Loss

10    $$\text{Insurer pays} = \frac{\$30,000}{80\% \times \$45,000} \times \$20,000 = \frac{\$30,000}{\$36,000} \times \$20,000$$

Property value

$$= \$16,666.666 \ldots = \$16,666.67 \text{ rounded}$$

Remember, the insurance company will never pay more than the loss.

Example: The Davis Company carries $40,000 in fire insurance with an 80% coinsurance clause. The value of the property is $48,000. The company sustained a $25,000 fire loss. Calculate the amount paid by the insurer.

$$\text{Insurer pays} = \frac{\$40,000}{80\% \times \$48,000} \times \$25,000 = \frac{\$40,000}{\$38,400} \times \$25,000$$
$$= \$26,041.67 ? \text{ rounded}$$

But $26,041.67 is more than the total loss! In this case the insurer pays the entire loss of $25,000. The insurance company will not pay more than the loss. In this case the Davis Company carries fire insurance for more than 80% of the property value.

Work a few more problems to be certain you understand the process.

1. The Rector Corporation carries $50,000 in fire insurance with an 80% coinsurance clause. The value of the property is $60,000. The corporation had a $15,000 fire loss. What is the amount paid by the insurer?
2. The Great Pacific Nut Company carries $70,000 in fire insurance with an 80% coinsurance clause. The value of the property is $100,000. The company sustained a $25,000 fire loss. Calculate the amount paid by the insurance company.

Check your calculations in **11**.

---

1. Insurer pays $= \dfrac{\$50,000}{80\% \times \$60,000} \times \$15,000 = \dfrac{\$50,000}{\$48,000} \times \$15,000$

$$= \$15,625 \text{ ? rounded}$$

But $15,625 is larger than the total loss. The insurer will pay only $15,000.

2. Insurer pays $= \dfrac{\$70,000}{80\% \times \$100,000} \times \$25,000 = \dfrac{\$70,000}{\$80,000} \times \$25,000$

$$= \$21,875$$

For a set of practice problems on fire insurance turn to 12.

Insurance

12   The answers are on page 599.

11.1   Fire Insurance   **A. Calculate the following premiums.**

	Amount of Insurance	Annual Rate per $100	Annual Premium	Two-Year Premium	Three-Year Premium
1.	$37,000	$0.37	——	——	——
2.	29,000	0.25	——	——	——
3.	45,000	0.29	——	——	——
4.	63,000	0.19	——	——	——
5.	58,000	0.42	——	——	——
6.	49,500	0.33	——	——	——
7.	39,000	0.47	——	——	——
8.	32,500	0.38	——	——	——
9.	48,000	0.27	——	——	——
10.	42,000	0.31	——	——	——
11.	125,000	0.52	——	——	——
12.	58,000	0.41	——	——	——
13.	62,000	0.27	——	——	——
14.	89,000	0.38	——	——	——
15.	38,000	0.47	——	——	——
16.	75,000	0.42	——	——	——

Date

Name

Course/Section

B. Calculate the following short-term premiums. Use the short-rate table on page 421.

	Amount of Insurance	Annual Rate per $100	Length of Policy in Days	Premium
1.	$73,000	$0.33	30	_____
2.	26,000	0.47	180	_____
3.	32,000	0.38	45	_____
4.	45,000	0.27	90	_____
5.	37,000	0.31	75	_____
6.	52,000	0.42	105	_____
7.	19,000	0.29	114	_____
8.	39,000	0.19	237	_____
9.	47,500	0.37	142	_____
10.	56,000	0.25	183	_____
11.	82,000	0.39	90	_____
12.	75,000	0.41	125	_____

C. For the following cancelled policies, calculate the time the policy was in force, the amount of premium retained by the insurer, and the refund to the insured. Use the short-rate table on page 421 where necessary. (Assume it's not a leap year.)
Note that the policies are sometimes cancelled by the insurer and sometimes by the insured.

	Annual Premium	Cancelled by	Policy Starting Date	Policy Cancellation Date	Time	Amount Retained by Insurer	Amount Refunded
1.	$450	Insurer	3/8	6/15	_____	_____	_____
2.	375	Insured	9/5	11/15	_____	_____	_____
3.	250	Insurer	4/5	9/30	_____	_____	_____
4.	462	Insurer	8/3	10/1	_____	_____	_____
5.	547	Insured	3/25	9/5	_____	_____	_____
6.	212	Insured	2/16	8/17	_____	_____	_____
7.	198	Insurer	6/15	11/15	_____	_____	_____
8.	367	Insured	3/18	6/21	_____	_____	_____
9.	405	Insured	5/23	12/1	_____	_____	_____
10.	396	Insurer	5/7	11/1	_____	_____	_____
11.	582	Insurer	5/17	8/12	_____	_____	_____
12.	755	Insured	6/12	12/18	_____	_____	_____

D. For the following multiple insurers, calculate the amount of the loss paid by each insurance company.

	Face Value of Policy Company A	Face Value of Policy Company B	Face Value of Policy Company C	Amount of Loss	Amount Paid by Company A	Amount Paid by Company B	Amount Paid by Company C
1.	$15,000	$20,000	$25,000	$36,000	_____	_____	_____
2.	10,000	15,000	20,000	35,000	_____	_____	_____
3.	20,000	30,000	25,000	82,000	_____	_____	_____
4.	35,000	25,000	10,000	55,000	_____	_____	_____
5.	10,000	15,000	10,000	22,000	_____	_____	_____
6.	15,000	20,000	20,000	63,000	_____	_____	_____
7.	12,000	15,000	20,000	5,000	_____	_____	_____
8.	35,000	50,000	40,000	14,000	_____	_____	_____
9.	60,000	55,000	25,000	78,000	_____	_____	_____
10.	5,000	10,000	5,000	22,000	_____	_____	_____
11.	15,000	20,000	15,000	52,000	_____	_____	_____
12.	20,000	25,000	30,000	80,000	_____	_____	_____

E. For the following problems, assume a coinsurance clause of 80%. Calculate the amount of the loss paid by the insurer.

	Value of Property	Face Value of Policy	Amount of Loss	Amount Paid by Insurer
1.	$73,000	$50,000	$12,500	_____
2.	32,000	24,000	17,000	_____
3.	37,000	30,000	5,000	_____
4.	39,000	29,000	30,000	_____
5.	19,000	12,000	2,000	_____
6.	48,000	35,000	15,000	_____
7.	39,500	34,000	12,000	_____
8.	58,000	48,000	30,000	_____
9.	65,000	40,000	32,000	_____
10.	82,000	55,000	22,000	_____
11.	95,000	70,000	52,000	_____
12.	44,000	36,000	12,000	_____

When you have had the practice you need, either return to the Preview for this unit on page 414 or continue in frame 13 with the study of automobile insurance.

SECTION 2: AUTOMOBILE INSURANCE

<table>
<tr><td>13</td><td>

OBJECTIVE

Determine driver classifications and calculate insurance premiums for liability, medical payments, collision and comprehensive coverage. Determine what portion of damages are paid by the insurance company in the cases of bodily injury and auto damage.

</td></tr>
</table>

Most families in the United States own at least one automobile. Because of the large number of accidents, automobile insurance is practically a necessity. Most states require businesses and individuals to provide financial responsibility in case of an auto accident. One method of providing financial responsibility is to obtain insurance.

There are many factors used to determine auto insurance premiums, including territory or region of residence, age and sex of drivers, driving record, whether the car is used in business or driven to work, and age, cost, and size of the auto.

The first step in calculating the insurance premiums is to determine the driver classification. The following is an abbreviated list of classes.

DRIVER CLASSIFICATIONS

1A: No business use; no driver under 25; not driven to work.
1B: No business use; no driver under 25; less than 10 miles to work one way.
1C: No business use, no driver under 25; more than 10 miles to work one way.
2A: Male under 25 not owner or principal driver; or married male under 25.
2C: Male under 25 unmarried owner and/or principal driver.
2F: Female driver under 25.
3: Business use.

Find the driver classifications for the following individuals.

1. Allen Klein, age 32, drives his car to work 7 miles one way. He does not use his car in business.
2. Karen Bishop, age 21, drives her car to work 12 miles one way. She does not use her car in business.
3. Sam Biggs, age 52, uses his automobile in his business.
4. Mary Cervantes, age 45, uses her car for pleasure only.

Check your answers in **14**.

---

14   1. Class 1B: No business use, no driver under 25, and only 7 miles to work.
      2. Class 2F: A female driver under 25.
      3. Class 3:   Used in business.
      4. Class 1A: No business use, no driver under 25, and not driven to work.

LIABILITY INSURANCE

One of the most important segments of automobile insurance is *personal liability insurance*. Liability insurance covers medical and property damage to persons other than the insured. Although property damage is usually limited to the value of an automobile and its contents, it is very difficult to place a dollar value on an individual's injuries and resulting pain and suffering. Oftentimes the price of an injury is many times the property damage. Personal liability insurance is a virtual necessity!

Bodily injury and property damage are usually written together into an insurance specification, such as 10/20/5. All three amounts are thousands of dollars. The first two figures 10/20 denote the maximum amounts for bodily injury. The first number, 10, or $10,000, is the maximum amount payable by the insurer to any one individual sustaining injuries in an accident. The second amount, 20, or $20,000, is the maximum total amount payable to all individuals in a single accident for injuries. The third figure, 5 or $5000, is the maximum amount payable for property damage in an accident.

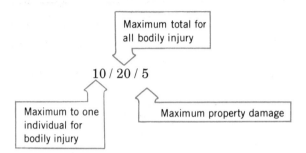

There are a variety of bodily injury amounts, usually including 5/10, 10/20, 15/30, 20/40, 50/100, 100/300, and in even larger amounts. Property damage usually includes amounts of $5000, $10,000, $25,000, $50,000 or larger.

Example: Jim Williams' auto ran into Lloyd Lusk's car, causing injuries to Mr. Lusk and his wife. Jim Williams carries 10/20/5 liability insurance. Mr. Lusk was awarded $8000 and his wife $12,500 for injuries. Damage to the Lusk's car was $4500. (a) What portion of the settlement is paid by Mr. Williams' insurance company? (b) What portion (if any) remains to be paid by Mr. Williams?

Since $10,000 is the maximum payment by the insurance company to an individual for injuries, Mrs. Lusk is paid $10,000. Mr. Lusk received $8000 for injuries. The insurer pays $4500 for the automobile.

Mr. Williams must pay the remaining $2500 for Mrs. Lusk's injury.

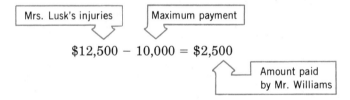

Work the following problem.

Myrtle Mesa's auto ran into Ruby Hudson's car, causing injuries to Mrs. Hudson and her husband. Myrtle Mesa carries 5/10/5 liability insurance. The court awarded Ruby Hudson $6000, and her husband $4500 for injuries. Damage to the Hudson's car was $5950. (a) What portion of the settlement is paid by Ms. Mesa's insurance company? (b) What portion must be paid by Ms. Mesa?

Check your answer in 15.

___

(a)   Paid by insurer:

Injury to Mrs. Hudson :	$5,000	($5000 maximum)
Injury to Mr. Hudson :	4,500	
Auto damage	5,000	($5000 maximum)
TOTAL	$14,500	

(b) Paid by Ms. Mesa:
Injury to Mrs. Hudson : $6000 - 5000 = $1000
Injury to Mr. Hudson :   0
Auto damage   : $5950 - 5000 =   950
                              TOTAL $1950

Calculating
Auto Insurance
Premiums

The moral of this problem is always carry adequate auto insurance.

The annual *premium* for liability coverage is easily calculated by using the correct driver classification and finding the amount of coverage in the "Automobile Liability Insurance" table. The premium tables vary from region to region.

Find the proper driver class in the left-hand column and read across the row to the appropriate bodily injury and property damage amounts.

AUTOMOBILE LIABILITY INSURANCE: ANNUAL PREMIUM

Driver Class	Bodily Injury			Property Damage	
	10/20	20/40	50/100	$5000	$10,000
1A	$68	88	112	51	56
1B	72	94	119	54	59
1C	77	101	127	59	65
2A	156	203	257	135	149
2C	182	237	301	148	163
2F	95	124	157	80	88
3	78	102	129	83	91

Example: Calculate the annual liability insurance premium for 10/20/10 for driver class 1C.

Bodily injury 10/20   : $77
Property damage $10,000 :   65
                      TOTAL $142   Annual premium

Calculate the liability insurance premium for the following:

1. 20/40/5; Driver class 2A   2. 50/100/10; Driver class 2F

Check your work in **16**.

---

16 | 1. Bodily injury   20/40: $203   2. Bodily injury   50/100: $157
   | Property damage $5,000:   135   Property damage $10,000:   88
   | TOTAL $338   TOTAL $245

MEDICAL PAYMENTS

Liability insurance covers injuries to others, *not* the insured. *Medical payments* pay for injuries to the insured. Driver classification may or may not affect the medical payments premium, depending on the state. The face value of insurance is the maximum amount paid by the insurance company for an accident.

AUTOMOBILE MEDICAL PAYMENTS

Amount	Annual Premium
$1,000	$16
2,000	18
5,000	23

Example: Calculate the annual automobile insurance premium for liability coverage 10/20/10 and $5000 medical payments for driver classification 3.

(Class 3)   Bodily injury 10/20 : $78
(Class 3)   Property damage $10,000 :   91
           Medical payments $5,000 :   23
                  TOTAL $192   Annual premium

Calculate the annual insurance premium for liability coverage 20/40/10 and medical payment $2000 for driver class 1B.

Check your work in **17**.

If I have three traffic tickets, will it increase my auto insurance?

Oops. Then I better not mention the time I backed into the garage wall.

You bet your backup lights it will. Once your basic premium is calculated, an additional amount will be added for your bad driving record. Get enough of those tickets and you can buy a bicycle.

17   Bodily injury 20/40 : $94
    Property damage $10,000 :   59
    Medical payments $2,000 :   18
             TOTAL $171   Annual premium

COLLISION AND COMPREHENSIVE

The property damage portion of the liability insurance covers damage to property belonging to others, *not* the insured. If you wish to insure your car and its contents, you must obtain *collision* and *comprehensive* coverage.

Collision covers damages to the insured car if the insured is at fault, or if the driver of the other car is at fault and unable to pay the damages. Most collision clauses carry a deductible provision. With a $100 *deductible* clause, the insured must pay the first $100 while the insurance company pays the rest of the damages. Collision coverage usually has $50, $100, $250 or higher deductibles.

Try the following problem.

Don Eagle was in an auto accident in which he was at fault. His collision insurance coverage has a $250 deductible clause. The damages to his car were $987. How much of the damages are paid by (a) Mr. Eagle and (b) the insurance company?

Turn to **18** to check your answer.

18   (a) **Mr. Eagle** must pay the deductible amount—$250. (b) The insurance company pays the damages less the deductible amount, $987 − 250 = $737.

Collision coverage depends on the driver classification, the auto's price and age, and the deductible.

The following is an abbreviated table. The auto's price group is determined by its original price. The auto's age group is determined by the age of the car.

To calculate the collision premium, first find the proper row using the auto's price and age groups. Next, move right to the proper driver class.

AUTOMOBILE COLLISION AND COMPREHENSIVE $100 DEDUCTIBLE

Price Group	Age Group	Collision Class							Compre-hensive
		1A	1B	1C	2A	2C	2F	3	
D	1	$109	$115	$119	$156	$212	$140	$132	$45
	2	92	98	103	148	201	132	125	39
	3	78	85	91	135	185	119	117	32
E	1	120	127	131	185	255	172	163	51
	2	103	109	104	168	231	161	155	45
	3	85	92	96	153	212	149	142	40
F	1	137	144	150	198	280	195	189	59
	2	121	128	133	182	263	180	176	52
	3	102	110	114	168	245	163	160	44

Example: Calculate the premium for a $100 deductible collision clause with price group E, age group 2, and driver class 1C.

First, find the row for price group E, age group 2. Read across to class 1C. The premium is $104.

It's easy; just look it up in the table.

Work the following problem.

Calculate the premium for a $100 deductible collision clause with price group F, age group 3, and driver class 2A.

Check your answer in 19.

---

19   The collision premium is $168.

Included in the collision table is a column for comprehensive coverage. Comprehensive pays for damages caused by fire, theft, vandalism, and similar events. Comprehensive, like collision coverage, usually has a deductible amount. Only the auto's price and age groups are used to determine the premium.

To calculate the comprehensive premium, use the automobile collision and comprehensive table on page 437. The comprehensive is given in the rightmost column.

Example: Calculate the premium for $100 deductible collision and comprehensive coverage with price group D, age group 1, and driver class 2F.

First, find the row for price group D, age group 1. Read across the row to class 2F. The collision is $140.

Next, read across the same row to the rightmost column. The comprehensive is $45.

$$
\begin{array}{rr}
\text{Collision:} & \$140 \\
\text{Comprehensive:} & \underline{45} \\
& \$185
\end{array}
$$

Your turn.

Calculate the premium for $100 deductible collision and comprehensive with price group E, age group 3 and driver classification 3.

Check your work in **20**.

20

$$
\begin{array}{rr}
\text{Collision:} & \$142 \\
\text{Comprehensive:} & \underline{40} \\
& \$182
\end{array}
$$

In addition to liability, medical payments, collision and comprehensive, you can obtain insurance to include: uninsured motorists, towing, road service, and disability and death benefits.

Are you ready to put it all together? Good, calculate the premium in the following problem.

What is the total annual premium for an automobile policy with 20/40/10 liability, $5000 medical payments, and $100 deductible collision and comprehensive for an auto with price group F, age group 1, and driver classification 1B?

Turn to **21** to check your work.

21

$$
\begin{array}{rll}
\text{Bodily injury 20/40:} & \$94 & \\
\text{Property damage \$10,000:} & 59 & \\
\text{Medical payments \$5,000:} & 23 & \\
\text{Collision:} & 144 & \\
\text{Comprehensive:} & \underline{59} & \\
\text{TOTAL} & \$379 & \text{Annual premium}
\end{array}
$$

It's easy if you work it carefully, step by step.

Need some more practice? Work another one.

Debbie Britton, age 39, drives her car to work, 17 miles one way. Calculate her automobile insurance premium for 10/20/5 liability, $2000 medical payments, and $100 deductible collision and comprehensive. Her car is in price group F, age group 1.

Check your work in **22**.

---

22 Driver class 1C.

$$
\begin{array}{rll}
\text{Bodily injury 10/20:} & \$77 & \\
\text{Property damage \$5000:} & 59 & \\
\text{Medical payments \$2000:} & 18 & \\
\text{Collision:} & 150 & \\
\text{Comprehensive:} & \underline{59} & \\
\text{TOTAL} & \$363 & \text{Annual premium}
\end{array}
$$

For a set of practice problems on automobile insurance turn to **23**.

Insurance

The answers are on page 599.

A. Calculate the following automobile premiums. Use the tables in the text.

1. Calculate the premium for an automobile policy with 10/20/10 liability, $2000 medical payments, and $100 deductible collision and comprehensive for an auto with price group E, age group 3, and driver classification 3.

2. Calculate the premium for an automobile policy with 50/100/10 liability, $5000 medical payments, and $100 deductible collision and comprehensive for an automobile with price group D, age group 2, and driver classification 1B.

3. Calculate the premium for an automobile policy with 20/40/10 liability, $2000 medical payments, and $100 deductible collision and comprehensive for an auto with price group F, age group 2, and driver classification 2F.

4. Calculate the premium for an automobile policy with 10/20/5 liability, $1000 medical payments, and $100 deductible collision and comprehensive for an auto with price group F, age group 1, and driver classification 2C.

5. Nancy Palmer, age 21, drives her car to work 9 miles one way. Calculate her automobile insurance premium for 20/40/10 liability, $5000 medical payments, and $100 deductible collision and comprehensive. Her car is in price group F, age group 2.

6. Robert Killian, age 23, drives his car to work, 19 miles one way. Calculate his automobile insurance premium for 20/40/5 liability, $2000 medical payments, and $100 deductible collision and comprehensive. His car is price group D, age group 1.

7. Mike Dominguez, age 45, drives his car to work 12 miles one way. Calculate his automobile insurance premium for 50/100/10 liability, $5000 medical payments, and $100 deductible collision and comprehensive. His car is in price group F, age group 1.

8. Jane Goben, age 39, uses her car in her business. Calculate her automobile insurance premium for 20/40/5 liability, $2000 medical payments, and $100 deductible collision and comprehensive. Her car is in price group F, age group 2.

Date

Name

Course/Section

9. George Sanchez, age 45, drives his car to work 15 miles one way. Calculate his automobile insurance premium for 20/40/10 liability, $5000 medical payments, and $100 deductible collision and comprehensive. His car is price group D, age group 2.

_____

10. Valeria Garrett, age 32, drives her car to work 8 miles one way. Calculate her automobile insurance premium for 50/100/10 liability, $5000 medical payments, and $100 deductible collision and comprehensive. Her car is price group F, age group 3.

_____

11. Michael Ball, age 24, owns his own car. Calculate the premium for 10/20/5 liability, $2000 medical payments, and $100 deductible collision and comprehensive. His car is price group E, age group 1.

_____

12. William French, age 56, drives his car to work 12 miles one way. Calculate his automobile insurance premium for 20/40/10 liability, $5000 medical payments, and $100 deductible collision and comprehensive. His car is price group D, age group 1.

_____

13. Scott Brown uses his car in his business. Calculate his automobile insurance premium for 10/20/5 liability, $1000 medical payments, and $100 deductible collision and comprehensive. His car is price group F, age group 1.

_____

14. Wanda Brayton, age 24, drives her car 5 miles to work one way. Calculate her automobile insurance premium for 20/40/10 liability, $2000 medical payments, and $100 deductible collision and comprehensive. Her car is price group F, age group 3.

_____

15. Ali Hurson, age 35, drives his car to work 18 miles one way. Calculate his automobile insurance premium for 20/40/10 liability, $5000 medical payments, and $100 deductible collision and comprehensive. His car is price group D, age group 2.

_____

B. **Work the following problems.**

1. Raymond Walker's auto ran into Ruth Sims' car, causing injuries to Ms. Sims. Raymond Walker carries 15/30/5 liability insurance. Ms. Sims was awarded $14,000 for injuries. Damage to Ms. Sims' car was $5400. (a) What portion of the settlement is paid by Mr. Walker's insurance company? (b) What portion (if any) remains to be paid by Mr. Walker?

(a) _____

(b) _____

2. Cynthia Bronson's auto ran into George Proctor's car, causing injuries to Mr. Proctor and his daughter. Cynthia Bronson carries 10/20/5 liability insurance. Mr. Proctor was awarded $9000 and his daughter $12,000 for injuries. Damage to the Proctors' car was $4200. (a) What portion of the settlement is paid by Ms. Bronson's insurance company? (b) What portion (if any) remains to be paid by Ms. Bronson?

   (a) _____

   (b) _____

3. Walter McLennon's auto ran into Ginger Holt's car, causing injuries to her and her husband. Walter McLennon carries 50/100/25 liability insurance. Mrs. Holt was awarded $25,000 and her husband $21,000 for injuries. Damage to the Holts' car was $6500. (a) What portion of the settlement is paid by Mr. McLennon's insurance company? (b) What portion (if any) remains to be paid by Mr. McLennon?

   (a) _____

   (b) _____

4. Myrtle Newby's auto ran into Felix Boddillo's car, causing injuries to him, his wife, and son. Myrtle carries 15/30/10 liability insurance. Mr. Boddillo was awarded $14,000, his wife $17,500, and his son $11,500 for injuries. Damage to the Boddillo's car was $4750. (a) What portion of the settlement is paid by Ms. Newby's insurance company? (b) What portion (if any) remains to be paid by Ms. Newby?

   (a) _____

   (b) _____

5. Ralph Story was in an accident in which he was at fault. His collision insurance coverage has a $250 deductible clause. The damages to his car were $1578. How much of the damages are paid by (a) Mr. Story and (b) the insurance company?

   (a) _____

   (b) _____

6. Cindy Prier was in an automobile accident in which she was at fault. Her automobile collision coverage has a $50 deductible clause. The damages to her car amounted to $2843. How much of the damages are paid by (a) Ms. Prier and (b) Ms. Prier's insurance company?

   (a) _____

   (b) _____

7. Ron Coughman's auto ran into Dave French's car, causing injuries to Mr. French, his wife, and his two daughters. Mr. Coughman carries 10/20/5 liability insurance. Mr. French was awarded $22,000, his wife was awarded $15,000, and each daughter was awarded $9,000 for injuries. Damage to Mr. French's car was $6500. (a) What portion of the settlement is paid by Mr. Coughman's insurance company? (b) What portion (if any) remains to be paid by Mr. Coughman?

(a) _____

(b) _____

8. Richard Thomas was in an automobile accident in which he was at fault. His automobile collision coverage has a $250 deductible clause. The damage to his car amounted to $4127. How much of the damages are paid by (a) Mr. Thomas and (b) Mr. Thomas's insurance company?

(a) _____

(b) _____

When you have had the practice you need, either return to the Preview on page 414 or continue in frame 24 with the study of life insurance.

SECTION 3: LIFE INSURANCE

<table>
<tr><td>24</td><td>

OBJECTIVE

Calculate premiums and surrender options for life insurance policies.

</td></tr>
</table>

The primary purpose of life insurance is to continue family support after the death of a family income earner. The insurance policy has a face value that is usually in multiples of $1000. The face value of the policy is paid to the designated beneficiary upon the death of the insured. (Other options are available.)

There are four basic types of life insurance: term, ordinary (straight), limited-payment, and endowment policies. *Term* insurance is simply life insurance for a certain term or period of time, usually one to ten years. If the insured dies during the term, the beneficiary receives the face value of the policy. Term insurance provides only death benefits; no other benefits are usually provided. The rates are relatively inexpensive for younger individuals. The rate increases with age and is very expensive for older people. Term insurance is very popular with employers who provide life insurance as a fringe benefit to employees.

In addition to death benefits, ordinary, limited-payment, and endowment policies provide other additional benefits. These policies build a *cash value* (or *cash surrender value*) that increases with the length of the policy. If, at some future date, the policy is discontinued, the insured may "cash in" the policy and receive the current cash value. This benefit may provide retirement income, or for a dependent's education, etc. The cash value may also be used as collateral for low interest loans.

Ordinary life insurance provides both death benefits and builds a cash value. The premium is determined by the insured's age when initiating the policy. The premium remains the same and is paid each year until the insured dies.

Limited-payment life insurance provides death benefits and builds a cash value. The premium remains the same but is paid up after a limited number of years. The most common time period is 20 years, although other periods are available. Since the premiums are paid for only a limited time period, the rates are usually higher than ordinary life insurance.

An *endowment* policy is a form of limited-payment life insurance, but the premiums are generally higher than the other types of insurance. An endowment policy is designed to build a large cash value after a certain number of years.

For a discussion of calculating premiums, turn to 25.

Calculating Life
Insurance Premiums

25

Life insurance premiums depend on the age of the insured. The rate is usually stated per $1,000 of insurance. The following is a sample premium schedule for males. Because of their longer life expectancy, subtract 3 years from the ages of females and use the premium table for males.

ANNUAL LIFE INSURANCE PREMIUMS FOR MALES* PER $1,000

Age	10-Year Term	Ordinary Life	20-Payment Life	20-Year Endowment
18	$ 4.94	$13.07	$21.53	$41.17
20	5.02	13.72	22.60	41.23
22	5.10	14.42	23.75	41.36
24	5.16	15.18	24.94	41.57
26	5.21	16.09	25.91	41.72
28	5.29	17.02	27.36	41.89
30	5.42	17.93	29.17	42.06
35	5.98	21.52	32.28	43.23
40	7.72	25.61	36.42	44.42
45	11.26	30.46	41.50	46.95
50	17.02	37.48	48.92	50.18
55	26.12	49.47	57.43	57.42

* Subtract 3 years for female.

To calculate the annual premium, first find the age in the left-hand column. Read across the row to the desired life insurance type. Finally, multiply the rate in the table by the number of $1000's of insurance.

The following example will clarify the procedure.

Example: Calculate the premium for a $5000 ordinary life insurance policy started at age 24 (male).

From the table, the rate is $15.18 per $1000.

```
5 × $1000 policy        Rate per $1000
```

Premium = 5 × $15.18 = $75.90

```
        Annual premium
```

Your turn. Work the following problems.

1. Calculate the annual premium for a $15,000 20-payment life insurance policy for Calvin Pryor, age 30.
2. What is the premium for a $45,000 10-year term insurance for Keri Lawrence, age 29? (**Hint.** Remember to subtract 3 years from the age of a female.)

Turn to **26** to check your work.

Is the cheapest ordinary life insurance policy the best?

No, not necessarily. You must also consider the cash value of the policy. Remember an ordinary policy is a combination of life insurance and savings.

26
1. Premium = 15 × $29.17 = $437.55
2. Use age = 29 − 3 = 26
   Premium = 45 × $5.21 = $234.45

SURRENDER OPTIONS

If the insured discontinues an ordinary, limited-payment or endowment life insurance policy, he has several *surrender* or *nonforfeiture options*. Most policies have three options: cash surrender value, paid-up insurance, and extended term insurance. When the insured discontinues a policy, only one option can be chosen.

The *cash surrender value* is the amount of money the insured receives from the insurance company for "cashing in" or discontinuing his policy. The cash surrender value is usually stated per $1000 of insurance.

Paid-up insurance is a reduced amount of insurance on which no additional payments are due. The insurance is paid-up for the life of the insured; there are no more premiums. Paid-up insurance is usually stated per $1000 of insurance.

The third option is *extended term insurance*. This option provides insurance for the face value of the original policy for a given term or time period.

Each life insurance policy will contain a table of surrender options for the age of the insured when the policy was started. The amount of each option depends on the length of time the policy was in force. The following is an abbreviated table for a policy issued at age 20 for a male or age 23 for a female.

NONFORFEITURE OPTIONS FOR ORDINARY LIFE POLICY PER $1,000

| End of Year | Age 20 Male or Age 23 Female | | Extended Term Insurance | |
	Cash Surrender Value	Paid-up Insurance	Years	Days
1	$ 0	$ 0	0	0
2	0	0	0	0
3	8	19	1	50
4	15	35	2	232
5	22	49	3	159
10	82	198	12	341
15	129	291	17	118
20	201	412	21	47

Example: Calculate the three surrender options for a $15,000 ordinary life insurance policy issued at age 23 for a female and in force 10 years.

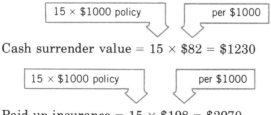

Cash surrender value = 15 × $82 = $1230

Paid-up insurance = 15 × $198 = $2970

Extended term insurance: 12 years, 341 days

Work the following problem.

John Strickland is discontinuing his $12,000 ordinary life insurance policy issued at age 20. The policy has been in force 5 years. Calculate the (a) cash surrender value, (b) paid-up insurance, and (c) extended term insurance.

Turn to **27** to check your work.

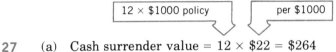

27 (a) Cash surrender value = 12 × $22 = $264

(b) Paid-up insurance = 12 × $49 = $588

(c) Extended term insurance: 3 years, 159 days

SETTLEMENT OPTIONS

The beneficiary of a life insurance policy has several options for receiving the proceeds when the insured dies. One option is to receive the face value of the policy in a one lump sum payment. Two additional options are usually provided.

The second option is monthly (or annual, semi-annual, or quarterly) payments to the beneficiary for a fixed number of years. The amount of each payment depends on the number of years selected by the beneficiary.

The third option is monthly payments for the life of the beneficiary. This option also contains a minimum guaranteed period of payments. The period is usually 10, 15, or 20 years. If the beneficiary dies during the guaranteed period, the beneficiary's heirs receive the payment for the remainder of the period. The amount of each payment depends on the age of the beneficiary and the length of the guaranteed period selected.

For a set of practice problems on life insurance, turn to **28**.

Insurance

The answers are on page 600.

A.   Calculate the annual premiums for the following life insurance policies. Use the table on page 444.

	Type of Insurance	Face Value	Age	Sex	Premium
1.	Ordinary life	$10,000	26	M	_____
2.	10-year term	15,000	21	F	_____
3.	20-payment life	5,000	22	M	_____
4.	Ordinary life	20,000	45	M	_____
5.	20-year endowment	30,000	27	F	_____
6.	20-payment life	25,000	33	F	_____
7.	Ordinary life	20,000	35	M	_____
8.	20-year endowment	15,000	38	F	_____
9.	10-year term	50,000	28	M	_____
10.	20-payment life	10,000	43	F	_____
11.	20-year endowment	25,000	29	F	_____
12.	10-year term	80,000	30	M	_____
13.	Ordinary life	18,000	35	M	_____
14.	Ordinary life	25,000	43	F	_____
15.	10-year term	40,000	26	M	_____
16.	10-year term	55,000	25	F	_____
17.	20-payment life	25,000	23	F	_____
18.	20-payment life	30,000	30	M	_____
19.	20-year endowment	45,000	40	M	_____
20.	20-year endowment	28,000	38	F	_____

Date _____

Name _____

Course/Section _____

B. Calculate the three nonforfeiture options for the following ordinary life insurance policies. Use the table on page 445.

	Starting Age	Sex	Years Policy in Force	Face Value	Cash Surrender Value	Paid-up Insurance	Extended Term Insurance
1.	23	F	5	$ 5,000	——	——	——
2.	20	M	10	10,000	——	——	——
3.	20	M	2	30,000	——	——	——
4.	23	F	15	20,000	——	——	——
5.	20	M	4	35,000	——	——	——
6.	23	F	3	25,000	——	——	——
7.	23	F	10	40,000	——	——	——
8.	20	M	3	20,000	——	——	——
9.	20	M	15	50,000	——	——	——
10.	23	F	20	15,000	——	——	——
11.	20	M	2	12,000	——	——	——
12.	20	M	4	25,000	——	——	——
13.	23	F	10	15,000	——	——	——
14.	20	M	5	8,000	——	——	——
15.	23	F	20	50,000	——	——	——
16.	23	F	3	35,000	——	——	——
17.	20	M	20	25,000	——	——	——
18.	23	F	15	35,000	——	——	——
19.	23	F	4	18,000	——	——	——
20.	20	M	10	14,000	——	——	——

When you have had the practice you need, turn to 29 for the Self-Test over this unit.

Insurance

29   Self-Test

The answers are on page 600.

Use the tables in Unit 11 when needed.

1. A building is insured for $56,000. The annual fire insurance rate is $0.47 per $100. Calculate the (a) annual premium, (b) two-year premium, and (c) three-year premium.

    (a) _____

    (b) _____

    (c) _____

2. A building is insured for $45,000. The annual rate is $0.38 per $100. Calculate the premium on a 120-day policy.

    _____

3. Mr. Shapiro insured his house for one year starting May 12 at a cost of $452. The policy was cancelled by the insurance company on October 15. Calculate the amount of the premium (a) retained by the insurer and (b) the refund to Mr. Shapiro.

    (a) _____

    (b) _____

4. Ms. Graves insured her house for one year starting June 27 at a cost of $527. She cancelled her policy on November 1. Calculate the amount of the premium (a) retained by the insurance company and (b) the refund to Ms. Graves.

    (a) _____

    (b) _____

5. Raymond's Rental Center carries three fire insurance policies: Company A, $15,000; Company B, $25,000; and Company C, $20,000. The building sustained a $30,000 fire loss. Calculate the amount paid by each company.

    _____

    _____

    _____

6. The Dorn Corporation carries two fire insurance policies: Company A $10,000 and Company B $15,000. The building sustained a $30,000 fire loss. Calculate the amount paid by each company.

    _____

    _____

Date

Name

Course/Section

7. The Rich Company carries a $40,000 fire insurance policy with an 80% coinsurance clause. The value of the property is $55,000. The company had a $20,000 fire loss. Calculate the amount paid by the insurer.

_____

8. Calculate the premium for an automobile policy with 20/40/10 liability, $5000 medical payments, and $100 deductible collision and comprehensive for an auto with price group E, age group 2, and driver classification 1B.

_____

9. Jeff Mann, age 20, drives his car to work, 12 miles one way. Calculate his automobile insurance premium for 10/20/5 liability, $1000 medical payments, and $100 deductible collision and comprehensive. His car is price group F, age group 3.

_____

10. Mark Rankin's auto ran into Laura Holloway's car causing injuries to Mrs. Holloway and her husband. Mr. Rankin carries 15/30/5 liability insurance. Mrs. Holloway was awarded $10,000 and her husband was awarded $16,500 for injuries. Damage to the Holloway's car was $5825. (a) What portion of the settlement is paid by Mr. Rankin's insurance company? (b) What portion (if any) remains to be paid by Mr. Rankin?

(a) _____

(b) _____

11. Bob Dalton was in an auto accident in which he was at fault. His collision insurance has a $250 deductible clause. The damages to his car were $1814. How much of the damages are paid by (a) Mr. Dalton and (b) the insurance company?

(a) _____

(b) _____

12. Calculate the premium for a $25,000 10-year term life insurance policy for Stan Koonce, age 35.

_____

13. Calculate the premium for a $15,000 ordinary life insurance policy for Wendy Wright, age 33.

_____

14. Carolyn Harp is discontinuing her $8000 ordinary life insurance policy issued at age 23. The policy has been in force 15 years. Calculate the (a) cash surrender value, (b) paid-up insurance, and (c) extended term insurance.

(a) _____

(b) _____

(c) _____

12 Depreciation

WHAT DO YOU THINK OF THIS UNIT?

I DEPRECIATE IT VERY MUCH!

Obviously, our friend doesn't understand exactly what "depreciation" is. Depreciation is the loss of value of an asset. Anyone who owns a car is all too familiar with this concept. The car is an asset, and as it is used and grows older its value decreases. The loss of value is approximated for business record keeping purposes with a depreciation schedule.

In this unit you will learn that several methods of calculating depreciation are used by businesses. If an asset loses $6000 in value over a three-year period, the depreciation is $6000. Using the *straight-line* method of depreciation, one-third of this amount would be recorded each year. Two other methods commonly used are the *declining-balance* and *sum-of-the-years' digits* methods. With each of these there is a larger depreciation in the first year and smaller depreciations in each following year. All of these methods are used for tax reporting.

In this unit you will also learn about a fourth method, the *units-of-production* method, in which the amount of depreciation is related directly to the use or production of the output during the year and the new ACRS method.

Ready? Pencil in hand? Good! Let's get ready to understand and "appreciate" depreciation . . . especially when it involves *your* assets.

UNIT OBJECTIVE

After you successfully complete this unit, you will be able to construct a depreciation schedule using the straight-line, units-of-production, declining-balance, sum-of-the-years' digits, and ACRS methods.

PREVIEW 12   This is a sample test for Unit 12. All of the answers are placed immediately after the test. Work as many of the problems as you can, then check your answers and look after the answers for further directions.

1. Carr's Athletic Supply Company purchased a new delivery truck for $6500. The total estimated mileage is 100,000 miles with a salvage value of $500. Calculate the depreciation by the units-of-production method for a year in which the truck was driven 32,157 miles.

2. The Anderson Vitamin Supply Company purchased a new delivery truck for $7800. It has a useful life of five years with a salvage value of $600. Construct a depreciation schedule using:
   (a) Straight-line method

   (b) Declining-balance method

   (c) Sum-of-the-years' digits method

3. The C. M. Corporation purchased a new machine on October 25, 1980 for $650. It has an estimated useful life of three years with a salvage value of $110. Construct a depreciation schedule using the sum-of-the years' digits method.

4. Construct a depreciation schedule using the ACRS method for a new light truck with a cost of $17,500.

Preview 12
  Answers

1. $1929.42

2. (a)

Year	Annual Depreciation	Accumulated Depreciation	Book Value
—	—	—	$7800
1	$1440	$1440	6360
2	1440	2880	4920
3	1440	4320	3480
4	1440	5760	2040
5	1440	7200	600

(b)

Year	Annual Depreciation	Accumulated Depreciation	Book Value
—	—	—	$7800.00
1	$3120.00	$3120.00	4680.00
2	1872.00	4992.00	2808.00
3	1123.20	6115.20	1684.80
4	673.92	6789.12	1010.88
5	404.35	7193.47	606.53

(c)

Year	Annual Depreciation	Accumulated Depreciation	Book Value
—	—	—	$7800
1	$2400	$2400	5400
2	1920	4320	3480
3	1440	5760	2040
4	960	6720	1080
5	480	7200	600

3.

Year	Annual Depreciation	Accumulated Depreciation	Book Value
—	—	—	$650
1980	$ 45	$ 45	605
1981	255	300	350
1982	165	465	185
1983	75	540	110

4.

Year	Depreciation	Accumulated Depreciation	Book Value
—	—	—	$17,500
1	$4375	$ 4,375	13,125
2	6650	11,025	6,475
3	6475	17,500	0

★ If you cannot work any of these problems, start with frame 1 on page 455.
★ If you missed some of the problems, turn to page numbers indicated.
★ If all of your answers were correct, you are probably ready to proceed to Unit 13. (If you would like more practice on Unit 12 before turning to Unit 13, try the Self-Test on page 499.)
★ Super-students—those who want to be certain they learn all of this material—will turn to frame 1 and begin work there.

SECTION 1: STRAIGHT-LINE AND UNITS OF PRODUCTION METHODS

1

> **OBJECTIVES**
>
> Given the cost, useful life, and salvage value, construct a depreciation schedule using the straight-line method.
>
> Given the cost, salvage value, estimated units of production, and number of units produced per year, construct a depreciation schedule using the units-of-production method.
>
> The depreciation schedules will include the annual depreciation, accumulated depreciation, and book value.

The cost of machinery, equipment, buildings, trucks, office furniture, and other business property with a useful life of more than one year is a capital expenditure and may not be deducted as an expense in the year of purchase. For example, when a business purchases a vehicle for $7000, it expects to get several years' use from the vehicle. Because of the time factor in the use of the vehicle, the business may not deduct the full $7000 as a business expense during the first year. The total loss of value of the vehicle must be spread out over the years the vehicle is to be used.

Since assets are used over a period of several years, a portion of the cost, called the *depreciation,* may be deducted each year. The amount deducted each year is determined by the depreciation method used.

All assets have three characteristics: an initial cost, a useful life, and a salvage value. The *initial cost* is the purchase price of the assets. The *useful life* is the estimated length of time the asset is to be used. The *salvage value* of an asset is its estimated value at the end of its useful life. The salvage value may be zero.

Both the useful life and salvage value are only *estimates,* made by the business based on such factors as usage, repair policy, replacement policy, and obsolescence. For income tax purposes, the Internal Revenue Service (IRS) has published guidelines to aid business in making these estimations.

The *total depreciation* is the total loss of value of an asset due to depreciation. The total depreciation is the difference between the initial cost and the salvage value.

> Total depreciation = initial cost − salvage value

Example 1:   A new deluxe steam cleaner for Storm's Steam Cleaning cost $4500. It has an estimated useful life of five years, and at the end of that time can be sold for about $360. What is the total depreciation?

Total depreciation = cost − salvage value
          = $4500 − $360 = $4140

The steam cleaner's total loss of value due to depreciation is $4140.

Your turn.

French's Valve Company purchased a new delivery truck for $7946. It has a useful life of five years with a salvage value of $150. Calculate the truck's total depreciation.

Check your work in 2.

2   Total depreciation = cost − salvage value
$$= \$7946 - 150 = \$7796$$

The truck's total loss of value is $7796. During the five years of the truck's useful life, it will depreciate $7796.

In calculating the company's income tax, how much depreciation will be deducted for each of the five years? The answer to this question depends on the depreciation method used.

STRAIGHT-LINE
METHOD

The straight-line method is the easiest method of depreciation. The straight-line method spreads the depreciation equally throughout the useful life of the asset. Hence, the depreciation is the same for each full year.

$$\text{Annual depreciation} = \frac{\text{total depreciation}}{\text{useful life}}$$

Example 2:   Using the straight-line method, what is the annual depreciation of the steam cleaner in Example 1? The original cost was $4500, useful life 5 years, and salvage value $360.

Total depreciation = cost − salvage value
$$= \$4500 - 360 = \$4140$$

$$\text{Annual depreciation} = \frac{\$4140}{5} = \$828$$

The depreciation is $828 per year for five years for a total of 5 × $828 = $4140

Now, you try one.

French's Valve Company purchased a new delivery truck for $7946. It has a useful life of five years, with a salvage value of $150.

As business manager of the company, you calculated the total depreciation at the end of frame **1**. Now, determine the annual depreciation using the straight-line method.

Check your answer in **3**.

---

3   Total depreciation = $7946 − 150 = $7796

$$\text{Annual depreciation} = \frac{\text{total depreciation}}{\text{useful life}}$$

$$= \frac{\$7796}{5} = \$1559.20$$

THE DEPRECIATION
SCHEDULE

Depreciation information is usually summarized in a depreciation schedule. The *depreciation schedule* contains the year, annual depreciation, accumulated depreciation, and the book value. The *accumulated depreciation* is the total depreciation to date. The book value is the current value of an asset. That is, the book value is the accounting value, but it is not necessarily the current market value. Don't confuse the book value and market value. The market value is the amount it can be sold for. The book value is simply the difference between the initial cost and the accumulated depreciation.

Book value = cost − accumulated depreciation

Example 3: Construct a depreciation schedule using the straight-line method for the steam cleaner of Examples 1 and 2.

Total depreciation = cost − salvage value
= \$4500 − \$360 = \$4140

$$\text{Annual depreciation} = \frac{\text{total depreciation}}{\text{useful life}}$$

$$= \frac{\$4140}{5} = \$828$$

Year	Annual Depreciation	Accumulated Depreciation	Book Value
0	——	——	\$4500
1	\$828	\$828	4500 − 828 = 3672
2	828	828 + 828 = 1656	4500 − 1656 = 2844
3	828	1656 + 828 = 2484	4500 − 2484 = 2016
4	828	2484 + 828 = 3312	4500 − 3312 = 1188
5	828	3312 + 828 = 4140	4500 − 4140 = 360 Salvage value

Note that with the straight-line method, the book value at the end of the useful life is *always* equal to the salvage value.

Test your understanding with this problem.

The Pekara Manufacturing Company purchased a machine for \$1800. It has an estimated useful life of six years with a salvage value of \$288. Construct a depreciation schedule using the straight-line method.

Turn to 4 to check your work.

Is the "salvage value" the same as the junk value?

Not necessarily. If the asset is in good condition when it is sold, the salvage value may be quite high. Many companies replace cars and trucks while they are still in very good operating condition.

4 Total depreciation = cost − salvage value

$$= \$1800 - 288 = \$1512$$

Annual depreciation $= \dfrac{\text{total depreciation}}{\text{useful life}}$

$$= \dfrac{\$1512}{6} = \$252$$

Year	Annual Depreciation	Accumulated Depreciation	Book Value
0	——	——	$1800
1	$252	$252	1800 − 252 = 1548
2	252	504	1800 − 504 = 1296
3	252	756	1800 − 756 = 1044
4	252	1008	1800 − 1008 = 792
5	252	1260	1800 − 1260 = 540
6	252	1512	1800 − 1512 = 288 Salvage value

Remember, the book value at the end of the useful life must be equal to the salvage value with the straight-line method. Always verify this; it's a good way to help check the problem.

UNITS-OF-PRODUCTION METHOD

The units-of-production method is often used with assets that are only occasionally used, such as a machine utilized in seasonal production. With the units-of-production method, rather than the useful life, we must estimate the total number of units to be produced by the asset. This is called the *estimated units of production.*

First, calculate the *depreciation per unit of production* that is the total depreciation divided by the estimated units of production.

Depreciation per unit of production $= \dfrac{\text{total depreciation}}{\text{estimated units of production}}$

The depreciation per year is the product of the number of units produced during the year times the depreciation per unit of production.

Annual depreciation = number of units × depreciation per unit
per year   of production

An example will help clarify this technique.

Example 4: A machine that originally cost \$1250 has an estimated production of 5500 units, and a salvage value of \$150. Calculate the depreciation per unit of production and the annual depreciation for a year in which 1573 units were produced.

Total depreciation = cost − salvage value
$$= \$1250 - 150 = \$1100$$

$$\text{Depreciation per unit of production} = \frac{\text{total depreciation}}{\text{estimated units of production}}$$

$$= \frac{\$1100}{5500} = \$0.20$$

Annual depreciation = number of units × depreciation per unit
$$= 1573 \times \$0.20 = \$314.60$$

During the year in which 1573 units were produced, we would claim a depreciation of \$314.60.

Try the following problem.

A cam construction machine for the Randolf Manufacturing Company cost \$4500. The estimated number of cams this machine is expected to produce in its lifetime is 8280. The salvage value of the machine is \$360. Calculate the depreciation per cam produced, and the annual depreciation for a year in which 1736 cams were produced.

Our solution is in **5**.

---

5   Total depreciation = cost − salvage value
$$= \$4500 - 360 = \$4140$$

$$\text{Depreciation per cam} = \frac{\text{total depreciation}}{\text{estimated cams of production}}$$

$$= \frac{\$4140}{8280} = \$0.50$$

Annual depreciation = number of cams × depreciation per cam
$$= 1736 \times \$0.50 = \$868$$

We can easily incorporate the units-of-production method in a depreciation schedule by including an additional column for the number of units produced each year.

Example 5: Lou's Auto Parts Company purchased a used delivery van for \$5900. On the basis of past experience, the total estimated mileage is 60,000 with a salvage value of \$500. The actual mileage driven for the years 1980 to 1983 is:

Year	Miles
1980	15,473
1981	19,857
1982	16,152
1983	8,518

Construct a depreciation schedule using the units-of-production method.

Total depreciation = cost − salvage value
$$= \$5900 - 500 = \$5400$$

$$\text{Depreciation per mile} = \frac{\text{total depreciation}}{\text{total miles}}$$

$$= \frac{\$5400}{60,000} = \$0.09$$

Year	Miles	Annual Depreciation	Accumulated Depreciation	Book Value
				$5900.00
1980	15,473	15,473 × $0.09 = $1392.57	$1392.57	4507.43
1981	19,857	19,857 × $0.09 = 1787.13	3179.70	2720.30
1982	16,152	16,152 × $0.09 = 1453.68	4633.38	1266.62
1983	8518	8518 × $0.09 = 766.62	5400.00	500.00

The book value at the end of the van's useful life, $500, is equal to the salvage value.

Now, it's your turn. Carefully work the following problem.

The Martin Factory purchased a widget-producing machine for $1800. The total estimated number of widgets to be produced is 4200, and the machine has a salvage value of $288. The actual production for the years 1979 to 1984 is as follows:

Year	Widgets Produced
1979	400
1980	550
1981	700
1982	800
1983	925
1984	825

Construct a depreciation schedule using the units-of-production method.

Check your work in 6.

---

6   Total depreciation = cost − salvage value
$$= \$1800 - 288 = \$1512$$

$$\text{Depreciation per widget} = \frac{\text{total depreciation}}{\text{total widgets}}$$

$$= \frac{\$1512}{4200} = \$0.36$$

Year	Units Produced	Annual Depreciation	Accumulated Depreciation	Book Value
—	—	—————	—	$1800
1979	400	$400 \times 0.36 = 144$	$144	1656
1980	550	$550 \times 0.36 = 198$	342	1458
1981	700	$700 \times 0.36 = 252$	594	1206
1982	800	$800 \times 0.36 = 288$	882	918
1983	925	$925 \times 0.36 = 333$	1215	585
1984	825	$825 \times 0.36 = 297$	1512	288

Remember the book value at the end of the useful life is equal to the salvage value.

Now, turn to 7 for a set of practice problems involving depreciation by the straight-line and units-of-production methods.

Is the "book value" and the "Blue Book" value of my car the same thing?

No. The book value we refer to in the text is the accounting value. It is the value of the asset as depreciated. The "Blue Book" value is the car's market value.

12 Depreciation

12.1 Straight-Line and Units-of-Production Methods

The answers are on page 600.

A. Solve the following.

1. The Wright Oil Company purchased a new copy machine for $6000. It has a useful life of five years with a salvage value of $600. Construct a depreciation schedule using the straight-line method.

2. The Zig-Zag Dress Company purchased a new Zinger Sewing machine for $650. It has a useful life of four years with a salvage value of $50. Construct a depreciation schedule using the straight-line method.

3. The Chartreuse Cab Company purchased a new automobile for $8400. It has a useful life of four years with a salvage value of $1200. Construct a depreciation schedule using the straight-line method.

4. Howard Huge purchased the Topple Tower Apartments for $800,000. He estimates that the building has a useful life of 25 years with a salvage value of $20,000. Construct a depreciation schedule for the first four years using the straight-line method.

Date

Name

Course/Section

5. Johnson's Tax Service purchased a new computer for $240,000. It has a useful life of 10 years with a salvage value of $55,200. Construct a depreciation schedule using the straight-line method.

B. Solve the following.

1. The Wright Oil Company purchased a new machine for $4200. The total estimated units-of-production are 18,000, with a salvage value of $600. The actual production for the years 1980 to 1984 is as follows:

Year	Units Produced
1980	2500
1981	3800
1982	4900
1983	3900
1984	2900

Construct a depreciation schedule using the units-of-production method.

2. The Chartreuse Cab Company purchased a new automobile for $8400. The total estimated mileage is 100,000 miles with a salvage value of $200. The actual mileage for the years 1980 to 1983 is as follows:

Year	Miles
1980	26,410
1981	28,550
1982	24,190
1983	20,850

Construct a depreciation schedule using the units-of-production method.

3. The Zig-Zag Dress Company purchased a new machine for $5000. The estimated units-of-production is 80,000 with no salvage value. The actual production of the years 1980 to 1985 is as follows:

Year	Units Produced
1980	15,000
1981	17,000
1982	19,000
1983	20,000
1984	6,000
1985	3,000

Construct a depreciation schedule using the units-of-production method.

4. The Trawick Company purchased a new machine for $6000. The total estimated units-of-production are 10,800 with a salvage value of $600. The actual production for the years 1980 through 1985 is as follows:

Year	Units Produced
1980	1576
1981	1652
1982	1856
1983	1885
1984	1842
1985	1989

Construct a depreciation schedule using the units-of-production method.

5. The Greenwell Novelty Company purchased a new balloon inflater for $650. The total estimated units-of-production is 2400, with a salvage value of $50. The actual production for the years 1980 to 1984 is as follows:

Year	Balloons Inflated
1980	403
1981	506
1982	553
1983	515
1984	423

Construct a depreciation schedule using the units-of-production method.

When you have had the practice you need, either return to the Preview on page 452 or continue in 8 with the study of declining-balance and sum-of-the years' digits methods of depreciation.

SECTION 2: DECLINING-BALANCE AND SUM-OF-THE-YEAR'S DIGITS METHODS

8

> **OBJECTIVE**
>
> Given the cost, useful life, and salvage value, construct a depreciation schedule using the declining-balance and sum-of-the-years' digits methods. The depreciation schedule will include annual depreciation, accumulated depreciation, and book value.

The straight-line method has the same depreciation each year. But many assets, such as cars, depreciate more at the beginning of their useful life and less at the end. The declining-balance and sum-of-the-years' digits methods have a larger initial depreciation and smaller depreciation each subsequent year.

DECLINING-BALANCE
METHOD

With the declining-balance method, we always use the balance or book value, rather than the total depreciation. The depreciation rate varies according to whether the asset is new or used.

$$(\text{New Asset}) \text{ rate} = \frac{2}{\text{useful life}}$$

This is a twice the straight-line rate.

Annual depreciation for a new asset is given by the following formula:

$$(\text{New Asset}) \text{ annual depreciation} = \underbrace{\frac{2}{\text{useful life}}}_{\text{Rate}} \times \text{book value}$$

For example, Storm's Steam Cleaning purchased a new deluxe steam cleaner for $4500. It has an estimated useful life of five years with a salvage value of $360.

The first year's depreciation by the declining-balance method is:

$$\text{First-year depreciation} = \frac{2}{\text{useful life}} \times \text{book value}$$

$$= \frac{2}{5} \times \$4500$$

The useful life is 5 years, and the book value at the beginning of the first year is equal to the original cost.

Complete the depreciation calculation; then turn to **9**.

---

9 First-year depreciation $= \frac{2}{5} \times \$4500 = \$1800$.

This depreciation is much larger than the $828 we obtained with the straight-line method in frame **2**, page 456. The first-year depreciation will always be greatest with the declining-balance method.

Before calculating the depreciation for the second year, the new book value must be computed. The book value will change each year and must be calculated before computing the next year's depreciation.

New book value = $4500 − $1800 = $2700

Calculate the second year's depreciation using the new book value. Check your work in 10.

---

10   Second-year depreciation = $\dfrac{2}{\text{useful life}} \times$ book value

$$= \frac{2}{5} \times \$2700 = \$1080$$

Using the same information as in the straight-line method, we can now summarize this material in a depreciation schedule.

Year	Annual Depreciation	Accumulated Depreciation	Book Value
0	————————	————	$4500.00
1	$\frac{2}{5} \times \$4500 = \$1800$	$1800.00	2700.00
2	$\frac{2}{5} \times 2700 = 1080$	2880.00	1620.00

Complete the entries for the third and fourth years. Remember to use the new book value each year.

Check your calculations in 11.

Can I change methods of depreciation during an asset's useful life? — use one method for the first few years and then switch to a different method later?

Yes. You can change from the declining-balance to the straight-line method easily enough, and it is legal.

11

Year	Annual Depreciation	Accumulated Depreciation	Book Value
0	————————	————	$4500.00
1	$\frac{2}{5} \times \$4500 = 1800$	$1800.00	2700.00
2	$\frac{2}{5} \times 2700 = 1080$	2880.00	1620.00
3	$\frac{2}{5} \times 1620 = 648$	3528.00	972.00
4	$\frac{2}{5} \times 972 = 388.80$	3916.80	583.20

When using the declining-balance method, *be careful!* Never depreciate below the salvage value. The IRS would frown on such a situation.

In our problem, the fifth-year calculation is:

$\frac{2}{5} \times 583.20 = \$233.28$

But this leads to trouble. Remember, at the start of this example we said that the salvage value was $360. A depreciation of $233.28 is too large because it results in a book value of $583.20 − 233.28 = $349.92, which is below the salvage value. In the fifth year, the asset can only be depreciated down to the salvage value. The depreciation will be the difference between the previous book value and the salvage value.

Our completed depreciation schedule looks like this:

Year	Annual Depreciation	Accumulated Depreciation	Book Value
0	———	———	$4500.00
1	$\frac{2}{5} \times 4500 = 1800$	$1800.00	2700.00
2	$\frac{2}{5} \times 2700 = 1080$	2880.00	1620.00
3	$\frac{2}{5} \times 1620 = 648$	3528.00	972.00
4	$\frac{2}{5} \times 972 = 388.80$	3916.80	583.20
5	223.20 to salvage value	4140.00	360.00

Now it's your turn.

A new machine was purchased for $14,000. It has an estimated life of six years with a salvage value of $1500. Construct a depreciation schedule using the declining balance method.

Our schedule is in 12.

12

Year	Annual Depreciation	Accumulated Depreciation	Book Value
0	———	———	$14,000.00
1	$\frac{2}{6} \times 14{,}000.00 = \$4{,}666.67$	$ 4,666.67	9,333.33
2	$\frac{2}{6} \times 9{,}333.33 = 3{,}111.11$	7,777.78	6,222.22
3	$\frac{2}{6} \times 6{,}222.22 = 2{,}074.07$	9,851.85	4,148.15
4	$\frac{2}{6} \times 4{,}148.15 = 1{,}382.72$	11,234.57	2,765.43
5	$\frac{2}{6} \times 2{,}765.43 = 921.81$	12,156.38	1,843.62
6	343.62 to salvage value	12,500.00	1,500.00

For the last year, the depreciation is only $343.62, although $\frac{2}{6} \times 1843.62 = 614.54$. A depreciation of $614.54 is too large. This would result in a book value below the salvage value, not to mention a very embarrassed accountant.

The declining-balance method is the only depreciation technique in which a different formula is used if the asset is not new. So far we have only worked with a new asset. For a used asset, the maximum rate is:

(Used Asset) $\dfrac{3}{2} \times \dfrac{1}{\text{useful life}}$

This is $1\frac{1}{2}$ times the straight-line rate.

The annual depreciation for a used asset by the declining-balance method is:

$$\text{(Used Asset) annual depreciation} = \frac{3}{2} \times \overbrace{\frac{1}{\text{useful life}}}^{\text{Rate}} \times \text{ book value}$$

The only difference between depreciation of a new or used asset with the declining-balance method is the rate.

The Pekara Manufacturing Company purchased a used machine for $1800. It has an estimated life of six years with a salvage value of $288. Calculate the depreciation for the first year using the declining-balance method.

If your answer is:

(a) $450   Go to **13**.
(b) $600   Go to **14**.
(c) I'm lost! Go to **15**.

13   You are correct.

$$\text{First-year depreciation} = \frac{3}{2} \times \overbrace{\frac{1}{\text{useful life}}}^{\text{Rate}} \times \text{ book value}$$

$$= \frac{3}{2} \times \frac{1}{6} \times \$1800 = \$450$$

It is usually easier to simplify the rate first.

$$\text{Rate} = \frac{\overset{1}{\cancel{3}}}{2} \times \frac{1}{\underset{2}{\cancel{6}}} = \frac{1}{4}$$

If you have trouble multiplying fractions, you may want to return to page 55 for a review.

Now complete the depreciation schedule. Don't forget to use the new book value for each year's calculation.

Our schedule is in **16**.

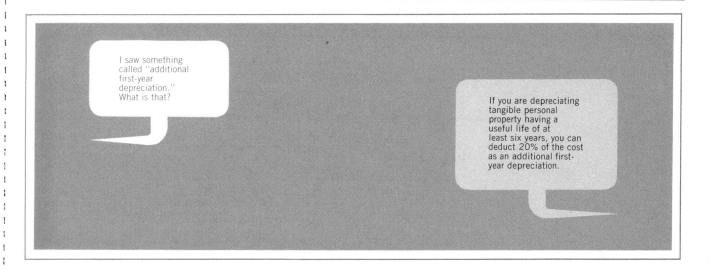

14   Your answer is incorrect. Remember the asset is used. You must apply the used rate.

$$\text{Used asset rate } = \frac{3}{2} \times \frac{1}{\text{useful life}} = \frac{\cancel{3}^{\,2}}{2} \times \frac{1}{\cancel{6}_{\,2}} = \frac{1}{4}$$

Complete the calculation. Then return to **12** and choose another answer.

15   Don't worry. Just apply the formula for the declining-balance method given on page 470.

$$\text{(Used asset) depreciation } = \frac{3}{2} \times \frac{1}{\text{useful life}} \times \text{book value}$$

$$= \frac{3}{2} \times \frac{1}{6} \times \$1800$$

The useful life is 6 years and the book value at the beginning of the first year is the purchase price.

Complete the calculation, then go to **13**.

Year	Annual Depreciation	Accumulated Depreciation	Book Value
0	——————	——	$1800.00
1	$\frac{1}{4} \times 1800.00 = 450.00$	$ 450.00	1350.00
2	$\frac{1}{4} \times 1350.00 = 337.50$	787.50	1012.50
3	$\frac{1}{4} \times 1012.50 = 253.13$	1040.63	759.37
4	$\frac{1}{4} \times$ 759.37 = 189.84	1230.47	569.53
5	$\frac{1}{4} \times$ 569.53 = 142.38	1372.85	427.15
6	$\frac{1}{4} \times$ 427.15 = 106.79	1479.64	320.36

In this problem, the book value at the end of the useful life is greater than the salvage value.

SUM-OF-THE-YEARS' DIGITS METHOD

Like the declining-balance method, the sum-of-the-years' digits method has a large initial depreciation and a smaller depreciation each subsequent year. In the sum-of-the-years' digits method, the book value at the end of the useful life is *always* equal to the salvage value.

The sum-of-the-years' digits method uses the total depreciation. Each year, the depreciation is the product of the total depreciation and a fraction that varies each year.

For example, if the useful life is four years, the numerators of the fractions are 4, 3, 2, and 1. The numerator is 4 for the first year, 3 for the second year, 2 for the third year, and 1 for the fourth year. The denominator is the same each year and is the sum of the numerators. For a useful life of four years, the denominator is $4 + 3 + 2 + 1 = 10$. This is where the sum-of-the-years' digits method get its name.

For a four-year depreciation the fractions are: $\frac{4}{10}, \frac{3}{10}, \frac{2}{10}$, and $\frac{1}{10}$, The depreciation will be:

First-year depreciation $= \dfrac{4}{10} \times$ total depreciation

Second-year depreciation $= \dfrac{3}{10} \times$ total depreciation

Third-year depreciation $= \dfrac{2}{10} \times$ total depreciation

Fourth-year depreciation $= \dfrac{1}{10} \times$ total depreciation

The fractions will be different for each useful life. The fractions will always add up to 1. In the above example, the fractions add up to 10/10 or 1.

What are the fractions for an asset with a useful life of seven years?

There will be seven of them. Turn to 17 after you have calculated them.

---

17   The numerators are 7, 6, 5, 4, 3, 2, and 1. The denominator is

$7 + 6 + 5 + 4 + 3 + 2 + 1 = 28.$

The fractions are $\frac{7}{28}, \frac{6}{28}, \frac{5}{28}, \frac{4}{28}, \frac{3}{28}, \frac{2}{28}$, and $\frac{1}{28}$.

It's really easy, but can be quite lengthy for assets with a long useful life. For an asset with a useful life of 50 years, the numerators will be 50, 49, 48, 47, . . . , 3, 2, 1. The denominator will be $50 + 49 + 48 + 47 . . . + 3 + 2 + 1$.

Luckily, there is a short-cut method to calculate the denominator using the following formula:

$$\text{Denominator} = \frac{N \times (N + 1)}{2} \quad \text{where } N = \text{useful life}$$

For a useful life of 50 years:

Useful life

$$\text{Denominator} = \frac{50 \times (50 + 1)}{2} = \frac{50 \times 51}{2} = 1275$$

That's much easier than adding all those numbers together. Now, you try it.

What are the first five fractions for an asset with a useful life of 35 years?

Check your work in 18.

---

18   The numerators are 35, 34, 33, 32, 31, and so on. The denominator is

$35 + 34 + 33 + 32 + . . . + 3 + 2 + 1$

or by the formula:

Useful life

$$\text{Denominator} = \frac{35 \times (35 + 1)}{2} = \frac{35 \times 36}{2} = 630$$

The fractions for the first five years of depreciation are $\frac{35}{630}$, $\frac{34}{630}$, $\frac{33}{630}$, $\frac{32}{630}$, and $\frac{31}{630}$.

Now, let's complete a sum-of-the-years' digits method depreciation problem.

Storm's Steam Cleaning Co. purchased a new deluxe steam cleaner for $4500. It has an estimated life of five years with a salvage value of $360.

First, calculate the fractions necessary for the sum-of-the-years' digits depreciation method.

Check your fractions in 19.

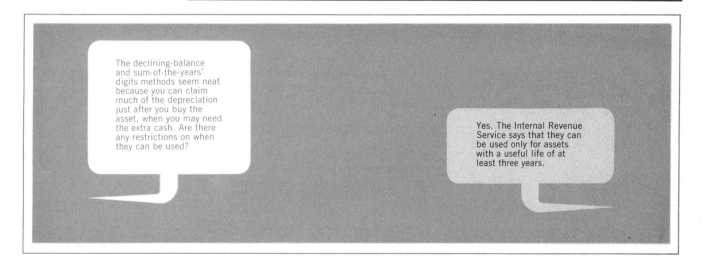

The declining-balance and sum-of-the-years' digits methods seem neat because you can claim much of the depreciation just after you buy the asset, when you may need the extra cash. Are there any restrictions on when they can be used?

Yes. The Internal Revenue Service says that they can be used only for assets with a useful life of at least three years.

19 The numerators are 5, 4, 3, 2, and 1. The denominator is

$5 + 4 + 3 + 2 + 1 = 15.$

The fractions are $\frac{5}{15}, \frac{4}{15}, \frac{3}{15}, \frac{2}{15}$, and $\frac{1}{15}$.

Next, calculate the total depreciation.

Total depreciation = cost − salvage value
= $4500 − 360 = 4140

The depreciation each year will be the product of the total depreciation and the proper fraction.

First-year Depreciation $= \dfrac{5}{15} \times 4140$

Second-year Depreciation $= \dfrac{4}{15} \times 4140$

Third-year Depreciation $= \dfrac{3}{15} \times 4140$

Fourth-year Depreciation $= \dfrac{2}{15} \times 4140$

Fifth-year Depreciation $= \dfrac{1}{15} \times 4140$

Complete the calculations and construct the depreciation schedule.

Our schedule is in **20**.

20

Year	Annual Depreciation	Accumulated Depreciation	Book Value
0	————	———	$4500
1	$\frac{5}{15} \times \$4140 = 1380$	$1380	3120
2	$\frac{4}{15} \times \ 4140 = 1104$	2484	2016
3	$\frac{3}{15} \times \ 4140 = 828$	3312	1188
4	$\frac{2}{15} \times \ 4140 = 552$	3864	636
5	$\frac{1}{15} \times \ 4140 = 276$	4140	360 Salvage value

Remember ▷ Note that the book value at the end of the fifth year is the same as the salvage value. The book value at the end of the useful life is *always* the same as the salvage value. This is a good way to help check your calculation.

Now for practice, work the following problem.

The Pekara Manufacturing Company purchased a machine for $1800. It has an estimated life of six years with a salvage value of $288. Construct a depreciation schedule using the sum-of-the-years' digits method.

Check your work in **21**.

21 The numerators will be 6, 5, 4, 3, 2, and 1. The denominator will be

$6 + 5 + 4 + 3 + 2 + 1 = 21$

The fractions are $\frac{6}{21}, \frac{5}{21}, \frac{4}{21}, \frac{3}{21}, \frac{2}{21}$, and $\frac{1}{21}$.

Total depreciation = cost − salvage value
= $1800 − 288 = $1512

Year	Annual Depreciation	Accumulated Depreciation	Book Value
0	———————	——	$1800
1	$\frac{6}{21} \times 1512 = 432$	$432	1368
2	$\frac{5}{21} \times 1512 = 360$	792	1008
3	$\frac{4}{21} \times 1512 = 288$	1080	720
4	$\frac{3}{21} \times 1512 = 216$	1296	504
5	$\frac{2}{21} \times 1512 = 144$	1440	360
6	$\frac{1}{21} \times 1512 = 72$	1512	288 Salvage value

Be sure to check that the book value at the end of the sixth year is equal to the salvage value.

Need more practice? Try this one.

The TALL Building cost $1,000,000. It has a useful life of 20 years with a salvage value of $160,000. Construct a depreciation schedule for the first five years using the sum-of-the-years' digits method.

Our work is in **22**.

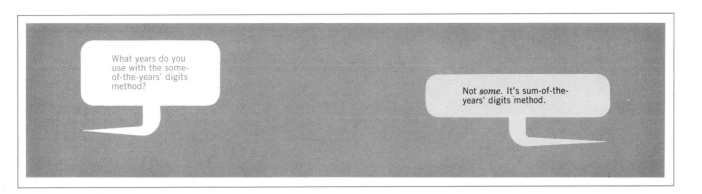

22 The numerators for the first five years are 20, 19, 18, 17, and 16. The denominator is $20 + 19 + 18 + \ldots + 3 + 2 + 1$ or by the formula

Useful life

$$\text{Denominator} = \frac{20 \times (20 + 1)}{2} = \frac{20 \times 21}{2} = 210$$

The fractions are $\frac{20}{210}, \frac{19}{210}, \frac{17}{210}$, and $\frac{16}{210}$.

Total depreciation = cost − salvage value
= $1,000,000 − 160,000 = $840,000

Year	Annual Depreciation	Accumulated Depreciation	Book Value
0	———————	———	$1,000,000
1	$\frac{20}{210} \times 840{,}000 = 80{,}000$	$80,000	920,000
2	$\frac{19}{210} \times 840{,}000 = 76{,}000$	156,000	844,000
3	$\frac{18}{210} \times 840{,}000 = 72{,}000$	228,000	772,000
4	$\frac{17}{210} \times 840{,}000 = 68{,}000$	296,000	704,000
5	$\frac{16}{210} \times 840{,}000 = 64{,}000$	360,000	640,000

Now, turn to **23** for a set of practice problems involving depreciation by the declining-balance and sum-of-the-years' digits methods.

12 Depreciation

PROBLEM SET 2

23

12.2 Declining-Balance and Sum-of-the-Years' Digits Methods

The answers are on page 603.

A. Construct a depreciation schedule using the declining-balance method for the following problems.

1. The Wright Oil Company purchased a new copy machine for $6000. It has a useful life of five years with a salvage value of $600.

2. The Zig-Zag Dress Company purchased a new Zinger Sewing machine for $650. It has a useful life of four years with a salvage value of $50.

3. The Chartreuse Cab Company purchased a new automobile for $8400. It has a useful life of four years with a salvage value of $1200.

4. Howard Huge purchased the Topple Tower (used) for $800,000. The building has a useful life of 25 years with a salvage value of $20,000. Complete the depreciation schedule for the first 4 years only.

Date

Name

Course/Section

5. Johnson's Tax Service purchased a new computer for $240,000. It has a useful life of 10 years with a salvage value of $55,200.

B. Construct a depreciation schedule using the sum-of-the-years' digits method for the following problems.

1. The Wright Oil Company purchased a new copy machine for $6000. It has a useful life of five years with a salvage value of $600.

2. The Zig-Zag Dress Company purchased a new Zinger sewing machine for $650. It has a useful life of four years with a salvage value of $50.

3. The Chartreuse Cab Company purchased a new automobile for $8400. It has a useful life of four years with a salvage value of $1200.

4. Howard Huge purchased the Topple Tower (new) for $800,000. The building has a useful life of 25 years with a salvage value of $20,000. Complete the depreciation schedule for the first 4 years only.

5. Johnson's Tax Service purchased a new computer for $240,000. It has a useful life of 10 years with a salvage value of $55,200.

When you have had the practice you need, either return to the Preview on page 452 or continue in **24** with the study of partial-year depreciation.

24

OBJECTIVE

Given the cost, useful life, salvage value, and purchase date (partial year), construct a depreciation schedule using the straight-line, declining-balance, and sum-of-the-years' digits methods. The depreciation schedule will include the annual depreciation, accumulated depreciation, and the book value.

In the previous two sections, depreciation was calculated only for entire years. This can be used only for assets purchased on January 1. Assets not purchased on January 1 must be depreciated for only the portion of the calendar year the asset is owned by the business. This is called *partial-year depreciation.*

Partial-year depreciation works on a monthly basis, rounding each purchase date to the "nearest" month. If an asset is purchased on or before the fifteenth of a month, we round the date to the first of the month. If the asset is purchased after the fifteenth, the date is rounded to the first day of the next month. For example, a machine purchased July 14, 1980 would be rounded to July 1, 1980.

Round a purchase date of September 16, 1980.

Check your answer in **25**.

---

25 Since the sixteenth is after the fifteenth, round to the first of the next month. September 16, 1980 is rounded to October 1, 1980.

STRAIGHT-LINE
METHOD FOR PARTIAL
YEARS

With the straight-line method, the depreciation for a calendar year is the product of the annual depreciation and the fractional part of the year the business has the asset.

Let's work an example.

Storm's Steam Cleaning purchased a new deluxe steam cleaner on April 15, 1980 for $4500. It has a useful life of five years with a salvage value of $360.

The purchase date is rounded to April 1, 1980.

Total depreciation = cost − salvage value
= $4500 − 360 = $4140

$$\text{Annual depreciation} = \frac{\text{total depreciation}}{\text{useful life}}$$

$$= \frac{\$4140}{5} = \$828$$

The annual depreciation of $828 is for an entire calendar year. For 1980, Storm's Steam Cleaning had the machine from April 1 on, that is, for 9 months or $\frac{9}{12}$ of the year. Therefore, the 1980 depreciation will be:

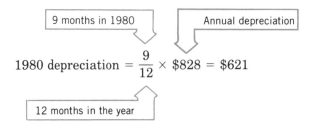

1980 depreciation $= \dfrac{9}{12} \times \$828 = \$621$

Since the business has the machine for the entire year in 1981, 1982, 1983, and 1984, the depreciation will be $828 each year.

Storm's Steam Cleaning has the machine until April 1, 1985; that is, for 3 months—January, February, and March. The 1985 depreciation will be:

3 months in 1985

1985 depreciation $= \dfrac{3}{12} \times \$828 = \$207$

Now that the depreciation has been calculated, the depreciation schedule can be constructed.

Complete the depreciation schedule. Then turn to 26.

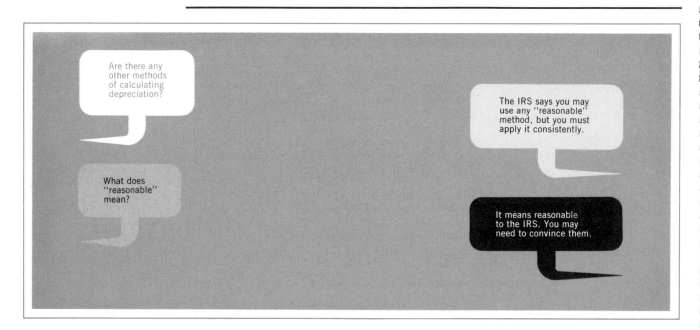

		Accumulated	
Year	Annual Depreciation	Depreciation	Book Value
——	————	——	$4500
1980	$\frac{9}{12} \times 828 = 621$	$ 621	3879
1981	828	1449	3051
1982	828	2277	2223
1983	828	3105	1395
1984	828	3933	567
1985	$\frac{3}{12} \times 828 = 207$	4140	360 Salvage value

Remember, the book value at the end of the useful life is equal to the salvage value with the straight-line method.

Partial-year calculations must be made only for the first and last years of the useful life when using the straight-line method.

Use the same procedure to solve the following problem.

The Pekara Manufacturing Company purchased a machine on August 18, 1980 for $1800. It has a useful life of six years, with a salvage value of $288. Construct a depreciation schedule using the straight-line method.

Our schedule is in **27**.

---

27   Total depreciation = cost − salvage value
$$= \$1800 - 288 = \$1512$$

$$\text{Annual depreciation} = \frac{\text{total depreciation}}{\text{useful life}}$$

$$= \frac{\$1512}{6} = \$252$$

The purchase date is rounded to September 1, 1980. For 1980, the company had the machine for 4 months or $\frac{4}{12}$ of the year.

> 4 months in 1980   Annual depreciation

$$1980 \text{ depreciation} = \frac{4}{12} \times \$252 = \$84$$

For each year from 1981 to 1985, there is a full year of depreciation. For the last year, 1986, the company has the machine for 8 months or $\frac{8}{12}$ of the year (until September).

> 8 months in 1986

$$1986 \text{ depreciation} = \frac{8}{12} \times \$252 = \$168$$

Year	Annual Depreciation	Accumulated Depreciation	Book Value	
——	————	——	$1800	
1980	$\frac{4}{12} \times 252 =$ 84	$ 84	1716	
1981	252	336	1464	
1982	252	588	1212	
1983	252	840	960	
1984	252	1092	708	
1985	252	1344	456	
1986	$\frac{8}{12} \times 252 =$ 168	1512	288	Salvage value

The book value for 1986 is the salvage value.

DECLINING-BALANCE METHOD FOR PARTIAL YEARS

With the declining-balance method, the depreciation for a calendar year is the product of the annual depreciation and the fractional part of the year the business has the asset.

Let's look at an example.

Storm's Steam Cleaning purchased a new deluxe steam cleaner on April 15, 1980 for $4500. It has a useful life of five years with a salvage value of $360.

Since the asset is new, the depreciation rate is

$$\text{Rate} = \frac{2}{\text{useful life}} = \frac{2}{5}$$

The purchase date is rounded to April 1, 1980.

For 1980, Storm's Steam Cleaning had the asset for 9 months, or $\frac{9}{12}$ of the year. The 1980 depreciation is:

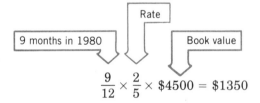

$$\frac{9}{12} \times \frac{2}{5} \times \$4500 = \$1350$$

For each of the years from 1981 to 1984, the machine is depreciated for the entire year. No partial-year calculations are necessary for these years.

Complete the calculations for 1981 to 1984 and construct the depreciation schedule for 1980 to 1984.

Our work is in **28**.

28			

Year	Annual Depreciation	Accumulated Depreciation	Book Value
———	———————————	———	$4500.00
1980	$\frac{9}{12} \times \frac{2}{5} \times 4500.00 = 1350.00$	$1350.00	3150.00
1981	$\frac{2}{5} \times 3150.00 = 1260.00$	2610.00	1890.00
1982	$\frac{2}{5} \times 1890.00 = 756.00$	3366.00	1134.00
1983	$\frac{2}{5} \times 1134.00 = 453.60$	3819.60	680.40
1984	$\frac{2}{5} \times 680.40 = 272.16$	4091.76	408.24

Partial-year calculations are necessary for the first year, 1980, and the last year, 1985. In 1985, the machine is used for 3 months or $\frac{3}{12}$ of the year.

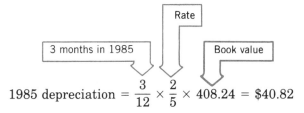

$$1985 \text{ depreciation} = \frac{3}{12} \times \frac{2}{5} \times 408.24 = \$40.82$$

The depreciation of $40.82 leaves a book value of $408.24 − 40.82 = $367.42, which is larger than the salvage value. Remember, always check the book value. The book value may *never* be below the salvage value.

The last line of our depreciation looks like this:

Year	Annual Depreciation	Accumulated Depreciation	Book Value
1985	$\frac{3}{12} \times \frac{2}{5} \times 408.24 = 40.82$	4132.58	367.42

Your turn.

The Pekara Manufacturing Company purchased a *used* machine on August 18, 1980 for $1800. It has a useful life of six years with a salvage value of $288. Construct a depreciation schedule using the declining-balance method.

Remember to apply the used rate. Our work is in **29**.

---

29   Since the machine is used, the rate is:

$$\text{Rate} = \frac{3}{2} \times \frac{1}{\text{useful life}} = \frac{\overset{1}{\cancel{3}}}{2} \times \frac{1}{\underset{2}{\cancel{6}}} = \frac{1}{4}$$

The purchase date is rounded to September 1, 1980. For 1980, the machine is depreciated for 4 months, or $\frac{4}{12}$ of the year. For each of the years from 1981 to 1985, the machine is depreciated for the entire year. For 1986, the machine is depreciated for 8 months, or $\frac{8}{12}$ of the year.

Year	Annual Depreciation	Accumulated Depreciation	Book Value
———	—————————	———	$1800.00
1980	$\frac{4}{12} \times \frac{1}{4} \times 1800.00 = 150.00$	$ 150.00	1650.00
1981	$\frac{1}{4} \times 1650.00 = 412.50$	562.50	1237.50
1982	$\frac{1}{4} \times 1237.50 = 309.38$	871.88	928.12
1983	$\frac{1}{4} \times\ \ 928.12 = 232.03$	1103.91	696.09
1984	$\frac{1}{4} \times\ \ 696.09 = 174.02$	1277.93	522.07
1985	$\frac{1}{4} \times\ \ 522.07 = 130.52$	1408.45	391.55
1986	$\frac{8}{12} \times \frac{1}{4} \times\ \ 391.55 =\ \ 65.26$	1473.71	326.29

SUM-OF-THE-YEARS' DIGITS METHOD FOR PARTIAL YEARS

With the straight-line and declining-balance methods, partial-year calculations must be made only for the first and last years of the useful life. Partial-year depreciation with the sum-of-the years' digits method affects every year throughout the useful life.

Let's work through the Storm's Steam Cleaning problem using the sum-of-the-years' digits method.

Storm's Steam Cleaning purchased a new deluxe steam cleaner on April 15, 1980 for $4500. It has an estimated useful life of five years with a salvage value of $360.

Total depreciation = cost − salvage value
= $4500 − 360 = $4140

The numerators of the sum-of-the-years' digits' fractions are 5, 4, 3, 2, and 1. The denominator is $5 + 4 + 3 + 2 + 1 = 15$. The fractions are $\frac{5}{15}$, $\frac{4}{15}$, $\frac{3}{15}$, $\frac{2}{15}$, and $\frac{1}{15}$.

The purchase date is rounded to April 1, 1980.

For 1980, the company had the machine for 9 months, or $\frac{9}{12}$ of the year.

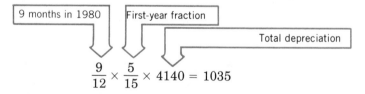

$$\frac{9}{12} \times \frac{5}{15} \times 4140 = 1035$$

That's not very hard, but the next year's calculation is more involved. Let's work through it carefully.

For 1981, the first three months (until April 1, 1981) are part of the first year the company had the machine.

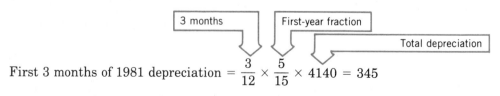

First 3 months of 1981 depreciation $= \frac{3}{12} \times \frac{5}{15} \times 4140 = 345$

Be very careful here. All 12 months of the first-year depreciation fraction, $\frac{5}{15}$, must be used before starting with the second-year fraction, $\frac{4}{15}$. Since we used only the first 9 months in 1980, we must use the last 3 months at the start of 1981.

Now the last 9 months of 1981 begin the second year the company has the machine.

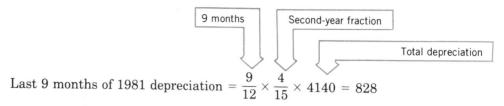

Last 9 months of 1981 depreciation $= \dfrac{9}{12} \times \dfrac{4}{15} \times 4140 = 828$

The depreciation for the entire 1981 year is the sum of the first 3 months depreciation and the last 9 months.

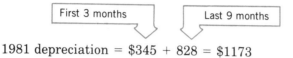

1981 depreciation $= \$345 + 828 = \$1173$

This technique of breaking each calendar year into two parts of 3 months and 9 months must be continued through the useful life.

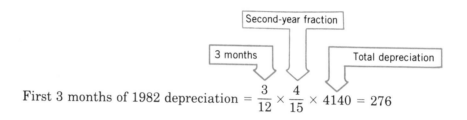

First 3 months of 1982 depreciation $= \dfrac{3}{12} \times \dfrac{4}{15} \times 4140 = 276$

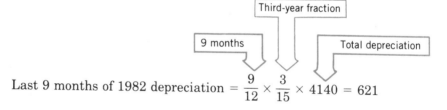

Last 9 months of 1982 depreciation $= \dfrac{9}{12} \times \dfrac{3}{15} \times 4140 = 621$

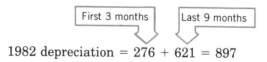

1982 depreciation $= 276 + 621 = 897$

Calculate the depreciation for 1983 and 1984. Then check your work in 30.

---

30   1983 depreciation $= \dfrac{3}{12} \times \dfrac{3}{15} \times 4140 + \dfrac{9}{12} \times \dfrac{2}{15} \times 4140$

$= 207 + 414 = 621$

$$\text{1984 depreciation} = \overbrace{\frac{3}{12} \times \frac{2}{15} \times 4140}^{\text{First 3 months}} + \overbrace{\frac{9}{12} \times \frac{1}{15} \times 4140}^{\text{Last 9 months}}$$

$$= 138 + 207 = 345$$

For the last year 1985, the company has the machine for only 3 months.

$$\text{1985 depreciation} = \frac{3}{12} \times \frac{1}{15} \times 4140 = 69$$

Complete the problem by constructing the depreciation schedule. Our schedule is in 31.

Is it really necessary to do all of those partial-year calculations when you use the sum-of-the-years' digits method?

Yes, it is necessary, unless you want to be audited by the IRS.

31

Year	Annual Depreciation	Accumulated Depreciation	Book Value
——	——	——	$4500
1980	$1035	$1035	3465
1981	1173	2208	2292
1982	897	3105	1395
1983	621	3726	774
1984	345	4071	429
1985	69	4140	360 Salvage value

Note that the book value at the end of the useful life is the same as the salvage value.

Now try one on your own.

The Pekara Manufacturing Company purchased a machine on August 18, 1980 for $1800. It has a useful life of six years with a salvage value of $288. Construct a depreciation schedule using the sum-of-the-years' digits method.

Our work is in 32.

32 Total depreciation = cost − salvage value
$$= \$1800 - 288 = \$1512$$

The numerators of the fractions are 6, 5, 4, 3, 2, and 1. The denominator is
$$6 + 5 + 4 + 3 + 2 + 1 = 21.$$

The fractions are $\frac{6}{21}, \frac{5}{21}, \frac{4}{21}, \frac{3}{21}, \frac{2}{21},$ and $\frac{1}{21}$.

The purchase date is rounded to September 1, 1980.

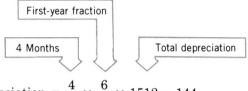

$$1980 \text{ depreciation} = \frac{4}{12} \times \frac{6}{21} \times 1512 = 144$$

$$1981 \text{ depreciation} = \frac{8}{12} \times \frac{6}{21} \times 1512 + \frac{4}{12} \times \frac{5}{21} \times 1512$$
$$= 288 + 120 = 408$$

$$1982 \text{ depreciation} = \frac{8}{12} \times \frac{5}{21} \times 1512 + \frac{4}{12} \times \frac{4}{21} \times 1512$$
$$= 240 + 96 = 336$$

$$1983 \text{ depreciation} = \frac{8}{12} \times \frac{4}{21} \times 1512 + \frac{4}{12} \times \frac{3}{21} \times 1512$$
$$= 192 + 72 = 264$$

$$1984 \text{ depreciation} = \frac{8}{12} \times \frac{3}{21} \times 1512 + \frac{4}{12} \times \frac{2}{21} \times 1512$$
$$= 144 + 48 = 192$$

$$1985 \text{ depreciation} = \frac{8}{12} \times \frac{2}{21} \times 1512 + \frac{4}{12} \times \frac{1}{21} \times 1512$$
$$= 96 + 24 = 120$$

$$1986 \text{ depreciation} = \frac{8}{12} \times \frac{1}{21} \times 1512 = 48$$

Year	Annual Depreciation	Accumulated Depreciation	Book Value
————	————	————	$1800
1980	$144	$144	1656
1981	408	552	1248
1982	336	888	912
1983	264	1152	648
1984	192	1344	456
1985	120	1464	336
1986	48	1152	288 Salvage value

Whew, that's a lot of work, but because the sum-of-the-years' digits is a commonly used method, it is important that you know it. It has a larger initial depreciation and smaller depreciation each subsequent year. Also, the book value at the end of the useful life is always the same as the salvage value.

UNITS OF
PRODUCTION METHOD
FOR PARTIAL YEARS

Since the units-of-production method is dependent only on the number of units produced during the year, rather than the number of months, a partial-year procedure is not necessary.

Now turn to 33 for a set of practice problems involving partial-year depreciation.

12 Depreciation

33

12.3 Partial-Year
Depreciation

The answers are on page 605.

A. Construct a depreciation schedule using the straight-line method for the following problems.

1. The Wright Oil Company purchased a new copy machine on September 27, 1980 for $6000. It has a useful life of five years with a salvage value of $600.

2. The Zig-Zag Dress Company purchased a new Zinger Sewing Machine on March 7, 1980 for $650. It has a useful life of four years with a salvage value of $50.

3. The Chartreuse Cab Company purchased a new automobile on November 13, 1980 for $8400. It has a useful life of four years with a salvage value of $1200.

4. Howard Huge purchased the Topple Tower (used) on June 15, 1980 for $800,000. The building has a useful life of 25 years with a salvage value of $20,000. Construct this depreciation schedule for 1980 through 1983.

Date

Name

Course/Section

5. Johnson's Tax Service purchased a new computer on May 3, 1980 for $240,000. It has a useful life of 10 years with a salvage value of $55,200.

B. Construct a depreciation schedule using the declining-balance method for the following problems.

1. The Wright Oil Company purchased a new copy machine on September 27, 1980 for $6000. It has a useful life of five years with a salvage value of $600.

2. The Zig-Zag Dress Company purchased a new Zinger sewing machine on March 7, 1980 for $650. It has a useful life of four years with a salvage value of $50.

3. The Chartreuse Cab Company purchased a new automobile on November 13, 1980 for $8400. It has a useful life of four years with a salvage value of $1200.

4. Howard Huge purchased the Topple Tower (used) on June 15, 1980 for $800,000. The building has a useful life of 25 years with a salvage value of $20,000. Construct this depreciation schedule for 1980 through 1983.

5. Johnson's Tax Service purchased a new computer on May 3, 1980 for $240,000. It has a useful life of 10 years with a salvage value of $55,200.

C. **Construct a depreciation schedule using the sum-of-the-years' digits method for the following problems.**

1. The Wright Oil Company purchased a new copy machine on September 27, 1980 for $6000. It has a useful life of five years with a salvage value of $600.

2. The Zig-Zag Dress Company purchased a new Zinger Sewing Machine on March 7, 1980 for $650. It has a useful life of four years with a salvage value of $50.

3. The Chartreuse Cab Company purchased a new automobile on November 13, 1980 for $8400. It has a useful life of four years with a salvage value of $1200.

4. Howard Huge purchased the Topple Tower (new) on June 15, 1980 for $800,000. The building has a useful life of 25 years with a salvage value of $20,000. Construct this depreciation schedule for 1980 through 1983.

5. Johnson's Tax Service purchased a new computer on May 3, 1980 for $240,000. It has a useful life of 10 years with a salvage value of $55,200.

When you have had the practice you need, continue in **34** with the study of the ACRS method.

34

OBJECTIVES

Given the cost and recovery period, construct a depreciation schedule using the ACRS method. The depreciation schedule will include the annual depreciation, accumulated depreciation, and the book value.

The Accelerated Cost Recovery System (ACRS) is a new depreciation method used on all assets purchased after December 31, 1980. This method has several advantages over the older techniques.

First, the ACRS is an accelerated depreciation method. This permits the business to quickly write off the expense for its taxes.

There are no partial year calculations with the ACRS method. If an asset is purchased anytime during the year, even on December 31, depreciation for a whole year is allowed.

Another benefit when using the ACRS method is that the depreciation calculations are very easy. Each year, the depreciation is simply the product of the given depreciation rate (a percent) times the cost.

Another advantage is that the time period, called the recovery period, is shorter than the time period in the older methods. There are only 4 different recovery periods: 3, 5, 10, and 15 years.

Assets eligible for a 3-year recovery include automobiles, light trucks, and machinery with a useful life of 7 years or less. The depreciation rates for each year are given in the following table:

3-year ACRS Recovery	
	Rate
1st year	25%
2nd year	38%
3rd year	37%

Assets eligible for a 5-year recovery include heavy duty trucks, most machinery and equipment, and most office equipment. The depreciation rates for each year are given in the following table:

5-year ACRS Recovery	
	Rate
1st year	15%
2nd year	22%
3rd through 5th year	21%

Assets eligible for a 10-year recovery include manufactured homes.

10-year ACRS Recovery

	Rate
1st year	8%
2nd year	14%
3rd year	12%
4th through 6th year	10%
7th through 10th year	9%

Assets eligible for a 15-year recovery include most real property with a life of more than 25 years.

15-year ACRS Recovery

	Rate
1st year	5%
2nd year	10%
3rd year	9%
4th year	8%
5th and 6th year	7%
7th through 15th year	6%

The depreciation for any year is the product of the depreciation rate and the cost.

Depreciation = Rate × Cost

Example: Calculate the first-year depreciation on a new car with an original cost of $12,500.

A new car has a 3-year recovery.

1st year rate   cost

First-year depreciation = 25% × $12,500
                        = 0.25 × $12,500
                        = $3125

The first-year depreciation is $3125. Now use the 3-year ACRS recovery table to complete the second and third years' depreciation.

Check your work in 35.

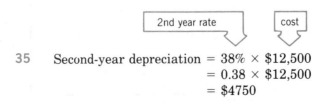

35   Second-year depreciation = 38% × $12,500
                               = 0.38 × $12,500
                               = $4750

Third-year depreciation = 37% × $12,500
= 0.37 × $12,500
= $4625

Now that you have checked your depreciation calculations, complete the depreciation calculations and complete the depreciation schedule. Remember to include the accumulated depreciation and book value for each year.

Our answer is in 36.

36

Year	Depreciation	Accumulated Depreciation	Value
—	—	—	$12,500
1	$3125	$3125	9375
2	4750	7875	4625
3	4625	12,500	0

For some additional practice, complete a description schedule for a heavy duty truck that cost $38,900.

Check your work in 37.

37   For a heavy duty truck, use a 5-year ACRS recovery.

First-year depreciation = 15% × $38,900
= 0.15 × $38,900 = $5835

Second-year depreciation = 22% × $38,900
= 0.22 × $38,900 = $8558

Third through fifth years' depreciation = 21% × $38,900
= 0.21 × $38,900 = $8169

Year	Depreciation	Accumulated Depreciation	Book Value
—	—	—	$38,900
1	$5835	$ 5,835	33,065
2	8558	14,393	24,507
3	8169	22,562	16,338
4	8169	30,731	8169
5	8169	38,900	0

The purchase date during the year does not affect the depreciation, because the ACRS method has no partial-year calculations.

Now turn to 38 for a set of practice problems involving the ACRS method.

12 Depreciation

PROBLEM SET 4

38

12.4 ACRS
METHOD

The answers are on page 609.

Construct a depreciation schedule using the ACRS method for the following problems.

1. A business bought a new automobile for $17,800.

2. A business purchased a new heavy duty truck for $32,900.

3. A small business purchased new office furniture for $2850.

4. A business purchased a light truck for $14,600.

5. A business purchased a manufactured residential home for $58,500.

6. A business purchased some machinery eligible for a 3-year recovery for $8250.

When you have had the practice you need, turn to **39** for the Self-Test on depreciation.

Date

Name

Course/Section

SECTION 4: ACRS METHOD **499**

12 Depreciation

39

Self-Test A

The answers are on page 611.

This Self-Test does not include partial-year depreciation. For a Self-Test covering partial-year depreciation, turn to page 503.

Analla's Cabinet Shop purchased a new lathe for $6000. It has a useful life of five years with a salvage value of $600. Construct a depreciation schedule using:

1. Straight-line method.

2. Declining-balance method.

3. Sum-of-the-years' digits method.

Date

Name

Course/Section

4. The Fleming Company purchased a new machine for $9700. The estimated units-of-production is 50,000 with a salvage value of $400. The actual production is as follows:

Year	Units Produced
1980	12,560
1981	17,890
1982	13,120
1983	6,430

Construct a depreciation schedule using the units-of-production method.

5. Construct a depreciation schedule using the ACRS method for a new automobile that costs $13,200.

12 Depreciation

Self-Test B

The answers are on page 612.

This Self-Test covers partial-year depreciation.

Analla's Cabinet Shop purchased a new lathe on August 21, 1980 for $4200. It has a useful life of four years with a salvage value of $600. Construct a depreciation schedule using:

1. Straight-line method.

2. Declining-balance method.

3. Sum-of-the-years' digits method.

4. The Fleming Company purchased a new machine for $9700. The estimated units-of-production is 50,000 with a salvage value of $400. The actual production is as follows:

Year	Units Produced
1980	12,560
1981	17,890
1982	13,120
1983	6,430

Date

Name

Course/Section

Construct a depreciation schedule using the units-of-production method.

5. Construct a depreciation schedule using the ACRS method for a new automobile that costs $13,200.

13 Financial Statement Analysis

THAT FINANCIAL STATEMENT WOULD LOOK BETTER IF YOU TOOK YOUR GLASSES OFF !

With or without glasses, a person in business cannot be successful unless he keeps proper records. His business, to be successful, must not operate on a day-to-day basis; in other words, he cannot expect a day's income to pay that day's debts.

Financial records must be carefully maintained. These records generally include summary statements such as balance sheets and income statements. Proper analysis of these important summaries allows a business manager to make effective financial decisions.

This unit is designed to acquaint you with the analysis of business financial statements.

To complete this unit successfully, you will need the following: (1) several sharp pencils, (2) a "charged up" calculator, (3) lots of scratch paper, and (4) a determination to master this unit!

UNIT OBJECTIVE

When you complete this unit successfully, you will be able to complete a horizontal and vertical analysis of two successive balance sheets or income statements. You will also be able to calculate current ratio, acid-test ratio, owner's rate or return, and inventory turnover.

This is a sample test for Unit 13. All the answers are placed immediately after the test. Work as many of the problems as you can, then check your answers and look after the answers for further directions.

Where to Go for Help

Page   Frame

1. Complete the horizontal analysis of the following balance sheets. Round all percents to the nearest 0.1%. Use an * to denote any decreases.

511   1

J. NICHOLS, INC.
COMPARATIVE BALANCE SHEETS
DECEMBER 31, 1983 AND 1984

Assets	1983	1984	Increase or Decrease Amount	Percent
Current Assets				
Cash	$ 12,500	$ 13,000		
Merchandise inventory	45,000	48,000		
Accounts receivable	30,000	35,000		
Total current assets	$ 87,500	$ 96,000		
Plant and Equipment				
Land	$ 8,000	$ 8,000		
Buildings	12,000	17,000		
Equipment	5,000	4,500		
Total plant and equipment	$ 25,000	$ 29,500		
Total assets	$112,500	$125,500		
Liabilities and Capital				
Current Liabilities				
Accounts payable	$ 33,000	$ 34,500		
Wages payable	5,000	5,500		
Total current liabilities	$ 38,000	$ 40,000		
Long-Term Liabilities				
Mortgage payable	15,000	18,000		
Total liabilities	$ 53,000	$ 58,000		
J. Nichols, Capital	59,500	67,500		
Total liabilities and capital	$112,500	$125,500		

2. Complete the vertical analysis of the following income statements. Round all percents to the nearest 0.1%.

529   12

S. KAMM, INC. COMPARATIVE INCOME STATEMENTS FOR THE YEARS ENDED DECEMBER 31, 1983 AND 1984				
	1983		1984	
	Amount	Percent	Amount	Percent
Income				
Sales	$79,000		$85,000	
Less sales returns	4,000		3,500	
Net sales	$75,000		$81,500	
Cost of goods sold:				
Inventory, January 1	$24,000		$27,000	
Purchases	52,000		55,000	
Goods available for sale	$76,000		$82,000	
Less inventory, December 31	27,000		30,000	
Cost of goods sold	$49,000		$52,000	
Gross profit	$26,000		$29,500	
Operating expenses:				
Selling	$ 9,000		$10,000	
Rent	4,000		4,500	
Utilities	1,500		1,600	
Miscellaneous	500		400	
Total operating expenses	$15,000		$16,500	
Net income	$11,000		$13,000	

3. Calculate the current ratio as of December 31, 1983 for J. Nichols, Inc., of Problem 1.

_____

543   20

4. Calculate the inventory turnover in 1983 for S. Kamm, Inc., in Problem 2. Round to the nearest tenth.

_____

543   20

Answers

1. See the following.
2. See page 509.
3. 2.3:1
4. 1.9:1

★ If you cannot work any of these problems, start with frame 1 on page 511.
★ If you missed some of the problems, turn to the page numbers indicated.
★ If all of your answers were correct, turn to page 549 for the Self-Test.
★ Super-students—those who want to be certain they learn all of this material—will turn to frame 1 and begin work there.

1.

J. NICHOLS, INC.
COMPARATIVE BALANCE SHEETS
DECEMBER 31, 1983 AND 1984

	1983	1984	Increase or Decrease*	
			Amount	Percent
Assets				
Current Assets				
Cash	$ 12,500	$ 13,000	$ 500	4.0%
Merchandise inventory	45,000	48,000	3,000	6.7%
Accounts receivable	30,000	35,000	5,000	16.7%
Total current assets	$ 87,500	$ 96,000	$8,500	9.7%
Plant and Equipment				
Land	$ 8,000	$ 8,000	0	0%
Buildings	12,000	17,000	5,000	41.7%
Equipment	5,000	4,500	500*	10.0%*
Total plant and equipment	$ 25,000	$ 29,500	$ 4,500	18.0%
Total assets	$112,500	$125,500	$13,000	11.6%
Liabilities and Capital				
Current Liabilities				
Accounts payable	$ 33,000	$ 34,500	$ 1,500	4.5%
Wages payable	5,000	5,500	500	10.0%
Total current liabilities	$ 38,000	$ 40,000	$ 2,000	5.3%
Long-Term Liabilities				
Mortgage payable	15,000	18,000	3,000	20.0%
Total liabilities	$ 53,000	$ 58,000	$ 5,000	9.4%
J. Nichols, Capital	59,500	67,500	8,000	13.4%
Total liabilities and capital	$112,500	$125,500	$13,000	11.6%

* Asterisks denote a decrease.

2.

S. KAMM, INC. COMPARATIVE INCOME STATEMENTS FOR THE YEARS ENDED DECEMBER 31, 1983 AND 1984				
	1983		1984	
	Amount	Percent	Amount	Percent
Income:				
Sales	$79,000	105.3%	$85,000	104.3%
Less sales returns	4,000	5.3%	3,500	4.3%
Net sales	$75,000	100.0%	$81,500	100.0%
Cost of goods sold:				
Inventory, January 1	$24,000	32.0%	$27,000	33.1%
Purchases	52,000	69.3%	55,000	67.5%
Goods available for sale	$76,000	101.3%	$82,000	100.6%
Less inventory, December 31	27,000	36.0%	30,000	36.8%
Cost of goods sold	$49,000	65.3%	$52,000	63.8%
Gross profit	$26,000	34.7%	$29,500	36.2%
Operating expenses:				
Selling	$ 9,000	12.0%	$10,000	12.3%
Rent	4,000	5.3%	4,500	5.5%
Utilities	1,500	2.0%	1,600	2.0%
Miscellaneous	500	0.7%	400	0.5%
Total operating expenses	$15,000	20.0%	$16,500	20.2%
Net income	$11,000	14.7%	$13,000	16.0%

SECTION 1: BALANCE SHEETS

<table>
<tr><td>1</td><td>**OBJECTIVE**

Complete a horizontal and vertical analysis of two successive balance sheets.</td></tr>
</table>

Balance Sheets   The *balance sheet* shows the financial condition of a business at a particular time. It is a stationary financial picture of the business showing the assets, liabilities, and capital or net worth.

Assets = values owned by the business
Liabilities = values owed to others
Capital = net worth of the business

All balance sheets must satisfy the fundamental accounting equation:

Assets = liabilities + capital

or, stated in its equivalent form, capital (net worth) of a business is the difference between its assets and its liabilities.

Capital = assets − liabilities

In addition to businesses, balance sheets may be used to present the financial position of an individual. For example, if you have $100 in cash (assets) and you owe $45 (liabilities), then your net worth or capital is

Capital = assets − liabilities
= $100 − 45 = $55

Now, this information may be summarized in an abbreviated balance sheet. The balance sheet is broken into two main parts: first—assets and second—liabilities and capital.

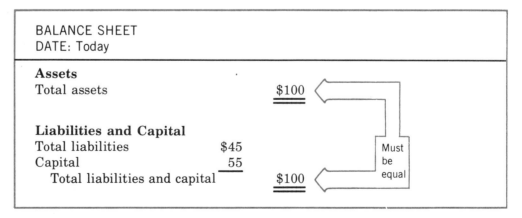

The above balance sheet satisfies the fundamental accounting equation:

Assets = liabilities + capital
$100 = $45 + $55

Balance sheets used in businesses are much more detailed than the last example. They contain additional categories including current assets, fixed assets, current liabilities, and long-term liabilities. For a discussion of these additional terms, turn to 2.

2   Before we look at a business' balance sheet, we must define some of the commonly used terms.

The *assets* of a business include the goods and property owned, and claims that have not been collected. Assets are usually broken into two categories: current assets and fixed assets.

Current assets are assets that will be used or converted into cash in a short period of time—usually within one year. Current assets include cash, merchandise on hand, and accounts receivable (sales for which payments have not been received).

Plant and equipment are items that will not be used or converted to cash within a year. Fixed assets include land, buildings, and equipment.

Liabilities are the amounts owed by the business to others. Liabilities are also divided into two categories: current liabilities and fixed or long-term liabilities.

Current liabilities are amounts owed by the business that must be paid in a short period of time. They include accounts payable, wages payable, utilities payable, federal income tax payable, and notes payable (due within one year).

Fixed or long-term liabilities are amounts that will be paid after a year. They include mortgages payable, bonds payable, and notes payable (due later than one year).

Capital is the difference between the assets and liabilities. Capital is also referred to as *net worth, net ownership,* or *stockholder's equity.*

Sound confusing? Just remember

Assets = values owned by the business
Liabilities = value owed to others
Capital = net worth of the business

Assets = liabilities + capital

Now we are ready to examine a typical business' balance sheet. Turn to 3.

3   Balance sheets are presented in a prescribed format. The balance sheet is divided into two main sections: first, assets, and second, liabilities and capital. The assets section is composed of current assets and fixed assets. The liabilities and capital section is composed of current liabilities, long-term liabilities, and capital. Examine the following balance sheet carefully.

```
┌─────────────────────────────────────────────────────────────────┐
│ R. TODD, INC.                                                     │
│ BALANCE SHEET                                                     │
│ DECEMBER 31, 1983                                                 │
├─────────────────────────────────────────────────────────────────┤
```

Assets

Current Assets

Cash	$ 3,000	
Merchandise inventory	5,000	
Accounts receivable	7,000	
Total current assets		$15,000

Plant and Equipment

Land	$ 8,000	
Buildings	17,000	
Equipment	10,000	
Total plant and equipment		$35,000
Total assets		$50,000

Liabilities and Capital

Current Liabilities

Accounts payable	$ 8,500	
Wages payable	3,000	
Total current liabilities		$11,500

Long-Term Liabilities

Mortgage payable		16,000
Total liabilities		$27,500
R. Todd, Capital		22,500
Total liabilities and capital		$50,000

Many businesses have an even more detailed balance sheet with many additional types of assets and liabilities.

Use the balance sheet to locate the following information.

(a) What is the value of the R. Todd factory buildings?
(b) What is the net worth or capital of the company?
(c) Is there enough cash on hand to pay wages and the immediate bills? If not, where will the additional money come from?

Turn to 4 to check your answers.

4   (a) The value of the buildings is $17,000.
     (b) The capital is $22,500.
     (c) The current liabilities are $11,500 whereas the cash is only $3000. The additional money should come from the accounts receivable and sales from the inventory.

The balance sheet presents a static picture of a business at a given time. It tells us nothing about how the business will change over a certain period of time. To obtain a dynamic picture, one which includes changes in the business' financial picture during a given time period, you must compare two successive balance sheets. An analysis of two balance sheets to show the amount and percent of change in each item is called a *horizontal analysis*.

In a horizontal analysis, first the increase or decrease (change) of each item is calculated. Next, the change in each item is compared to the **first balance sheet's amount** to obtain the percent of increase or decrease. This is equivalent to a %-type percent problems where the base is the first balance sheet's amount and the change is the percentage.

$$\% \text{ of change} = \frac{\text{change}}{\text{first amount}}$$

If you need a review of %-type percent problems, refresh your memory on page 136.

Before we look at the analysis of an entire balance sheet, let's first consider the analysis of a single item.

Example: Two successive annual balance sheets of R. Todd, Inc., show that the total assets for 1983 are $50,000 and for 1984 are $60,000. Complete a horizontal analysis of this item, finding (1) the increase or decrease and (2) the percent increase or decrease.

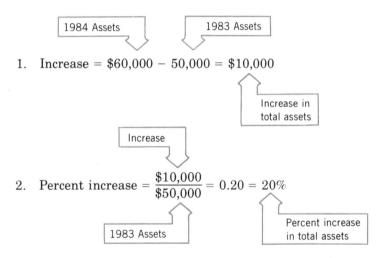

1. Increase = $60,000 − 50,000 = $10,000

2. Percent increase = $\dfrac{\$10,000}{\$50,000} = 0.20 = 20\%$

The increase in total assets is $10,000 and the percent of increase is 20%.

Careful! When calculating the percent of increase or decrease, always use the *first* balance sheet's amount (1983 in our example above) as the base.

It's your turn. Work the following problem.

Two balance sheets of R. Todd, Inc., show that the total liabilities for 1983 are $27,500 and for 1984 are $30,500. Complete a horizontal analysis showing (1) the increase or decrease, and (2) the percent increase or decrease. Round the percent to the nearest 0.1%.

Turn to **5** to check your answer.

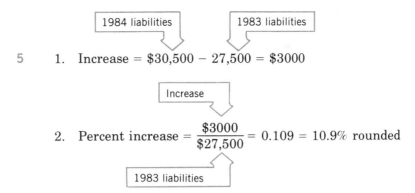

5   1.   Increase = $30,500 − 27,500 = $3000

2.   Percent increase = $\dfrac{\$3000}{\$27,500}$ = 0.109 = 10.9% rounded

Not all items on a balance sheet will be increases. Some items may decrease from one date to the next. In the following problem, calculate the amount and percent of decrease.

Two successive balance sheets of R. Todd, Inc., show that the mortgage payable for 1983 is $16,000 and for 1984 is $15,500. Complete a horizontal analysis of the mortgage payable. Round the percent to the nearest 0.1%.

Check your calculations in 6.

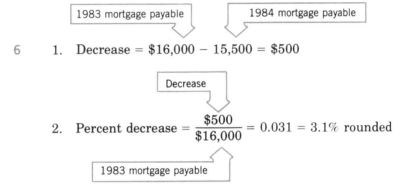

6   1.   Decrease = $16,000 − 15,500 = $500

2.   Percent decrease = $\dfrac{\$500}{\$16,000}$ = 0.031 = 3.1% rounded

On the balance sheet analysis, an asterisk (*) is used to denote decreases.

To calculate the change, simply subtract the smaller amount from the larger. Since the amount in 1984 is smaller than the 1983 amount, there was a decrease.

Whether there is an increase or decrease, the percent change is always the amount of change divided by the amount from the first balance sheet.

To complete a horizontal analysis of two balance sheets, simply perform the analysis item by item. For each item, calculate both the amount and percent of change.

Example: The following is a horizontal analysis of two balance sheets of R. Todd, Inc.. An * is used to denote any decreases.

R. TODD, INC.
COMPARATIVE BALANCE SHEETS
DECEMBER 31, 1983 AND 1984

	1983	1984	Increase or Decrease* Amount	Percent
Assets				
Current Assets				
Cash	$ 3,000	$ 3,500	$ 500	16.7%
Merchandise inventory	5,000	7,000	2,000	40.0%
Accounts receivable	7,000	8,500	1,500	21.4%
Total current assets	$15,000	$19,000	$ 4,000	26.7%
Plant and Equipment				
Land	$ 8,000	$ 9,000	$ 1,000	12.5%
Buildings	17,000	19,000	2,000	11.8%
Equipment	10,000	13,000	3,000	30.0%
Total plant and equipment	$35,000	$41,000	$ 6,000	17.1%
Total assets	$50,000	$60,000	$10,000	20.0%
Liabilities and Capital				
Current Liabilities				
Accounts payable	$ 8,500	$11,000	$ 2,500	29.4%
Wages payable	3,000	4,000	1,000	33.3%
Total current liabilities	$11,500	$15,000	$ 3,500	30.4%
Long-Term Liabilities				
Mortgage payable	16,000	15,500	500*	3.1%*
Total liabilities	$27,500	$30,500	$ 3,000	10.9%
R. Todd, Capital	$22,500	$29,500	$ 7,000	31.1%
Total liabilities and capital	$50,000	$60,000	$10,000	20.0%

* Asterisk denotes a decrease.

It's easy, just work one item at a time.

Ready to do a complete horizontal analysis on your own? Warm up your calculator, wind your mind, and try this one.

Complete a horizontal analysis for Gregory, Inc.

GREGORY, INC.
COMPARATIVE BALANCE SHEETS
DECEMBER 31, 1983 AND 1984

	1983	1984	Increase or Decrease* Amount	Percent
Assets				
Current Assets				
Cash	$ 36,000	$ 38,000		
Merchandise inventory	106,000	98,000		
Accounts receivable	58,000	52,500		
Total current assets	$200,000	$188,000		
Plant and Equipment				
Land	$ 25,000	$ 30,000		
Buildings	63,000	80,000		
Equipment	162,000	202,000		
Total plant and equipment	$250,000	$312,000		
Total assets	$450,000	$500,000		
Liabilities and Capital				
Current Liabilities				
Accounts payable	$ 78,000	$ 75,000		
Wages payable	29,000	30,000		
Total current liabilities	$107,000	$105,000		
Long-Term Liabilities				
Mortgage payable	100,000	90,000		
Total liabilities	$207,000	$195,000		
Gregory, Capital	243,000	305,000		
Total liabilities and capital	$450,000	$500,000		

Round all percents to the nearest 0.1%.

Turn to 7 to check your work.

GREGORY, INC. COMPARATIVE BALANCE SHEETS DECEMBER 31, 1983 AND 1984			Increase or Decrease*	
	1983	1984	Amount	Percent
Assets				
Current Assets				
Cash	$ 36,000	$ 38,000	$ 2,000	5.6%
Merchandise inventory	106,000	98,000	8,000*	7.5%*
Accounts receivable	58,000	52,000	6,000*	10.3%*
Total current assets	$200,000	$188,000	$12,000*	6.0%*
Plant and Equipment				
Land	$ 25,000	$ 30,000	$ 5,000	20.0%
Buildings	63,000	80,000	17,000	27.0%
Equipment	162,000	202,000	40,000	24.7%
Total plant and equipment	$250,000	$312,000	$62,000	24.8%
Total assets	$450,000	$500,000	$50,000	11.1%
Liabilities and Capital				
Current Liabilities				
Accounts payable	$ 78,000	$ 75,000	$ 3,000*	3.8%*
Wages payable	29,000	30,000	1,000	3.4%
Total current liabilities	$107,000	$105,000	$ 2,000*	1.9%*
Long-Term Liabilities				
Mortgage payable	100,000	90,000	10,000*	10.0%*
Total liabilities	$207,000	$195,000	$12,000*	5.8%*
Gregory, Capital	243,000	305,000	62,000	25.5%
Total liabilities and capital	$450,000	$500,000	$50,000	11.1%

Whew! That's a lot of work! The analysis of two balance sheets is a long process, but if you work each item carefully, it's easy!

As we have seen, the balance sheet is divided into two main categories: (1) assets and (2) liabilities and capital. *Vertical analysis* determines what percent each item of a category is of the total. For each asset, the percent of the total assets is calculated. For each liability or capital, the percent of the total liabilities and capital is determined. This calculation is a %-type percent problem where the total assets or total liabilities and capital is the base or total, and the amount of each item is the percentage.

$$\text{Percent} = \frac{\text{amount of individual item}}{\text{total}}$$

An example will clarify this procedure.

Example: On the December 31, 1983 balance sheet for R. Todd, Inc., cash is $3000 and total assets is $50,000. Complete the vertical analysis by determining what percent of the total assets is cash.

$$\text{Percent} = \frac{\$3000}{\$50,000} = 0.06 = 6\%$$

Your turn again. Work the following problem.

On the December 31, 1983 balance sheet for R. Todd, Inc., the merchandise inventory is $5000, the accounts receivable is $7000, and the total assets is $50,000. Complete the vertical analysis for the merchandise inventory and accounts receivable. Find the merchandise inventory as a percent of total assets, and the accounts receivable as a percent of total assets.

Check your work in **8**.

---

8 $$\text{Merchandise Inventory Percent} = \frac{\$5000}{\$50,000} = 0.10 = 10\%$$

$$\text{Accounts receivable percent} = \frac{\$7000}{\$50,000} = 0.14 = 14\%$$

The vertical analysis of a complete balance sheet is simply performed item by item. Each percent calculation is the same as the above examples.

Example: The following example shows a vertical analysis of the December 31, 1983 balance sheet of R. Todd, Inc.

R. TODD, INC. BALANCE SHEET DECEMBER 31, 1983	Amount	Percent
Assets		
Current Assets		
Cash	$ 3,000	6%
Merchandise inventory	5,000	10%
Accounts receivable	7,000	14%
Total current assets	$15,000	30%
Plant and Equipment		
Land	$ 8,000	16%
Buildings	17,000	34%
Equipment	10,000	20%
Total plant and equipment	$35,000	70%
Total assets	$50,000	100%
Liabilities and Capital		
Current Liabilities		
Accounts payable	$ 8,500	17%
Wages payable	3,000	6%
Total current liabilities	$11,500	23%
Long-Term Liabilities		
Mortgage payable	16,000	32%
Total liabilities	$27,500	55%
R. Todd, Capital	22,500	45%
Total liabilities and capital	$50,000	100%

Each percent is calculated by finding what percent each amount is of the total assets or total liabilities and capital. If the percent doesn't work out evenly, round to the nearest tenth of a percent.

The percents may be totaled. In our example, cash is 6%, merchandise is 10%, and accounts receivable is 14%. The total current assets is 6% + 10% + 14% or 30%.

In some cases, the total percent will be more or less than the correct percent because of rounding. The error may be only a couple of tenths of a percent. Any larger error is due to an incorrect calculation.

By comparing these percent differences in the separate balance sheets, management personnel may find financial trends.

Complete the vertical analysis for the 1984 balance sheet of R. Todd, Inc. (The 1983 analysis was completed in the previous example.) Round all percents to the nearest 0.1%.

R. TODD, INC.
COMPARATIVE BALANCE SHEETS
DECEMBER 31, 1983 AND 1984

	1983		1984	
	Amount	Percent	Amount	Percent
Assets				
Current Assets				
Cash	$ 3,000	6.0%	$ 3,500	
Merchandise inventory	5,000	10.0%	7,000	
Accounts receivable	7,000	14.0%	8,500	
Total current assets	$15,000	30.0%	$19,000	
Plant and Equipment				
Land	$ 8,000	16.0%	$ 9,000	
Buildings	17,000	34.0%	19,000	
Equipment	10,000	20.0%	13,000	
Total plant and equipment	$35,000	70.0%	$41,000	
Total assets	$50,000	100.0%	$60,000	
Liabilities and Capital				
Current Liabilities				
Accounts payable	$ 8,500	17.0%	$11,000	
Wages payable	3,000	6.0%	4,000	
Total current liabilities	$11,500	23.0%	$15,000	
Long-Term Liabilities				
Mortgage payable	16,000	32.0%	15,500	
Total liabilities	$27,500	55.0%	$30,500	
R. Todd, Capital	22,500	45.0%	29,500	
Total liabilities and capital	$50,000	100.0%	$60,000	

Turn to **9** to check your work.

9

R. TODD, INC.
COMPARATIVE BALANCE SHEETS
DECEMBER 31, 1983 AND 1984

	1983		1984	
	Amount	Percent	Amount	Percent
Assets				
Current Assets				
Cash	$ 3,000	6.0%	$ 3,500	5.8%
Merchandise inventory	5,000	10.0%	7,000	11.7%
Accounts receivable	7,000	14.0%	8,500	14.2%
Total current assets	$15,000	30.0%	$19,000	31.7%
Plant and Equipment				
Land	$ 8,000	16.0%	$ 9,000	15.0%
Buildings	17,000	34.0%	19,000	31.7%
Equipment	10,000	20.0%	13,000	21.7%
Total plant and equipment	$35,000	70.0%	$41,000	68.3%
Total assets	$50,000	100.0%	$60,000	100.0%
Liabilities and Capital				
Current Liabilities				
Accounts payable	$ 8,500	17.0%	$11,000	18.3%
Wages payable	3,000	6.0%	4,000	6.7%
Total current liabilities	$11,500	23.0%	$15,000	25.0%
Long-Term Liabilities				
Mortgage payable	$16,000	32.0%	15,500	25.8%
Total liabilities	$27,500	55.0%	$30,500	50.8%
R. Todd, Capital	22,500	45.0%	29,500	49.2%
Total liabilities and capital	$50,000	100.0%	$60,000	100.0%

For 1984, the amount of each item is divided by the total assets or total liabilities and capital, $60,000, to calculate the percent.

Notice that the long-term liability dropped from 32% to 25.8%. There was also a considerable increase in capital from 45% to 49.2%.

Note that in 1984, the percent for land is 15.0%, building is 31.7%, and equipment is 21.7%. The total of 15.0% + 31.7% + 21.7% is 68.4%. But the total plant and equipment percent is 68.3%. The 0.1% error is due to the rounding.

Use the following table to complete a vertical analysis for the December 31, 1983 and 1984 balance sheets for Gregory, Inc.

GREGORY, INC.
COMPARATIVE BALANCE SHEETS
DECEMBER 31, 1983 AND 1984

	1983		1984	
	Amount	Percent	Amount	Percent
Assets				
Current Assets				
Cash	$ 36,000		$ 38,000	
Merchandise inventory	106,000		98,000	
Accounts receivable	58,000		52,000	
Total current assets	$200,000		$188,000	
Plant and Equipment				
Land	$ 25,000		$ 30,000	
Buildings	63,000		80,000	
Equipment	162,000		202,000	
Total plant and equipment	$250,000		$312,000	
Total assets	$450,000		$500,000	
Liabilities and Capital				
Current Liabilities				
Accounts payable	$ 78,000		$ 75,000	
Wages payable	29,000		30,000	
Total current liabilities	$107,000		$105,000	
Long-Term Liabilities				
Mortgage payable	100,000		90,000	
Total liabilities	$207,000		$195,000	
Gregory, Capital	243,000		305,000	
Total liabilities and capital	$450,000		$500,000	

Check your work in 10.

10

GREGORY, INC.
COMPARATIVE BALANCE SHEETS
DECEMBER 31, 1983 AND 1984

	1983		1984	
	Amount	Percent	Amount	Percent
Assets				
Current Assets				
Cash	$ 36,000	8.0%	$ 38,000	7.6%
Merchandise inventory	106,000	23.6%	98,000	19.6%
Accounts receivable	58,000	12.9%	52,000	10.4%
Total current assets	$200,000	44.4%	$188,000	37.6%
Plant and Equipment				
Land	$ 25,000	5.6%	$ 30,000	6.0%
Buildings	63,000	14.0%	80,000	16.0%
Equipment	162,000	36.0%	202,000	40.4%
Total plant and equipment	$250,000	55.6%	$312,000	62.4%
Total assets	$450,000	100.0%	$500,000	100.0%
Liabilities and Capital				
Current Liabilities				
Accounts payable	$ 78,000	17.3%	$ 75,000	15.0%
Wages payable	29,000	6.4%	30,000	6.0%
Total current liabilities	$107,000	23.8%	$105,000	21.0%
Long-Term Liabilities				
Mortgage payable	100,000	22.2%	90,000	18.0%
Total liabilities	$207,000	46.0%	$195,000	39.0%
Gregory, Capital	243,000	54.0%	305,000	61.0%
Total liabilities and capital	$450,000	100.0%	$500,000	100.0%

Horizontal and vertical analysis of two successive balance sheets is a detailed procedure composed of many easy calculations. Just remember to work each item carefully.

For a set of practice problems on horizontal and vertical analysis of balance sheets, turn to 11.

13 Financial Statement Analysis

PROBLEM SET 1

11   The answers are on page 614.

13.1   Balance Sheets

A. Complete the horizontal analysis of the following balance sheets. Round all percents to the nearest 0.1%. Use an * to denote any decreases.

1.

PEKARA MANUFACTURING COMPANY COMPARATIVE BALANCE SHEETS DECEMBER 31, 1983 AND 1984			Increase or Decrease	
	1983	1984	Amount	Percent
Assets				
Current Assets				
Cash	$ 27,000	$ 30,000		
Merchandise inventory	78,000	85,000		
Accounts receivable	75,000	76,000		
Total current assets	$180,000	$191,000		
Plant and Equipment				
Land	$ 20,000	$ 20,000		
Buildings	72,000	72,000		
Equipment	85,000	92,000		
Total plant and equipment	$177,000	$184,000		
Total assets	$357,000	$375,000		
Liabilities and Capital				
Current Liabilities				
Accounts payable	$ 72,000	$ 75,000		
Wages payable	25,000	27,000		
Total current liabilities	$ 97,000	$102,000		
Long-Term Liabilities				
Mortgage payable	80,000	78,000		
Total liabilities	$177,000	$180,000		
Pekara, Capital	180,000	195,000		
Total liabilities and capital	$357,000	$375,000		

Date _____

Name _____

Course/Section _____

2.

BRAKEBILL LAB SUPPLY, INC. COMPARATIVE BALANCE SHEETS JUNE 30, 1983 AND 1984			Increase or Decrease	
	1983	1984	Amount	Percent
Assets				
Current Assets				
Cash	$ 5,000	$ 6,000		
Merchandise inventory	16,000	20,000		
Accounts receivable	9,000	9,500		
Total current assets	$30,000	$35,500		
Plant and Equipment				
Land	$ 4,000	$ 4,000		
Buildings	15,000	16,000		
Equipment	16,000	18,000		
Total plant and equipment	$35,000	$38,000		
Total assets	$65,000	$73,500		
Liabilities and Capital				
Current Liabilities				
Accounts payable	$ 7,000	$ 8,500		
Wages payable	5,000	5,500		
Total current liabilities	$12,000	$14,000		
Long-Term Liabilities				
Mortgage payable	13,000	12,500		
Total liabilities	$25,000	$26,500		
Brakebill, Capital	40,000	47,000		
Total liabilities and capital	$65,000	$73,500		

B. Complete the vertical analysis of the following balance sheets. Round all percents to the nearest 0.1%.

1.

PEKARA MANUFACTURING COMPANY
COMPARATIVE BALANCE SHEETS
DECEMBER 31, 1983 AND 1984

	1983		1984	
	Amount	Percent	Amount	Percent
Assets				
Current Assets				
Cash	$ 27,000		$ 30,000	
Merchandise inventory	78,000		85,000	
Accounts receivable	75,000		76,000	
Total current assets	$180,000		$191,000	
Plant and Equipment				
Land	$ 20,000		$ 20,000	
Buildings	72,000		72,000	
Equipment	85,000		92,000	
Total plant and equipment	$177,000		$184,000	
Total assets	$357,000		$375,000	
Liabilities and Capital				
Current Liabilities				
Accounts payable	$ 72,000		$ 75,000	
Wages payable	25,000		27,000	
Total current liabilities	$ 97,000		$102,000	
Long-Term Liabilities				
Mortgage payable	80,000		78,000	
Total liabilities	$177,000		$180,000	
Pekara, Capital	180,000		195,000	
Total liabilities and capital	$357,000		$375,000	

2.

	1983		1984	
	Amount	Percent	Amount	Percent

BRAKEBILL LAB SUPPLY, INC.
COMPARATIVE BALANCE SHEETS
JUNE 30, 1983 AND 1984

	1983 Amount	1983 Percent	1984 Amount	1984 Percent
Assets				
Current Assets				
Cash	$ 5,000		$ 6,000	
Merchandise inventory	16,000		20,000	
Accounts receivable	9,000		9,500	
Total current assets	$30,000		$35,500	
Plant and Equipment				
Land	$ 4,000		$ 4,000	
Buildings	15,000		16,000	
Equipment	16,000		18,000	
Total plant and equipment	$35,000		$38,000	
Total assets	$65,000		$73,500	
Liabilities and Capital				
Current Liabilities				
Accounts payable	$ 7,000		$ 8,500	
Wages payable	5,000		5,500	
Total current liabilities	$12,000		$14,000	
Long-Term Liabilities				
Mortgage payable	13,000		12,500	
Total liabilities	$25,000		$26,500	
Brakebill, Capital	40,000		47,000	
Total liabilities and capital	$65,000		$73,500	

When you have had the practice you need, either return to the Preview on page 506 or continue in 12 with the study of income statements.

SECTION 2: INCOME STATEMENTS

12 | **OBJECTIVE**

Complete a horizontal and vertical analysis of two successive income statements.

Income Statements

The *income statement* is a summary of all income and expenses for a certain period of time, such as a month or year. The main purpose of an income statement is to show the profitability of a business over a certain time period. Unlike the balance sheet, which presents a stationary financial picture or summary, the income statement reflects the business' operation over a given period of time.

The basic accounting formula for income statements is

Net profit = gross profit − operating expenses

For example, if your business had an income of $100, and expenses of $60 this month, the net profit is

Net profit = $100 − 60 = $40

This information may be summarized in an abbreviated income statement that is divided into two main parts: income and operating expenses.

INCOME STATEMENT
FOR THE MONTH ENDED OCTOBER 31, 1984

Income:
   Total income $100
Operating expenses:
   Total operating expenses   60
Net profit $ 40

As you probably guessed, a typical business' income statement will have many more items than the above example. For a look at en expanded income statement and a discussion of the additional items, turn to 13.

13   The following is a typical income statement

R. ALLEN, INC.
INCOME STATEMENT
FOR THE YEAR ENDED DECEMBER 31, 1983

Income:		
Sales		$40,000
Less sales returns		−1,500
(1) Net sales		$38,500
Cost of goods sold:		
Inventory, January 1, 1983	$10,000	
Purchases	25,000	
Goods available for sale	$35,000	
Less inventory, December 31, 1983	−12,000	
(2) Cost of goods sold [⟹] [Subtract⟹]		− 23,000
(3) *Gross profit*		$15,500
Operating expenses		
Selling	$ 1,500	
Rent	2,000	
Utilities	1,800	
Miscellaneous	1,500	
(4) Total operating expenses [⟹] [Add.⟹]		6,800
(5) *Net profit*		$ 8,700

This income statement contains five major summary items: net sales, cost of goods sold, gross profit, total operating expenses and, finally, the net profit.

(1)   The *net sales* is the net income from the sale of goods or services. The net sales is the difference between the total sales and the amount of sales returns and allowances.

Net sales = sales − sales returns
      = $40,000 − 1500 = $38,500

Sales returns and allowances result from incorrect or damaged merchandise shipped to customers.

(2)   Next, the *cost of goods sold* is calculated. The beginning inventory is added to the purchases for the time period to obtain the total goods available for sale. To calculate the cost of the goods sold for the year, we must subtract the value of the ending inventory from the total goods available for sale.

Goods available for sale = beginning inventory + purchases
          = $10,000 + 25,000 = $35,000

Cost of goods sold = goods available for sale − ending inventory
          = $35,000 − 12,000 = $23,000

(3)   The *gross profit* is the difference between the net sales and the cost of goods sold.

Gross profit = net sales − cost of goods sold
        = $38,500 − 23,000 = $15,500

$\textcircled{4}$ The *total operating expenses* is the sum of all the operating expenses. Operating expenses may include selling expense, rent, utilities, advertising, taxes, wages, depreciation, and insurance. In our example, the total operating expenses is $6800.

$\textcircled{5}$ Finally, the *net profit* is calculated using our basic accounting equation for income statements. The net profit is the difference between the gross profit and the total operating expenses.

Net profit = gross profit −total operating expenses
= $15,500 − 6800 = $8700

Turn to **14** for a discussion of horizontal analysis of income statements.

HORIZONTAL
ANALYSIS

14   Horizontal and vertical analysis may be performed on both income statements and balance sheets. The horizontal analysis of two income statements will show the increase or decrease in profit, along with changes in the various expense categories.

In a horizontal analysis, first calculate the increase or decrease of each item. Next, the percent of increase or decrease is computed using the first income statement's amount as the base.

$$\% \text{ of change} = \frac{\text{change}}{\text{first amount}}$$

Example: The income statements of R. Allen, Inc., show the gross profit for 1983 is $15,500 and for 1984 is $20,000. Complete a horizontal analysis of this item finding (1) the increase or decrease and (2) the percent of increase or decrease. Round the percent to the nearest 0.1%.

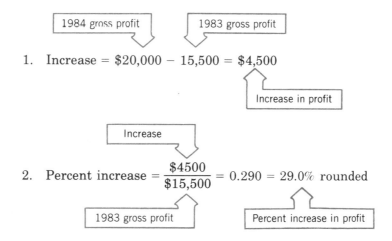

1.   Increase = $20,000 − 15,500 = $4,500

2.   Percent increase = $\dfrac{\$4500}{\$15,500}$ = 0.290 = 29.0% rounded

Work the following problem.

The income statements of R. Allen, Inc., show the net sales for 1983 is $38,500 and for 1984 is $48,000. Complete a horizontal analysis of the net sales finding (1) the increase or decrease and (2) the percent of increase or decrease. Round the percent to the nearest 0.1%.

Turn to **15** to check your work.

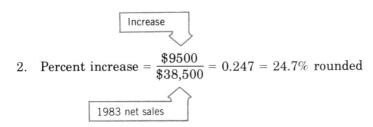

1984 net sales   1983 net sales

15   1. Increase = $48,000 − 38,500 = $9500

Increase

2. Percent increase = $\dfrac{\$9500}{\$38,500}$ = 0.247 = 24.7% rounded

1983 net sales

Careful! Always use the first income statement's amount as the base when calculating the percent change in a horizontal analysis.

The complete horizontal analysis of two income statements is simply performed item by item. Calculate both the change and percent of change for each item.

Example: The following table shows a horizontal analysis of two income statements for R. Allen, Inc. An * is used to denote a decrease.

R. ALLEN, INC.
COMPARATIVE INCOME STATEMENTS
FOR THE YEARS ENDED DECEMBER 31, 1983 AND 1984

	1983	1984	Increase or Decrease* Amount	Percent
Income:				
Sales	$40,000	$50,000	$10,000	25.0%
Less sales returns	1,500	2,000	500	33.3%
Net sales	$38,500	$48,000	$ 9,500	24.7%
Cost of goods sold:				
Inventory, January 1	$10,000	$12,000	$ 2,000	20.0%
Purchases	25,000	30,000	5,000	20.0%
Goods available for sale	$35,000	$42,000	$ 7,000	20.0%
Less inventory, December 31	12,000	14,000	2,000	16.7%
Cost of goods sold	$23,000	$28,000	$ 5,000	21.7%
Gross profit	$15,500	$20,000	$ 4,500	29.0%
Operating expenses:				
Selling	$ 1,500	$ 2,500	$ 1,000	66.7%
Rent	2,000	2,500	500	25.0%
Utilities	1,800	2,000	200	11.1%
Miscellaneous	1,500	1,400	100*	6.7%*
Total operating expenses	$ 6,800	$ 8,400	$ 1,600	23.5%
Net Profit	$ 8,700	$11,600	$ 2,900	33.3%

It's easy; calculate each item separately, row by row. The same technique is used for both income statements and balance sheets.

For practice in this kind of analysis, turn to the next page, and complete a horizontal analysis for Dorn's.

DORN'S
COMPARATIVE INCOME STATEMENTS
FOR THE YEARS ENDED DECEMBER 31, 1983 AND 1984

	1983	1984	Increase or Decrease* Amount	Percent
Income:				
Sales	$150,000	$182,000		
Less sales returns	3,000	2,500		
Net sales	$147,000	$179,500		
Cost of goods sold:				
Inventory, January 1	$ 42,000	$ 48,000		
Purchases	108,000	129,000		
Goods available for sale	$150,000	$177,000		
Less inventory, December 31	48,000	55,500		
Cost of goods sold	$102,000	$121,500		
Gross profit	$ 45,000	$ 58,000		
Operating expenses:				
Selling	$ 18,000	$ 22,000		
Rent	5,000	5,500		
Utilities	2,000	2,500		
Miscellaneous	3,000	2,500		
Total operating expenses	$ 28,000	$ 32,500		
Net Profit	$ 17,000	$ 25,500		

Check your work in 16.

DORN'S
COMPARATIVE INCOME STATEMENTS
FOR THE YEARS ENDED DECEMBER 31, 1983 AND 1984

	1983	1984	Increase or Decrease* Amount	Percent
Income:				
Sales	$150,000	$182,000	$32,000	21.3%
Less sales returns	3,000	2,500	500*	16.7%*
Net sales	$147,000	$179,500	$32,500	22.1%
Cost of goods sold:				
Inventory, January 1	$ 42,000	$ 48,000	$ 6,000	14.3%
Purchases	108,000	129,000	21,000	19.4%
Goods available for sale	$150,000	$177,000	$27,000	18.0%
Less inventory, December 31	48,000	55,500	7,500	15.6%
Cost of goods sold	$102,000	$121,500	$19,500	19.1%
Gross profit	$ 45,000	$ 58,000	$13,000	28.9%
Operating expenses:				
Selling	$ 18,000	$ 22,000	$ 4,000	22.2%
Rent	5,000	5,500	500	10.0%
Utilities	2,000	2,500	500	25.0%
Miscellaneous	3,000	2,500	500*	16.7%*
Total operating expenses	$ 28,000	$ 32,500	$ 4,500	16.1%
Net Profit	$ 17,000	$ 25,500	$ 8,500	50.0%

Careful!

In a horizontal analysis, the percents cannot be added or subtracted. In this case, the individual items are compared "horizontally" or across, from one income statement to another. They are percent changes from one year to the next. The percent numbers *cannot* be added or subtracted *vertically* in a *horizontal* analysis.

Vertical Analysis

In vertical analysis, only items within a given year's income statement are compared. Each individual item is compared to the net sales. For each item the percent of net sales is calculated. This is a %-type percent problem where the amount of each item is the percentage and the net sales is the base.

$$\text{Percent} = \frac{\text{amount of individual item}}{\text{net sales}}$$

Example: On the 1983 income statement for R. Allen, Inc., the total operating expenses are $6800 and the net sales are $38,500. Complete a vertical analysis on this item. Round to the nearest 0.1%.

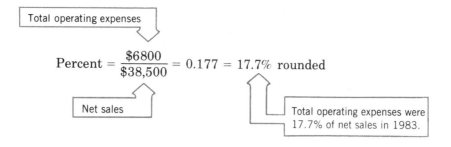

$$\text{Percent} = \frac{\$6800}{\$38,500} = 0.177 = 17.7\% \text{ rounded}$$

The vertical analysis of income statements uses the same procedure as the balance statement, with net sales as the base.

Work the following problem.

On the 1983 income statement for R. Allen, Inc., the cost of goods sold is $23,000 and the net sales are $38,500. Complete a vertical analysis on this item. Round the percent to the nearest 0.1%.

Turn to **17** to check your work.

17 $$\text{Percent} = \frac{\$23,000}{\$38,500} = 0.5974 = 59.7\%$$

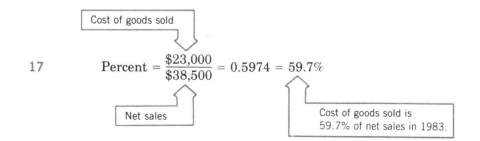

The vertical analysis of a complete income statement is simply performed item by item. Each percent calculation is performed like the above problem.

As in analysis of balance sheets, the vertical analysis of income statements is usually performed on two successive sheets. In this case, management can look for trends by analyzing any significant changes in percents from one income statement to the other.

Example: The following is a complete vertical analysis of the income statements of R. Allen, Inc, for 1983 and 1984.

R. ALLEN, INC. COMPARATIVE INCOME STATEMENTS FOR THE YEARS ENDED DECEMBER 31, 1983 AND 1984	1983		1984	
	Amount	Percent	Amount	Percent
Income:				
Sales	$40,000	103.9%	$50,000	104.2%
Less sales returns	1,500	3.9%	2,000	4.2%
Net sales	$38,500	100.0%	$48,000	100.0%
Cost of goods sold:				
Inventory, January 1	$10,000	26.0%	$12,000	25.0%
Purchases	25,000	64.9%	30,000	62.5%
Goods available for sale	$35,000	90.9%	$42,000	87.5%
Less inventory, December 31	12,000	31.2%	14,000	29.2%
Cost of goods sold	$23,000	59.7%	$28,000	58.3%
Gross profit	$15,500	40.3%	$20,000	41.7%
Operating expenses:				
Selling	$ 1,500	3.9%	$ 2,500	5.2%
Rent	2,000	5.2%	2,500	5.2%
Utilities	1,800	4.7%	2,000	4.2%
Miscellaneous	1,500	3.9%	1,400	2.9%
Total operating expenses	$ 6,800	17.7%	$ 8,400	17.5%
Net Profit	$ 8,700	22.6%	$11,600	24.2%

Remember, all items on the 1983 income sheet are compared to the net sales for 1983. All items on the 1984 income sheet are compared to the net sales for 1984.

Notice that there was a decrease in the percent of purchases and amount of inventory. In other words, R. Allen, Inc., has had a slight percent decrease in costs causing a slight percent increase in gross profit.

Complete a vertical analysis for the 1983 and 1984 income sheets for Dorn's.

DORN'S
COMPARATIVE INCOME STATEMENTS
FOR THE YEARS ENDED DECEMBER 31, 1983 AND 1984

	1983		1984	
	Amount	Percent	Amount	Percent
Income:				
Sales	$150,000		$182,000	
Less sales returns	3,000		2,500	
Net sales	$147,000		$179,500	
Cost of goods sold:				
Inventory, January 1	$ 42,000		$ 48,000	
Purchases	108,000		129,000	
Goods available for sale	$150,000		$177,000	
Less inventory, December 31	48,000		55,500	
Cost of goods sold	$102,000		$121,500	
Gross profit	$ 45,000		$ 58,000	
Operating Expenses:				
Selling	$ 18,000		$ 22,000	
Rent	5,000		5,500	
Utilities	2,000		2,500	
Miscellaneous	3,000		2,500	
Total operating expenses	$ 28,000		$ 32,500	
Net income	$ 17,000		$ 25,500	

Turn to **18** to check your work.

DORN'S
COMPARATIVE INCOME STATEMENTS
FOR THE YEARS ENDED DECEMBER 31, 1983 AND 1984

	1983		1984	
	Amount	Percent	Amount	Percent
Income:				
Sales	$150,000	102.0%	$182,000	101.4%
Less sales returns	3,000	2.0%	2,500	1.4%*
Net sales	$147,000	100.0%	$179,500	100.0%
Cost of goods sold:				
Inventory, January 1	$ 42,000	28.6%	$ 48,000	26.7%
Purchases	108,000	73.5%	129,000	71.9%
Goods available for sale	$150,000	102.0%	$177,000	98.6%
Less inventory, December 31	48,000	32.7%	55,500	30.9%
Cost of goods sold	$102,000	69.4%	$121,500	67.7%
Gross profit	$ 45,000	30.6%	$ 58,000	32.3%
Operating expenses:				
Selling	$ 18,000	12.2%	22,000	12.3%
Rent	5,000	3.4%	5,500	3.1%
Utilities	2,000	1.4%	2,500	1.4%
Miscellaneous	3,000	2.0%	2,500	1.4%
Total operating expenses	$ 28,000	19.0%	$ 32,500	18.1%
Net income	$ 17,000	11.6%	$ 25,500	14.2%

Although the analysis of financial statements is a long process, don't panic. Simply break the problem down and work item by item.

For a set of practice problems on horizontal and vertical analysis of income statements, turn to 19.

13 Financial Statement Analysis

PROBLEM SET 2

19   The answers are on page 618.

13.2   Income
Statements

A. Complete the horizontal analysis of the following income statements. Round all percents to the nearest 0.1%. Use an * to denote any decreases.

1.

TALL COMPANY COMPARATIVE INCOME STATEMENTS FOR THE YEARS ENDED DECEMBER 31, 1983 AND 1984			Increase or Decrease	
	1983	1984	Amount	Percent
Income:				
Sales	$95,000	$110,000		
Less sales returns	2,000	2,500		
Net sales	$93,000	$107,500		
Cost of goods sold:				
Inventory, January 1	$13,000	$ 12,000		
Purchases	62,000	71,000		
Goods available for sale	$75,000	$ 83,000		
Less inventory, December 31	12,000	14,000		
Cost of goods sold	$63,000	$ 69,000		
Gross profit	$30,000	$ 38,500		
Operating expenses:				
Selling	$12,000	$ 15,000		
Rent	3,000	3,500		
Utilities	500	1,000		
Miscellaneous	500	1,500		
Total operating expenses	$16,000	$ 21,000		
Net income	$14,000	$ 17,500		

Date

Name

Course/Section

2.

E. LESLIE, INC. COMPARATIVE INCOME STATEMENTS FOR THE YEARS ENDED DECEMBER 31, 1983 AND 1984				
			Increase or Decrease	
	1983	1984	Amount	Percent
Income:				
Sales	$460,000	$515,000		
Less sales returns	3,000	4,000		
Net sales	$457,000	$511,000		
Cost of goods sold:				
Inventory, January 1	$ 39,000	$ 42,000		
Purchases	359,000	402,000		
Goods available for sale	$398,000	$444,000		
Less inventory, December 31	42,000	41,000		
Cost of goods sold	$356,000	$403,000		
Gross profit	$101,000	$108,000		
Operating expenses:				
Selling	$ 32,000	$ 35,000		
Rent	15,000	16,000		
Utilities	7,000	7,500		
Miscellaneous	11,000	10,500		
Total operating expenses	$ 65,000	$ 69,000		
Net income	$ 36,000	$ 39,000		

B. Complete the vertical analysis of the following income statements. Round all percents to the nearest 0.1%.

1.

TALL COMPANY COMPARATIVE INCOME STATEMENTS FOR THE YEARS ENDED DECEMBER 31, 1983 AND 1984	1983		1984	
	Amount	Percent	Amount	Percent
Income:				
Sales	$95,000		$110,000	
Less sales returns	2,000		2,500	
Net sales	$93,000		$107,500	
Cost of goods sold:				
Inventory, January 1	$13,000		$ 12,000	
Purchases	62,000		71,000	
Goods available for sale	$75,000		$ 83,000	
Less inventory, December 31	12,000		14,000	
Cost of goods sold	$63,000		$ 69,000	
Gross profit	$30,000		$ 38,500	
Operating expenses:				
Selling	$12,000		$ 15,000	
Rent	3,000		3,500	
Utilities	500		1,000	
Miscellaneous	500		1,500	
Total operating expenses	$16,000		$ 21,000	
Net income	$14,000		$ 17,500	

2.

E. LESLIE, INC. COMPARATIVE INCOME STATEMENTS FOR THE YEARS ENDED DECEMBER 31, 1983 AND 1984	1983		1984	
	Amount	Percent	Amount	Percent
Income:				
Sales	$460,000		$515,000	
Less sales returns	3,000		4,000	
Net sales	$457,000		$511,000	
Cost of goods sold:				
Inventory, January 1	$ 39,000		$ 42,000	
Purchases	359,000		402,000	
Goods available for sale	$398,000		$444,000	
Less inventory, December 31	42,000		41,000	
Cost of goods sold	$356,000		$403,000	
Gross profit	$101,000		$108,000	
Operating expenses:				
Selling	$ 32,000		$ 35,000	
Rent	15,000		16,000	
Utilities	7,000		7,500	
Miscellaneous	11,000		10,500	
Total operating expenses	$ 65,000		$ 69,000	
Net income	$ 36,000		$ 39,000	

When you have had the practice you need, either return to the Preview on page 506 or continue in **20** with the study of ratio analysis.

B. Complete the vertical analysis of the following income statements. Round all percents to the nearest 0.1%.

1.

TALL COMPANY COMPARATIVE INCOME STATEMENTS FOR THE YEARS ENDED DECEMBER 31, 1983 AND 1984	1983		1984	
	Amount	Percent	Amount	Percent
Income:				
Sales	$95,000		$110,000	
Less sales returns	2,000		2,500	
Net sales	$93,000		$107,500	
Cost of goods sold:				
Inventory, January 1	$13,000		$ 12,000	
Purchases	62,000		71,000	
Goods available for sale	$75,000		$ 83,000	
Less inventory, December 31	12,000		14,000	
Cost of goods sold	$63,000		$ 69,000	
Gross profit	$30,000		$ 38,500	
Operating expenses:				
Selling	$12,000		$ 15,000	
Rent	3,000		3,500	
Utilities	500		1,000	
Miscellaneous	500		1,500	
Total operating expenses	$16,000		$ 21,000	
Net income	$14,000		$ 17,500	

2.

	1983		1984	
E. LESLIE, INC. COMPARATIVE INCOME STATEMENTS FOR THE YEARS ENDED DECEMBER 31, 1983 AND 1984	Amount	Percent	Amount	Percent
Income:				
Sales	$460,000		$515,000	
Less sales returns	3,000		4,000	
Net sales	$457,000		$511,000	
Cost of goods sold:				
Inventory, January 1	$ 39,000		$ 42,000	
Purchases	359,000		402,000	
Goods available for sale	$398,000		$444,000	
Less inventory, December 31	42,000		41,000	
Cost of goods sold	$356,000		$403,000	
Gross profit	$101,000		$108,000	
Operating expenses:				
Selling	$ 32,000		$ 35,000	
Rent	15,000		16,000	
Utilities	7,000		7,500	
Miscellaneous	11,000		10,500	
Total operating expenses	$ 65,000		$ 69,000	
Net income	$ 36,000		$ 39,000	

When you have had the practice you need, either return to the Preview on page 506 or continue in **20** with the study of ratio analysis.

SECTION 3: RATIO ANALYSIS

20

> **OBJECTIVE**
>
> Calculate the current ratio, acid-test ratio, owner's rate of return, and inventory turnover.

It is often necessary in business, industry, or everyday activities to compare two quantities. In some cases, we could compare them by subtracting to find their differences, but very often a better comparison is found by dividing to find the ratio of the size of one quantity to the other. A *ratio* is a comparison of the sizes of two like quantities found by dividing them. It is a single number commonly written as a fraction.

Example: My car weighs 2100 lb and I weigh 150 lb. What is the ratio of our weights?

$$\text{Ratio of weights} = \frac{\text{car's weight}}{\text{my weight}} = \frac{2100 \text{ lb}}{150 \text{ lb}} = \frac{210}{15} = \frac{14}{1}$$

The ratio may be written as a fraction $\frac{210}{15}$ or $\frac{14}{1}$ or as $14:1$. The ratio $\frac{14}{1}$ or $14:1$ is read "14 to 1."

Jack's height is 74 inches and Jill's height is 61 inches. Calculate the ratio of their heights.

Turn to **21** to check your answer.

21

$$\text{Ratio of heights} = \frac{\text{Jack's height}}{\text{Jill's height}} = \frac{74 \text{ in.}}{61 \text{ in.}}$$

$$= 1.21 \quad \text{or} \quad 1.2 \text{ to } 1 \text{ rounded}$$

Ratios are usually rounded to the nearest tenth.

CURRENT RATIO

One measure of the ability of a business to meet its current debts is the current ratio. The *current ratio* is the ratio of current assets to current liabilities. Both the current assets and liabilities may be obtained from the balance sheet.

> $$\text{Current ratio} = \frac{\text{current assets}}{\text{current liabilities}}$$

Most businesses are expected to maintain a current ratio of at least 2 to 1.

Example: A recent balance sheet for French Manufacturing, Inc., shows current assets of \$204,000, and current liabilities \$85,000. Calculate the current ratio.

$$\text{Current ratio} = \frac{\text{current assets}}{\text{current liabilities}} = \frac{\$204,000}{\$85,000} = \frac{2.4}{1} = 2.4:1$$

The current ratio is $2.4:1$.

In a recent balance sheet for the Anderson Vitamin Company, current assets are $35,150 and current liabilities are $18,500. Calculate the current ratio.

Check your answer in 22.

I run a lawn mower repair service, and my current ratio is 1.5 to 1. Is that bad?

Yep. If your current ratio is less than 2 to 1 you may have a difficult time borrowing money.

22   Current ratio $= \dfrac{\text{current assets}}{\text{current liabilities}} = \dfrac{\$35,150}{\$18,500} = \dfrac{1.9}{1} = 1.9:1$

This ratio is slightly less than the desired current ratio of $2.0:1$.

ACID-TEST RATIO

Not all current assets, such as inventories, can be readily converted into cash. *Liquid assets* are current assets that may be easily converted into cash. Liquid assets include accounts receivable, short-term notes receivable, and cash.

The *acid-test ratio*, or *quick ratio*, is the ratio of liquid assets to current liabilities. The liquid assets and current liabilities are obtained from the balance sheet.

$$\text{Acid-test ratio} = \frac{\text{liquid assets}}{\text{current liabilities}}$$

The acid-test ratio measures the ability of a company to pay off its liabilities in a brief period of time.

Normally, a business is expected to maintain an acid-test ratio of at least $1:1$.

Example: The H. B. Company's recent balance sheet showed cash $30,000, accounts receivable $51,000, and current liabilities $90,000. Calculate the liquid assets and acid-test ratio.

Cash   Accounts receivable

Liquid assets = $30,000 + 51,000 = $81,000

Acid-test ratio $= \dfrac{\text{liquid assets}}{\text{current liabilities}} = \dfrac{\$81,000}{\$90,000} = \dfrac{0.9}{1} = 0.9:1$

The acid-test ratio is $0.9:1$, below the desired $1:1$ ratio.

Platt's Speed Shop's recent balance sheet showed cash $5000, accounts receivable $10,000, and current liabilities $12,500. Calculate the acid-test ratio.

Check your calculations in 23.

23 Liquid assets = $5000 + $10,000 = $15,000

$$\text{Acid-test ratio} = \frac{\text{liquid assets}}{\text{current liabilities}} = \frac{\$15,000}{\$12,500} = \frac{1.2}{1} = 1.2 : 1$$

The acid-test ratio is $1.2 : 1$.

OWNER'S RATE OF RETURN

The *owner's rate of return,* or *rate of return on ownership capital,* shows the profitability of the owner's investment capital. It is also called ROI or return on investment. This gives the rate of return on the investment. In the case of a stock company, the rate of return on ownership capital is a major guide for potential investors. The rate should at least be equal to the current interest rate.

The owner's rate of return is the ratio of net profit or income to the owner's capital. This figure is usually expressed as a percent rather than a ratio number. The net profit is obtained from the income statement and the owner's capital is from the balance sheet.

$$\text{Owner's rate of return} = \frac{\text{net profit}}{\text{owner's capital}}$$

Example: The Hutchinson Corporation's capital is $150,000 and its net profit is $21,000. Calculate the owner's rate of return.

$$\text{Owner's rate of return} = \frac{\text{net profit}}{\text{owner's capital}} = \frac{\$21,000}{\$150,000}$$

$$= \frac{0.14}{1} = 0.14 = 14\%$$

The owner's rate of return is 14%. This is a better rate of return than most savings accounts.

The L. Blackman Company's capital is $76,000 and its net profit is $6000. Calculate the owner's rate of return. Round to the nearest 0.1%.

Turn to **24** to check your answer.

---

24 $$\text{Owner's rate of return} = \frac{\text{net profit}}{\text{owner's capital}} = \frac{\$6000}{\$76,000}$$

$$= \frac{0.079}{1} = 0.079 = 7.9\% \text{ rounded}$$

This is about the same rate of return as some passbook savings accounts or similar investments.

INVENTORY TURNOVER

The *inventory turnover* is a measure of the volume of goods sold as compared to the amount of goods in the inventory. The inventory turnover is the ratio of the cost of goods sold to the average inventory.

$$\boxed{\text{Inventory turnover} = \frac{\text{cost of goods sold}}{\text{average inventory}}}$$

If the inventory turnover is equal to 1, the company sold its entire inventory during the period covered by the income statement. If the inventory turnover is 2, they sold twice their average inventory during this time.

The cost of goods sold is obtained from the income statement. The average inventory may be calculated using the beginning and ending inventories from the income statement.

$$\boxed{\text{Average inventory} = \frac{\text{beginning inventory} + \text{ending inventory}}{2}}$$

The inventory turnover varies widely among different types of businesses.

Example: The recent income statement of the Grumman Corporation showed a beginning inventory of $8000, ending inventory of $10,000, and cost of goods sold $72,000. Calculate the inventory turnover.

First, calculate the average inventory.

$$\text{Average inventory} = \frac{\text{beginning inventory} + \text{ending inventory}}{2}$$

$$= \frac{\$8000 + 10,000}{2} = \frac{\$18,000}{2} = \$9000$$

Next, calculate the inventory turnover.

$$\text{Inventory turnover} = \frac{\text{cost of goods sold}}{\text{average inventory}}$$

$$= \frac{\$72,000}{\$9000} = \frac{8}{1} = 8:1$$

The inventory turnover is 8 to 1. This means that the Grumman Corporation "sold out" its inventory eight times during the time period covered by the income statement. It is a low inventory, fast turnover business.

A recent income statment of M. Flynn, Inc., showed a beginning inventory of $23,000, ending inventory $27,000, and cost of goods sold $180,000. Calculate the inventory turnover.

Check your calculations in **25.**

---

25 $$\text{Average inventory} = \frac{\text{beginning inventory} + \text{ending inventory}}{2}$$

$$= \frac{\$23,000 + 27,000}{2} = \frac{\$50,000}{2} = \$25,000$$

$$\text{Inventory turnover} = \frac{\text{cost of goods sold}}{\text{average inventory}}$$

$$= \frac{\$180,000}{\$25,000} = \frac{7.2}{1} = 7.2:1$$

The inventory turnover is 7.2 to 1. The Flynn, Inc., "sold out" its inventory about 7 times during the year.

For a set of practice problems on ratio analysis, turn to **26.**

---

13 Financial Statement Analysis

PROBLEM SET 3

26   The answers are on page 621.

13.3   Ratio Analysis

A. Calculate the current ratio from the given information. Round to the nearest tenth.

	Current Assets	Current Liabilities	Current Ratio
1.	$ 52,900	$ 23,000	_____
2.	87,150	41,500	_____
3.	55,100	29,000	_____
4.	34,600	17,300	_____
5.	230,000	97,000	_____
6.	142,000	78,500	_____
7.	420,000	156,000	_____
8.	530,000	250,000	_____

B. Calculate the acid-test ratio from the given information. Round to the nearest tenth.

	Liquid Assets	Current Liabilities	Acid-Test Ratio
1.	$104,400	$ 87,000	_____
2.	12,600	14,000	_____
3.	30,800	28,000	_____
4.	63,000	42,000	_____
5.	280,000	236,000	_____
6.	29,000	32,500	_____
7.	30,000	18,000	_____
8.	190,000	147,500	_____

Date

Name

Course/Section

C. Calculate the owner's rate of return from the given information. Round to the nearest 0.1%.

	Owner's Capital	Net Profit	Owner's Rate of Return
1.	$ 27,000	$ 2,160	_____
2.	12,000	2,400	_____
3.	260,000	39,000	_____
4.	94,000	4,700	_____
5.	56,000	4,760	_____
6.	45,000	10,800	_____
7.	150,000	14,000	_____
8.	89,500	11,000	_____

D. Calculate the average inventory and the inventory turnover from the given information. Round to the nearest tenth.

	Cost of Goods Sold	Beginning Inventory	Ending Inventory	Average Inventory	Inventory Turnover
1.	$101,400	$12,000	$14,000	_____	_____
2.	163,400	23,000	20,000	_____	_____
3.	636,500	45,000	50,000	_____	_____
4.	231,650	27,000	29,500	_____	_____
5.	260,000	18,000	20,000	_____	_____
6.	370,000	56,500	52,000	_____	_____
7.	192,000	8,000	7,500	_____	_____
8.	210,000	18,700	17,300	_____	_____

When you have had the practice you need, turn to **27** for the Self-Test on this unit.

13 Financial Statement Analysis

27 Self-Test

The answers are on page 622.

1. Complete the vertical analysis of the following balance sheets. Round all percent to the nearest 0.1%.

J. NICHOLS, INC.
COMPARATIVE BALANCE SHEETS
DECEMBER 31, 1983 AND 1984

	1983 Amount	1983 Percent	1984 Amount	1984 Percent
Assets				
Current Assets				
Cash	$ 12,000		$ 13,000	
Merchandise inventory	45,000		48,000	
Accounts receivable	30,000		35,000	
Total current assets	$ 87,000		$ 96,000	
Plant and Equipment				
Land	$ 8,000		$ 8,000	
Buildings	12,000		17,000	
Equipment	5,000		4,500	
Total plant and equipment	$ 25,000		$ 29,500	
Total assets	$112,000		$125,500	
Liabilities and Capital				
Current Liabilities				
Accounts payable	$ 33,000		$ 34,500	
Wages payable	5,000		5,500	
Total current liabilities	$ 38,000		$ 40,000	
Long-Term Liabilities				
Mortgage payable	15,000		18,000	
Total liabilities	$ 53,000		$ 58,000	
J. Nichols, Capital	59,000		67,500	
Total liabilities and capital	$112,000		$125,500	

Date _____

Name _____

Course/Section _____

2. Complete the horizontal analysis of the following income statements. Round all percents to the nearest 0.1%. Use an * to denote any decreases.

S. KAMM, INC.
COMPARATIVE INCOME STATEMENTS
FOR THE YEARS ENDED DECEMBER 31, 1983 AND 1984

	1983	1984	Increase or Decrease Amount	Percent
Income:				
Sales	$79,000	$85,000		
Less sales returns	4,000	3,500		
Net sales	$75,000	$81,500		
Cost of goods sold:				
Inventory, January 1	$24,000	$27,000		
Purchases	52,000	55,000		
Goods available for sale	$76,000	$82,000		
Less inventory, December 31	27,000	30,000		
Cost of goods sold	$49,000	$52,000		
Gross profit	$26,000	$29,500		
Operating expenses:				
Selling	$ 9,000	$10,000		
Rent	4,000	4,500		
Utilities	1,500	1,600		
Miscellaneous	500	400		
Total operating expenses	$15,000	$16,500		
Net income	$11,000	$13,000		

3. Calculate the current ratio as of December 31, 1984 for J. Nichols, Inc., of Problem 1.

_____

4. Calculate the acid-test ratio as of December 31, 1984 for J. Nichols, Inc. of Problem 1.

_____

5. Calculate the inventory turnover in 1984 for S. Kamm, Inc., of Problem 2.

_____

6. The Waggoner Company's capital is $58,000 and its net profit is $8700. Calculate the owner's rate of return.

_____

Appendix A **The Electronic Calculator**

The hand-held electronic calculator is not a fad. In a very short time it has become an essential tool for people in every occupation, from engineers to shoppers, from clerks to bank presidents. Just as large-scale computers have revolutionized society, the inexpensive pocket calculator has changed personal calculating. For millions of people arithmetic will never be the same again.

It is reasonable to ask: if calculators are available why study arithmetic at all? The answer is that while a calculator can be a very useful tool, it must not become a crutch. A calculator will help you to make mathematical computations more quickly, but it will not tell you *what* to do or *how* to do it. Because the calculator allows you to make very long and difficult calculations quickly, the basic mathematical skills are more important than ever. The gadget itself can't make a mistake, but you can. If you know the basic mathematics skills, you will recognize when an answer must be wrong.

To use a calculator intelligently and effectively in your work, you must be able to:

(a) Multiply and add one-digit numbers quickly and correctly, almost like an instant reflex.
(b) Read and write any number, very large or very small, decimal, whole number, or fraction.
(c) Work with fractions and percents.
(d) Estimate answers and check your calculations.

We have included a very careful review of these basic skills in Units 1 to 4 in this textbook. If you understand these basic skills, the calculator will be a useful tool. If you do not understand these basic skills, you will find that a calculator is simply a lightning-fast way of arriving at a wrong answer.

This appendix is designed to help you to understand the calculator and to use it correctly and effectively.

There are dozens of electronic calculators available. They differ in size, shape, color, cost, and, most important, in how they work and what they can do. In this brief introduction we assume that the calculator you are using has the following characteristics.

★ It has at least an 8-digit display.
★ It performs at least the four basic arithmetic functions: addition (+), subtraction (−), multiplication (×), and division (÷).
★ It uses floating-point arithmetic, so that the position of the decimal point is given automatically in any answer.
★ It uses algebraic logic. (This is by far the most popular type of calculator in use today. If your calculator has a key marked $\boxed{+ =}$, it is probably an "arithmetic logic" device, and some of this appendix will not be very helpful to you.)

Your calculator may have a great many additional features, of course, including square root $\boxed{\sqrt{}}$ or square $\boxed{x^2}$, trigonometric or more advanced functions, one or more memories, and perhaps a programming capability. We'll consider only the simple calculations here. In the following examples, the displays shown for our "average" calculator may differ slightly from what you find with your calculator, but the differences will not be confusing.

THE CALCULATOR

First, check out the machine. You'll find an on-off switch somewhere, a display, and a keyboard with both *numerical* (0, 1, 2, 3, . . . 9) and basic *function* (+, −, ×, ÷, =, .) keys usually arranged like this:

$\boxed{7}\boxed{8}\boxed{9}\boxed{÷}$

$\boxed{4}\boxed{5}\boxed{6}\boxed{×}$

$\boxed{1}\boxed{2}\boxed{3}\boxed{−}$

$\boxed{0}\boxed{.}\boxed{=}\boxed{+}$

In addition, you will find a clear $\boxed{C}$ key somewhere on the keyboard, and perhaps other keys will appear such as $\boxed{\%}$ $\boxed{M+}$ $\boxed{MC}$ $\boxed{\sqrt{}}$ and so on.

In order to understand the operation of the calculator completely, it helps if you know a little about what goes on inside the case. As the following diagram shows, all information, numbers, or arithmetic instructions that are entered on the keyboard go immediately to an *operator register* or Op. Reg., where they are stored until they are used. The number in the operation register is also shown in the display.

All arithmetic calculations or other operations are performed in the *Arithmetic/Logic Unit* or ALU, the heart of the calculator.

ENTERING NUMBERS

The *clear* key $\boxed{C}$ "clears" all entries from the Op. Reg. and ALU, and causes the number zero to appear in the display. Pressing the clear key means that you want to start the entire calculation over or begin a new calculation.

Keyboard   **Display**

$\boxed{C}$ 0.

The *clear entry* key $\boxed{CE}$ "clears" the display only, leaves the Op. Reg. and ALU unchanged, and causes a zero to appear in the display. Pressing the clear entry key $\boxed{CE}$ means that you wish to remove the last entry only.

Keyboard	Display
CE	0.

Every number is entered into the calculator digit by digit, left to right. For example, to enter the number 37.4 the sequence looks like this:

Keyboard	Display
3	3
7	37
.	37.
4	37.4

Don't worry about leading zeros to the left of the decimal point (as in 0.5, 0.664, or 0.15627) or final zeros after the decimal point (as in 4.70, 32.500, or 8.25000). You need not enter these zeros; the calculator will automatically interpret the number correctly and display the zeros if they are needed. Of course no harm will be done if you do enter them.

	Keyboard	Display
	. 5	0.5
or	0 . 5	0.5
or	. 5 0	0.50

If you make an error in entering the number, press either the clear C or clear entry CE key. For example, if you wanted to enter the number 46 and made an error, the following sequence might result:

Keyboard	Display	
4	4	
7	47	an error
CE	0	Pressing the CE key clears the entire number. Start over.
4	4	
6	46	The correct entry

Fractions must be translated to decimal form before they can be entered. The fraction $4\frac{1}{4}$ is entered as 4.25. and $5\frac{1}{9}$ as 5.11111 . . .

Keyboard	Display
4 . 2 5	4.25
5 . 1 1 1 1	5.1111

Converting numbers from fraction to decimal form is covered in Chapter 3, pages 85 to 117 of this textbook.

Exercises A-1   Entering Numbers

Enter each of the following numbers into your calculator. Convert to decimal form if necessary.

1.   327   2.   46,002   3.   0.0120   4.   .4

5.	0.00053	6.	137,620.4	7.	3.1400	8.	seven tenths
9.	$6\frac{1}{2}$ lb	10.	3 dollars and 37¢ (in dollars)	11.	four and one one-quarter inches	12.	one and one-third
13.	$14.75	14.	16¢ (in dollars)	15.	three thousand twenty one	16.	two and one-tenth miles

Check your answers on page 623.

When you solve a problem using a calculator, your result will be a number that appears in the display. To find the actual answer from this display may require a bit of interpretation on your part. For example, if you solve a business problem where the answer is in dollars, the display

    37.6   means $37.60

    0.07   means $ .07 or 7¢

    14250.1   means $14,250.10   . . . and so on.

The calculator is capable of displaying an answer with eight numerical digits, so that in general you must round the displayed number to find the correct answer to your problem. For example, if the answer is in dollars, the display

    6.7912   means $6.79   rounding to the nearest cent,

and

    145.3086   means $145.31   rounding to the nearest cent.

Rounding is discussed in detail in section 2 on page 102.

Exercises A-2   Interpreting Numbers The answers are on page 623.

Interpret each of the following calculator displays as an amount of money.

1.	**4.2**	2.	**0.4**	3.	**0.02**
4.	**11.7426**	5.	**431.751**	6.	**3786506.1**
7.	**0.3881**	8.	**6.007143**	9.	**29214**

Round each of the following displays to the nearest tenth.

| 10. | **4.107** | 11. | **189.01** | 12. | **3278.5674** |
| 13. | **0.1987** | 14. | **4.25** | 15. | **5.719** |

Round each of the following displays to the nearest hundredth.

| 16. | **0.49021** | 17. | **3.6967** | 18. | **137.515** |
| 19. | **12.4849** | 20. | **6.845** | 21. | **4.0076** |

Notice that there are no commas in large numbers displayed on the calculator. The display in Problem 6 above **3786506.1** should be written as $3,786,506.10. You need to count over from the decimal point and insert the commas.

Every numerical entry is built up digit by digit. The calculator does not know if the number is 3 or 37 or 37.4 or 37.492116 until you stop entering numerical digits and press a function key: $+$, $-$, $\times$, $\div$ or $=$.

For example, to add 48 + 37 = ? use this sequence:

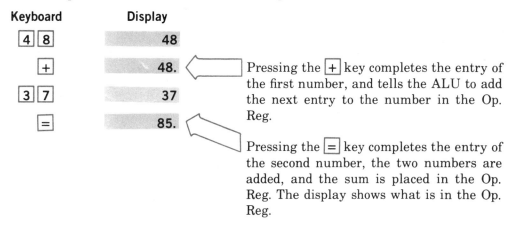

Keyboard	Display
4 8	48
+	48.
3 7	37
=	85.

Pressing the $+$ key completes the entry of the first number, and tells the ALU to add the next entry to the number in the Op. Reg.

Pressing the $=$ key completes the entry of the second number, the two numbers are added, and the sum is placed in the Op. Reg. The display shows what is in the Op. Reg.

Another example: $7.42 \times 3.5 = ?$

Keyboard	Display
7 . 4 2	7.42
$\times$	7.42
3 . 5	3.5
=	25.97

Pressing the $\times$ key completes the entry of the first number into the Op. Reg.

Pressing the $=$ key completes the entry of the second number. The two numbers are multiplied and the product is placed in the Op. Reg. The display shows this product.

Suppose that you made an error in entering one of the numbers to be multiplied. The sequence of actions might look like this:

Keyboard	Display
7 . 4 2	7.42
$\times$	7.42
3 . 6	3.6
CE	0
3 . 5	3.5
=	25.97

Oops, an error.

The correct multiplier.

Notice that pressing the **CE** key clears only the second entry (3.6) and does not disturb the earlier part of the calculation. Pressing the **C** key would clear the entire calculation from the calculator. The **CE** key clears only the display and does not change any previous entries in the Op. Reg.

ESTIMATING

In order to use a calculator most effectively and accurately, you need to develop the skill of *estimating*. Before beginning any calculation you should first estimate the answer by doing simple arithmetic. This estimate gives you a quick and easy check on the calculator answer.

For example, before calculating the addition

4820 + 1241 = ?

mentally round each number to the nearest thousand and add

5000 + 1000 = 6000

Your final calculator answer should be roughly 6000. The actual sum is 6061 in this case.

Estimating the answer in this way greatly reduces the possibility that you will make an error. The first law of effective calculating is

Never use the calculator until you have
a good estimate of the answer.

Here are a few examples of the process of estimating:

Problem	Simplified Problem	Estimate	Actual Answer
82.2 × 47.1	80 × 50	4000	3871.62
0.0912 × 0.615	0.09 × 0.6	0.054	0.056088
$117.92 + $6.37	$120 + $6	$126	$124.29
217.64 ÷ 41.62	200 ÷ 40	5	5.2292167

In each problem round the numbers to form a simplified problem that can be solved mentally—"in your head." Use this mental arithmetic estimate to check the answer obtained on the calculator.

Try the following simple problem.

Exercises A-3   Simple Operations

Perform the following calculations. The answers are on page 624.

1. 3569 + 407
2. 7.6 × 12.8
3. 333 + 777
4. 333 × 777
5. 94.7 − 65.9
6. 4.17 ÷ 3.05
7. 3675.42 − 1996.87
8. 4478.65 ÷ 0.073
9. 0.076 − 0.04
10. 14 ÷ 384.17
11. $42.76 + $5.48
12. $377 ÷ 23
13. Add "two and three-quarters" to "five and one-half."
14. Find the total cost of 6 shirts costing $9.75 each.
15. If hiking socks are on sale at 3 pairs for $6.59, what will one pair cost? (Divide by 3 and round to the nearest cent.)
16. A 23.5 ounce jar of jelly sells for $1.19. What is the cost per ounce? (Divide by 23.5 and round to the nearest cent.)
17. What is the cost of 135 sq ft of flooring at $8.75 per sq ft?

A calculator can only operate on two numbers at a time, but very often you need to use the result of one calculation in a second calculation or in a long string of calculations. Because the answer to any calculation remains in the Op. Reg., you can carry out a very complex chain of calculations without ever stopping the calculator to write down an intermediate step. For example, to add a list of numbers simply continue the addition step.

$4.1 + 0.72 + 12.68 + 3.2 = ?$ **Estimate:** $4 + 1 + 12 + 3 = 20.$

Keyboard	Display	
4 . 1	4.1	
+	4.1	
. 7 2	0.72	
+	4.82	The sum of the first two numbers
1 2 . 6 8	12.68	
+	17.5	The sum of the first three numbers
3 . 2	3.2	
=	20.7	The total sum of all four numbers

Each sum sits in the Op. Reg, and the next number entered is added to the Op. Reg.

Of course we can combine addition and subtraction in the same sequence of operations. For example,
$462 - 294 + 31 = ?$ **Estimate:** $460 - 300 + 30 = 190$

Keyboard	Display	
4 6 2	462	
−	462.	
2 9 4	294	
+	168.	The difference of the first two numbers
3 1	31	
=	199.	The answer

Multiplication of a list of factors is similar to addition and quite simple. For example,

$3.2 \times 4.1 \times 0.53 \times 2.6 = ?$ **Estimate:** $3 \times 4 \times 0.5 \times 3 = 18.$

Keyboard	Display	
3 . 2	3.2	
×	3.2	
4 . 1	4.1	
×	13.12	◁ Product of the first two numbers
. 5 3	0.53	
×	6.9536	◁ Product of the first three factors
2 . 6	2.6	
=	18.07936	◁ Final product of all four factors

Each product sits in the Op. Reg. and the next number multiples the number in the Op. Reg.

Another example: $2.1^3 = ?$ **Estimate:** $2^3 = 8.$

Keyboard	Display
2 . 1	2.1
×	2.1
2 . 1	2.1
×	4.41
2 . 1	2.1
=	9.261

We can easily combine multiplication and division operations. For example,
$2 \times 4 \div 5 = ?$

Keyboard	Display
2	2
×	2.
4	4.
÷	8.
5	5
=	1.6

The operations of multiplication and division can be combined with the operations of addition and subtraction using a calculator, but it may be necessary to write down the intermediate results. If your calculator has a memory, these intermediate results may be stored in the calculator and used in the rest of the calculation.

In general, you must be very careful when combining unlike arithmetic operations. For example, punch this problem into your calculator:

$\boxed{2}\ \boxed{+}\ \boxed{3}\ \boxed{\times}\ \boxed{4}\ \boxed{=}$?

If you press the keys in the order given and your calculator displays the answer as 20, then your calculator is doing the operations sequentially, left to right. It has interpreted these instructions as

$(2 + 3) \times 4$ or 5×4.

A different model calculator may display the answer 14. This second kind of calculator interprets the instructions as

$2 + (3 \times 4)$ or $2 + 12$.

The second calculator follows an order of operations where multiplications or divisions are done first, then additions and subtractions are done.

To be certain there is no misinterpretation, parentheses are used to show which calculation should be done first. For example,

$(3.1 \times 4.2) + 1.5 = ?$ **Estimate:** $(3 \times 4) + 1 = 12 + 1 = 13.$

Keyboard	Display	
$\boxed{3}\boxed{.}\boxed{1}$	**3.1**	
$\boxed{\times}$	**3.1**	
$\boxed{4}\boxed{.}\boxed{2}$	**4.2**	
$\boxed{=}$	**13.02**	Pressing the = key gives the product of the multiplication in the parentheses.
$\boxed{+}$	**13.02**	
$\boxed{1}\boxed{.}\boxed{5}$	**1.5**	
$\boxed{=}$	**14.52**	The final answer

Another example: $(6.16 + 3.75) \times (1.14 + 3.58)$

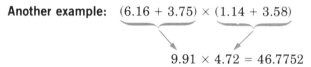

$9.91 \times 4.72 = 46.7752$

In this problem we must write down or store in memory the first sum (9.91) while the second sum is being performed. Once both sums are available, multiply them to find the answer.

Another example: $(37.01 - 14.65) \div (8.4 - 4.791)$

$22.36 \div 3.609 = 6.195622$

A fraction bar is also used to group numbers in a complex calculation. For example, in the following calculation the fraction bar tells you that the upper number $(2 + 4)$ is to be divided by the lower number $(3 + 7)$.

$$\frac{2 + 4}{3 + 7} = \frac{6}{10}$$ Divide 6 by 10 to get $6 \div 10 = 0.6.$

Another example: $\dfrac{3.15 + 1.77}{9.42 - 3.67} = \dfrac{4.92}{5.75}$ Divide: $4.92 \div 5.75 = 0.8556521$

Calculate and write down both numerator and denominator and then divide.

Another example: $\dfrac{(3.18 \times 1.4) + 0.66}{14.4 - 12.007} = \dfrac{5.112}{2.393}$

Dividing $5.112 \div 2.393$ gives 2.1362306.

Exercises A-4 Combined Operations

Perform the following calculations. The answers are on page 624.

1. $35 + 23 + 16 + 47$
2. $32 + 1.8 + 3.09 + 17$
3. $16.4 + 97.1 + 112 + 4.67$
4. $72 - 117 - 6.5 + 68$
5. $3468 - 3061 + 1109 - 702$
6. $1.06 - 0.005 + 0.91 - 0.07$
7. $\$6.24 + \$1.75 - 98¢ + 6¢ - \$2.15$
8. $\$112.41 - \$87.75 - \$31.98 + \$14.95$
9. $37¢ + \$1.14 - 79¢ - 43¢$
10. $6.25 \times 7.04 \times 13.67 \times 2$
11. 3.14×1.17^2
12. 8.04^2
13. $13.19 \div 175.15 \div 2$
14. $\dfrac{81.06}{39.9} \times 1.05$
15. $\dfrac{23.82 \times 31.4}{8.6}$
16. $\dfrac{3.14 \times 1.25^2}{4}$
17. $\dfrac{(4.1 \times 3.2) - 1.8}{1.20}$
18. $\dfrac{(5.92 + 3.08) \times 3.1}{19.1}$
19. $\dfrac{3.5 \times 7.5}{6} + 8.25$
20. $14.9 - \dfrac{3.14 \times 2.8^2}{4}$
21. $6.1 \div \left(\dfrac{2.5 - 1.78}{3}\right)$

22. An employee in the Acme Store is paid \$3.91 per hour and works 32 hours per week. How much does he earn in $5\frac{1}{2}$ weeks?

23. Use the formula $C = \dfrac{5 \times (F - 32)}{9}$ to find C when F is equal to 65. Round to the nearest tenth.

DIVISION WITH A REMAINDER

Division is easy with a calculator. The answer, or quotient, is given as a decimal number. The answer will either be exact or will be given to the limit of the 8-digit display. For example,

$17 \div 7 = 2.4285714$ to eight digits.

In many practical situations it is useful to write the answer to a division as a whole number quotient plus a remainder. For example,

$17 \div 7 = 2$ with remainder 3   In longhand division:

$$
\begin{array}{r}
2 \quad \leftarrow \text{quotient} \\
7\overline{)17} \\
\underline{14} \\
3 \quad \leftarrow \text{remainder}
\end{array}
$$

To do division with a remainder on your calculator follow this sequence of operations:

Keyboard	Display	
1 7	17	
÷	17.	
7	7	
=	2.4285714	The answer to 8 digits
−	2.4285714	
2	2	The quotient is 2.
=	0.4285714	
×	0.4285714	
7	7	
=	2.9999998	The remainder. Round it to 3.

Notice that you must usually round the final remainder to the nearest whole number. In some calculators the rounding is performed automatically and the answer displayed is exact.

Division with a remainder is useful in solving many problems involving units. For example, write the length measurement 191 inches in feet and inches.

Keyboard	Display	
1 9 1	191	
÷	191.	
1 2	12	
=	15.916666	
−	15.916666	
1 5	15	The quotient is 15 ft.
=	0.9166666	
×	0.9166666	
1 2	12	
=	10.999999	The remainder is 11 inches.

191 inches = 15 ft 11 in.

The following set of problems will provide you with a bit of practice in this kind of division.

Exercises A-5　Division with a Remainder

Find the quotient and remainder for each division. The answers are on page 624.

1. $41 \div 9$	2. $87 \div 6$	3. $106 \div 9$
4. $318 \div 13$	5. $999 \div 123$	6. $407 \div 365$
7. $1001 \div 333$	8. $5046 \div 1070$	9. $23781 \div 5107$
10. $1241 \div 12$	11. $1765 \div 60$	12. $207 \div 16$

13. Convert 112 hours to days and hours.
14. Convert 370 inches to feet and inches.
15. Convert 212 minutes to hours and minutes.

Why worry about the metric system? Why use it? First, about 95% of the world's people already measure, think, buy, and sell in metric units. The few who don't—Canada, Great Britain, and the United States—are switching.

Second, the United States already uses metric units in many ways: most electrical units such as volts, amperes, or watts are metric; cameras, film (8mm, 16mm, 35mm), lenses (50mm, 260mm), and other optical parts are measured in metric units; ball bearings, spark plugs, even skis and cigarettes come in metric sizes. Most important of all, the metric system is easier to use and simpler to learn. Working with metric units means there is much less use of fractions, fewer units to remember, less to memorize.

In a few years traffic signs, road maps, U.S. National Park information, grade school textbooks, automobiles, and machine parts may all be in metric units. A pinch of salt, a teaspoonful of sugar, or a cup of flour will probably stay the same in the kitchen and inchworms won't become centimeter worms, but commercial, industrial, and government measurements may be metric. It is important that you start learning to read, use, and think metric.

The most important common units in the metric system are those for length or distance, speed, weight or mass, volume, and temperature. Other metric units are used in science but are not generally needed in day-to-day activities. Time units—year, day, hour, minute, second—are the same in the metric as in the English system.

The basic unit of length in the metric system is the meter (pronounced *meet-ur* and abbreviated m). One meter is a little longer than a yard.

The basic metric unit for weight or mass is the gram (abbreviated g). One ounce is equivalent to about 28 grams; one pound is equal to 454 grams.

The basic metric unit of volume is the liter (pronounced *leet-ur* and abbreviated ℓ). A liter is a little larger than a quart.

One very important feature of the metric system is that there are no 2s, 3s, 4s, 12s, 16s, or 5280s involved when we change from one unit to another. All metric units are multiples of ten times the basic unit. Just as with money, metric units increase in steps of ten.

Money	Prefix	Multiplier
$1000	kilo (k)	1000 ×
$100	hecto (h)	100 ×
$10	deka (da)	10 ×
$1	unit	1 ×
10¢	deci (d)	0.1 ×
1¢	centi (c)	0.01 ×
$\frac{1}{10}$¢	milli (m)	0.001 ×

For example, a kilometer (abbreviated km) is 1000 × 1 meter; km = 1000 m. A centimeter (cm) is 0.01 × 1 meter, or 1 cm = 0.01 m. It is much easier to remember that 1 cm = 0.01 m, or 100 cm equals 1 meter than that 63,360 in. equals 1 mile.

Since the metric system is based on multiples of ten, conversion from metric units to other metric units is simple.

Example 1: Convert 7.3 m to cm.

1 cm = 0.01 m   or   100 cm = 1 m

$$7.3 \text{ m} = 7.3 \text{ m} \times \frac{100 \text{ cm}}{1 \text{ m}} = 7.3 \times 100 \text{ cm} = 730 \text{ cm}$$

Notice that since 100 cm = 1 m, then $\frac{100 \text{ cm}}{1 \text{ m}} = 1$.

When you multiply a number by 1, its value is unchanged. There are, however, two possible fractions: $\frac{100 \text{ cm}}{1 \text{ m}}$ and $\frac{1 \text{ m}}{100 \text{ cm}}$. We used $\frac{100 \text{ cm}}{1 \text{ m}}$ in order to cancel the m units leaving the cm units.

Example 2: Convert 9 cm to mm.

10 mm = 1 cm

$$9 \text{ cm} = 9 \text{ cm} \times \frac{10 \text{ mm}}{1 \text{ cm}} = 9 \times 10 \text{ mm} = 90 \text{ mm}$$

Example 3: Convert 775 dℓ to hℓ.

$$775 \text{ d}\ell = 775 \text{ d}\ell \times \frac{1 \text{ h}\ell}{1000 \text{ d}\ell} = \frac{775}{1000} \text{ h}\ell = 0.775 \text{ h}\ell$$

1 hℓ = 1000 dℓ

Example 4: Convert 8.72 hg to g.

$$8.72 \text{ hg} = 8.72 \text{ hg} \times \frac{100 \text{ g}}{1 \text{ hg}} = 8.72 \times 100 \text{ g} = 872 \text{ g}$$

$$\boxed{1 \text{ hg} = 100 \text{ g}}$$

Exercises A-6 Metric to Metric Conversions

Perform the following metric to metric conversions. The answers are on page 624.

1. 47.2 dag = _____ dg
2. 7842 mℓ = _____ cℓ
3. 3.72 km = _____ dam
4. 0.0037 hg = _____ mg
5. 5287 ℓ = _____ kℓ
6. 42.3 dm = _____ mm
7. 84 kg = _____ g
8. 7280 mℓ = _____ dkℓ
9. 45800 cm = _____ km
10. 0.75 hg = _____ cg

11. 342 dm = _____ dam
12. 0.82 g = _____ mg
13. 4300 dℓ = _____ hℓ
14. 350 dam = _____ km
15. 32 dag = _____ g
16. 42 dm = _____ mm
17. 10,000 m = _____ km
18. 85 cℓ = _____ mℓ
19. 85 cℓ = _____ ℓ
20. 2500 g = _____ kg

In order to become familiar with the metric system it is important to *think metric*. Most people in the United States know about how much a foot is in length, how much a pound is in weight, and how much a quart is in volume. But what is a meter, a gram, and a liter? The following will give you some idea of metric measurements.

LENGTH

1. The thickness of a dime is approximately one millimeter.
2. A king-size cigarette is about 100 millimeters.
3. An unsharpened pencil is approximately 18 centimeters long.
4. A meter is a little longer than a yard.
5. A mile is about 1.6 km.

WEIGHT

1. A paperclip weighs approximately 1 gram.
2. A nickel weighs 5 grams.
3. One kilogram is about 2.2 pounds.
4. A 100-pound barbell weighs about 45 kilograms.
5. A football player weighs about 120 kilograms.

VOLUME

1. One teaspoon is approximately 5 milliliters.
2. One fluid ounce is about 30 milliliters.
3. A liter is a little larger than a quart.
4. A gallon is approximately 3.8 liters.

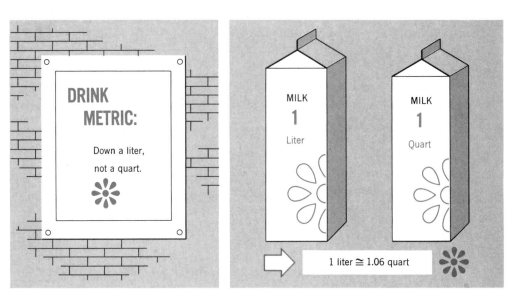

DRINK
METRIC:

Down a liter,
not a quart.

MILK
1
Liter

MILK
1
Quart

1 liter ≅ 1.06 quart

SPEED

1. Driving 55 miles per hour is about 90 kilometers per hour.
2. Driving 25 miles per hour is about 40 kilometers per hour.

Exercises A-7 Think Metric

State whether the following statements are likely or unlikely. The answers are on page 624.

1. The basketball player was 3 meters tall.
2. The bicycle was traveling 20 kilometers per hour.
3. He drank a liter of beer in one gulp.
4. The ballet dancer weighed 150 kilograms.
5. He was driving 80 kilometers per hour on the freeway and was arrested for speeding.
6. The pencil weighed 100 grams.
7. A cowboy was wearing a 35-liter hat.
8. He ate a 5-kilogram steak.
9. A football field is about 100 meters long.
10. She won the 10,000-kilometer race in 37:15 minutes.

Before all measurements are metric, we must be able to convert from one system to another. The following is a short table of English–metric conversions.

ENGLISH—METRIC CONVERSIONS

Length

1 inch = 2.54 cm exactly	0.0394 inch ≅ 1 mm
1 foot ≅ 30.5 cm	0.394 inch ≅ 1 cm
1 yard ≅ 91.4 cm	39.4 inch ≅ 1 m
1 mile ≅ 1610 m	3.28 feet ≅ 1 m
1 mile ≅ 1.61 km	1.09 yards ≅ 1 m
	0.621 mile ≅ 1 km

Weight

1 ounce ≅ 28.3 g	0.0353 ounce ≅ 1 g
1 pound ≅ 454 g	0.0022 pound ≅ 1 g
1 pound ≅ 0.454 kg	2.2 pounds ≅ 1 kg

Volume

1 gallon ≅ 3.79 ℓ	0.264 gallon ≅ 1 ℓ
1 quart ≅ 0.946 ℓ	1.06 quart ≅ 1 ℓ

≅ means "approximately equal"

Example 5: Convert 152 meters to yards.

$$152 \text{ m} = 152 \text{ m} \times \frac{1.09 \text{ yd}}{1 \text{ m}} = 152 \times 1.09 \text{ yd} = 165.68 \text{ yd}$$

1 m = 1.09 yd

Example 6: Convert 6 inches to centimeters.

$$6 \text{ in.} = 6 \text{ in.} \times \frac{2.54 \text{ cm}}{1 \text{ in.}} = 6 \times 2.54 \text{ cm} = 15.24 \text{ cm}$$

1 in. = 2.54 cm

Example 7: Convert 3 pounds to grams.

$$3 \text{ lb} = 3 \text{ lb} \times \frac{454 \text{ g}}{1 \text{ lb}} = 3 \times 454 \text{ g} = 1362 \text{ g}$$

1 lb = 454 g

Example 8: Convert 5 liters to gallons.

$$5 \ \ell = 5 \ \ell \times \frac{0.264 \text{ gal}}{1 \ \ell} = 5 \times 0.264 \text{ gal} = 1.32 \text{ gal}$$

1 ℓ = 0.264 gal

Example 9: Convert 25 kilograms to pounds.

$$25 \text{ kg} = 25 \text{ kg} \times \frac{2.2 \text{ lb}}{1 \text{ kg}} = 25 \times 2.2 \text{ lb} = 55 \text{ lb}$$

1 kg = 2.2 lb

Exercises A-8   English–Metric Conversions

Perform the following English–metric conversions. The answers are on page 624.

1. 5 in. = _____ cm
2. 1300 g = _____ lb
3. 3 gal = _____ ℓ
4. 2 m = _____ in.
5. 10 lb = _____ kg
6. 5 km = _____ mi
7. 6 qt = _____ ℓ
8. 5 lb = _____ kg
9. 10 km = _____ mi
10. 20 ℓ = _____ gal

11. 100 mm = _____ in.
12. 8 cm = _____ in.
13. 10 mi = _____ m
14. 5 oz = _____ g
15. 50 kg = _____ lb
16. 2 gal = _____ ℓ
17. 4 ℓ = _____ qt
18. 150 lb = _____ kg
19. 55 mi = _____ km
20. 5 yd = _____ m

Acid-test ratio. The acid-test ratio is the ratio of liquid assets to current liabilities.

Aliquot part. An aliquot part is any number that can be divided evenly into another number.

Amortization schedule. An amortization schedule is a listing of all principal and interest payments and principal balance on a loan.

APR. The APR or Annual Percentage Rate of interest is the true interest rate of a loan.

Automobile insurance. Automobile insurance is financial protection for automobiles and related injuries.

Balance sheet. A balance sheet is a business' financial condition summary at a particular time.

Bank discount. Bank discount is a method of lending money where the principal is discounted by the amount of interest.

Bank reconciliation. Bank reconciliation is a procedure of checking the bank's summary of checks, service fees, deposits, and balance against the checkbook register or check stubs.

Bond. A bond is a long-term note with a stated interest rate issued by companies to raise money.

Book value. The book value of an asset is its current accounting value as determined by a depreciation method.

Cancelling. Cancelling is the process of dividing by or eliminating common factors in a fraction.

Cash discounts. Cash discounts are discounts given for prompt payment.

Check. A check is a method of transferring money from one individual's account to another.

Check register. A check register is used to record checking account transactions.

Check stub. A check stub is attached to a check and is used to record checking transactions.

Coinsurance. Coinsurance is a method where the insured carries less than the required amount of insurance and shares a portion of any loss with the insurance company.

Commission. Commission is a payroll method in which each employee's wages are determined by his total sales.

Compound interest. Compound interest is a method of accumulating interest where the interest earned during each period is added to the principal.

Cost. The cost is the amount a business pays for merchandise.

Credit card accounts. A credit card account is an account with a store or lending institution where purchases may be "charged" to your account, and interest is calculated on the unpaid balance.

Current ratio. The current ratio is the ratio of current assets to current liabilities.

Decimal number. A decimal number is a fraction whose denominator is 10, 100, 1000, 10,000, and so on.

Denominator. The denominator of a common fraction is the bottom number.

Depreciation. Depreciation is the gradual decline in value of an asset.

Digits. Digits are the basic symbols used to construct numerals. The ten digits in our number system are 0, 1, 2, 3, 4, 5, 6, 7, 8, and 9.

Discount. A discount is an amount deducted from the list price.

Dividend. A dividend is a share of a company's profits distributed to the stockholder.

Exact time. Exact time is a method of counting the exact number of days between two dates. You do not count the first day.

Excise tax. Excise tax is a sales tax levied by the federal government on the manufacture, sale, or consumption of a commodity such as gasoline, tires, jewelry, or firearms.

FICA. FICA is an abbreviation for the Federal Insurance Contributions Act. It is a federal payroll deduction commonly known as Social Security.

Fire insurance. Fire insurance provides protection against fire loss and related damage.

Fraction. A common fraction is a number expressed as the quotient of two whole numbers.

Gram. A gram is the basic unit of weight in the metric system. One ounce is equivalent to about 28 grams.

Gross. The gross amount is before any deductions such as gross pay.

Horizontal analysis. A horizontal analysis shows the amount and percent of change in two successive financial statements.

Hourly. Hourly is a payroll method in which each employee's wages are determined by the number of hours worked.

Improper fraction. An improper fraction is a fraction greater than one.

Income statement. An income statement is a summary of all revenue and expenses for a certain period of time, such as a month or year.

Installment loan. An installment loan is a method of borrowing where regular equal payments are required.

Insurance. Insurance is financial protection against a loss.

Interest. Interest is the payment for the use of money.

Inventory valuation. Inventory valuation is the listing of individual items of merchandise in stock with their value.

Inventory turnover. The inventory turnover is the ratio of the cost of goods sold to the average inventory.

Invoice. An invoice is a record of merchandise ordered along with shipping and billing information.

Least common denominator. The least common denominator of several fractions is the smallest number evenly divisible by the fractions' denominators.

Life insurance. Life insurance provides family support after the death of a family income earner.

Liter. A liter is the basic unit of volume in the metric system. A liter is a little larger than a quart.

Markdown. Markdown is a reduction of the selling price.

Markup. Markup is the amount added to a business' cost to cover expenses and profit.

Maturity value. The maturity value of a loan is the total amount paid back, that is, principal plus interest.

Meter. A meter is the basic unit of length in the metric system. One meter is a little longer than a yard.

Metric system. The metric system is a system of measure based on ten.

Mill. A mill is one thousandths of a dollar; $0.001.

Mixed number. A mixed number is an improper fraction written as the sum of a whole number and a proper fraction.

Mortgage. A mortgage is a real estate loan.

Net. The net amount is the amount after deductions, such as net pay and net cost.

Numerator. The numerator of a common fraction is the top number.

Owner's rate of return. The owner's rate of return is the ratio of net profit or income to the owner's capital.

Payroll sheet. A payroll sheet is a summary of employees' earnings records.

Percent. A percent is a number expressed in hundredths.

Piecework. Piecework is a payroll method in which each employee's wage is determined by his production.

Place value system. In the place value system of naming numbers, the value of any digit depends on the place where it is located.

Points. Points are a percent of the principal charged the seller for FHA and VA loans.

Present value. The present value is the current value of a future sum.

Prime number. A prime number is a number greater than 1 and only divisible by itself and one.

Principal. The principal of a loan is the amount of money borrowed.

Proper fraction. A proper fraction is a number less than 1.

Proration. Proration is the process of dividing the cost of an expense over a period of time.

Ratio. A ratio is a comparison of the sizes of two like quantities found by dividing them.

Regulation Z. Regulation Z is the federal Truth-in-Lending Act requiring disclosures of credit costs.

Rounding. Rounding is a process of approximating a number.

Salary. A salary is a fixed amount of money paid to an employee for certain assigned duties.

Selling price. The selling price is a business' price to individual consumers. It is the sum of cost and markup.

Stock. Stock is issued by a corporation to raise money. Each share of stock represents a portion of the company's ownership.

Thirty-day-month time. Thirty-day-month time is a method of calculating the number of days between two dates that assumes each month has 30 days. You do not count the first day.

Trade discounts. A trade discount is an amount subtracted from the list price by the buyer. Trade discounts were originally given to members of various trades or businesses.

Vertical analysis. A vertical analysis of a financial statement shows what percent each item of a category is of the total.

UNIT 1: WHOLE NUMBERS

Problem Set 1: Page 7

A.
1. Eighty-two
2. Seventeen
3. One hundred fifty-six
4. Two hundred forty-seven
5. Eight hundred fifty-one
6. Four hundred thirty-nine
7. One thousand, eight hundred twenty-five
8. Seven thousand, six hundred twenty-one
9. Fourteen thousand, four hundred ninety-six
10. Fifty-seven thousand, four hundred eighteen
11. Two hundred thirty-nine thousand, one hundred fifty-three
12. One hundred fifty-six thousand, four hundred eighty-three
13. Ninety-two million, seven hundred forty-two thousand, seven
14. Four million, seven hundred eighty-five thousand, three hundred ten
15. One hundred eighty-six million, fourteen thousand, one hundred fifty-seven
16. Twenty-four million, one hundred sixty thousand, four hundred thirty-eight
17. Two million, seven hundred sixty-five thousand, nine hundred eighty-four
18. Five million, seven hundred one
19. Twenty-three billion, four hundred ninety-three million, six hundred thirty-one thousand, three hundred seventy-five
20. One hundred eighty-six billion, three hundred ninety-three million, six hundred thirteen thousand, seventy-four

B.
1.	97	2.	42
3.	652	4.	429
5.	24,000	6.	78,065
7.	9605	8.	747,489
9.	18,567,032	10.	83,903,651
11.	26,498	12.	28,406,531
13.	9,641,817	14.	98,005
15.	500,432	16.	39,809
17.	56,460,513,627	18.	14,821,401,052
19.	111,000,293,378	20.	2,400,000,186

Box, page 12

(a) 91, (b) 81, (c) 46, (d) 92, (e) 74, (f) 130

Problem Set 2: Page 15

A.
1.	70	2.	104	3.	65	4.	126	5.	80	6.	106
7.	112	8.	72	9.	131	10.	103	11.	105	12.	123
13.	103	14.	124	15.	100	16.	132	17.	52	18.	136

B.	1. 415	2. 393	3. 1113	4. 1003	5. 1530
	6. 1390	7. 1016	8. 831	9. 1262	10. 1009
	11. 824	12. 806	13. 1241	14. 861	15. 1001

C.	1. 5525	2. 9563	3. 9461	4. 2611	5. 9302
	6. 3513	7. 3702	8. 12,599	9. 7365	10. 10,122
	11. 6505	12. 11,428	13. 5781	14. 15,715	15. 9403
	16. 11,850				

D.	1. 25,717	2. 11,071	3. 70,251	4. 21,642	5. 14,711
	6. 89,211	7. 47,111	8. 175,728	9. 101,011	10. 180,197
	11. 110,102	12. 128,876			

E.	1. 1042	2. 5211	3. 2442	4. 6441	5. 7083
	6. 16,275	7. 6352	8. 7655	9. 6514	10. 9851

F.	1. $5251	2. $327	3. $1859	4. 371	5. $10,535
	6. $419				

7.

Branch	M	T	W	Th	F	S	Weekly Total
A	$567	$687	$592	$653	$751	$826	$4076
B	497	506	519	527	632	758	3439
Daily Totals	1064	1193	1111	1180	1383	1584	7515

8. 671   9. $8316   10. $353

11.

Employee	M	T	W	Th	F	Weekly Total
M. Schallhorn	$826	$921	$752	$950	$625	$4074
B. Coleman	675	680	723	850	752	3680
K. Wilson	515	758	652	725	432	3082
J. Mee	455	872	953	862	759	3901
E. Milliron	520	755	659	782	526	3242
Total	2991	3986	3739	4169	3094	17,979

12. 138,487

Problem Set 3: Page 23

A.	1. 6	2. 5	3. 7	4. 6	5. 2	6. 5
	7. 8	8. 0	9. 4	10. 1	11. 9	12. 8
	13. 3	14. 7	15. 0	16. 9	17. 3	18. 7
	19. 3	20. 4	21. 8	22. 1	23. 8	24. 4
	25. 9	26. 7	27. 9	28. 7	29. 9	30. 4
	31. 9	32. 6	33. 3	34. 1	35. 6	36. 6
	37. 8	38. 5	39. 5	40. 8	41. 7	42. 7
	43. 18	44. 7	45. 7	46. 3	47. 8	48. 9
	49. 0	50. 2				

B.	1. 13	2. 44	3. 29	4. 16	5. 12

	6. 57	7. 19	8. 17	9. 15	10. 28
	11. 36	12. 18	13. 22	14. 37	15. 25
	16. 26	17. 38	18. 85		

C.	1. 189	2. 458	3. 85	4. 877	5. 281
	6. 176	7. 154	8. 266	9. 273	10. 198
	11. 715	12. 51	13. 574	14. 45	15. 29
	16. 145				

D.	1. 2809	2. 7781	3. 5698	4. 28,842
	5. 12,518	6. 7679	7. 56,042	8. 37,328
	9. 4741	10. 9897	11. 9614	12. 26,807
	13. 47,593	14. 316,640	15. 22,422	16. 55,459
	17. 24,939	18. 7238	22 222	44 333

E.	1. $595	2. $89	3. 1546	4. $487	5. $3528
	6. $4048	7. $257	8. $23	9. $1896	10. 402

Problem Set 4: Page 31

A.	1. 42	2. 72	3. 56	4. 63	5. 48
	6. 54	7. 72	8. 42	9. 54	10. 63
	11. 56	12. 48			

B.	1. 87	2. 402	3. 576	4. 243	5. 282
	6. 792	7. 320	8. 259	9. 156	10. 294
	11. 290	12. 261	13. 564	14. 392	15. 153
	16. 161	17. 282	18. 424	19. 308	20. 324
	21. 720	22. 1505	23. 1728	24. 2736	25. 5040
	26. 2952	27. 7138	28. 1170	29. 1938	30. 2548
	31. 1650	32. 1349	33. 4484	34. 1458	35. 928
	36. 6232	37. 3822	38. 2030	39. 8930	40. 2752

C.	1. 37,515	2. 74,820	3. 375,750	4. 97,643
	5. 297,591	6. 384,030	7. 38,023	8. 108,486
	9. 378,012	10. 1,279,840	11. 41,064	12. 4,947,973
	13. 30,780	14. 225,852	15. 1,368,810	16. 31,152
	17. 397,584	18. 43,381	19. 60,241	20. 5,098,335

D.	1. $205	2. $338	3. $1196	4. $310	5. $4644
	6. $5035	7. $238	8. 24,440	9. $1224	10. 98,280
	11. $663	12. $11,304 (6 × 12 × $157)			

Problem Set 5: Page 39

A.	1. 9	2. 12	3. 11, remainder 4
	4. 7	5. Not defined	6. 13
	7. 7, remainder 2	8. 5	9. 10, remainder 1
	10. 7	11. 1	12. Not defined
	13. 8	14. 6	15. 4
	16. 7	17. 6	18. 9

B.	1. 35	2. 41	3. 23, remainder 6
	4. 42	5. 57	6. 46, remainder 2
	7. 51, remainder 4	8. 45	9. 112, remainder 1
	10. 44	11. 21	12. 27
	13. 52	14. 37	15. 88
	16. 125	17. 50, remainder 1	18. 67

C.	1. 23	2. 21	3. 20, remainder 2
	4. 31, remainder 4	5. 39	6. 50, remainder 2
	7. 25	8. 19, remainder 17	9. 9, remainder 1

10. 41	11. 53	12. 43
13. 22	14. 11, remainder 34	15. 12
16. 34	17. 9, remainder 6	18. 71, remainder 5

D.
1. 95, remainder 6   2. 104   3. 96
4. 208   5. 142, remainder 6   6. 107
7. 222, remainder 2   8. 170, remainder 10   9. 32
10. 1000   11. 305, remainder 5   12. 311, remainder 8
13. 84, remainder 41   14. 100, remainder 5   15. 119
16. 61   17. 102, remainder 98   18. 81

E.
1. 25   2. $265   3. 67, yes   4. $29   5. 24
6. $189   7. $2875   8. $19   9. 1530   10. $572

Self-Test 1: Page 41

1. Two million, eight hundred forty-seven thousand, three hundred ninety-one
2. 48,561,072   3. 17,085   4. $909   5. $13,066   6. $247
7. 2,078,909   8. 2,376   9. $359   10. $276

UNIT 2: FRACTIONS

Problem Set 1: Page 53

A.
1. $\frac{7}{3}$,   2. $\frac{22}{5}$,   3. $\frac{15}{2}$,   4. $\frac{94}{7}$,   5. $\frac{35}{4}$,
6. $\frac{4}{1}$,   7. $\frac{5}{3}$,   8. $\frac{35}{6}$,   9. $\frac{31}{8}$,   10. $\frac{13}{5}$,
11. $\frac{161}{10}$,   12. $\frac{535}{9}$,   13. $\frac{481}{40}$,   14. $\frac{170}{11}$,   15. $\frac{113}{3}$

B.
1. $8\frac{1}{2}$,   2. $7\frac{2}{3}$,   3. $1\frac{3}{5}$,   4. $4\frac{3}{4}$,   5. $6\frac{1}{6}$,
6. $9\frac{1}{3}$,   7. $4\frac{5}{8}$,   8. $4\frac{1}{7}$,   9. $1\frac{9}{25}$,   10. $5\frac{2}{9}$,
11. $52\frac{3}{4}$,   12. $7\frac{9}{23}$,   13. $4\frac{3}{10}$,   14. $20\frac{5}{6}$,   15. $9\frac{4}{15}$

C.
1. $\frac{13}{15}$,   2. $\frac{4}{5}$,   3. $\frac{4}{5}$,   4. $\frac{1}{2}$,   5. $\frac{1}{8}$
6. $\frac{2}{5}$,   7. $\frac{1}{6}$,   8. $\frac{8}{9}$,   9. $\frac{1}{3}$,   10. $\frac{3}{8}$
11. $\frac{7}{20}$,   12. $\frac{3}{8}$,   13. $\frac{1}{6}$,   14. $\frac{4}{7}$,   15. $\frac{5}{9}$

D.
1. $\frac{14}{16}$,   2. $\frac{27}{45}$,   3. $\frac{9}{12}$,   4. $\frac{145}{60}$,   5. $\frac{7}{63}$
6. $\frac{45}{35}$,   7. $\frac{20}{32}$,   8. $\frac{140}{25}$,   9. $\frac{39}{78}$,   10. $\frac{34}{51}$
11. $\frac{363}{44}$,   12. $\frac{82}{14}$,   13. $\frac{66}{72}$,   14. $\frac{185}{50}$,   15. $\frac{516}{54}$

Problem Set 2: Page 59

A.
1. $\frac{1}{8}$,   2. $\frac{1}{9}$,   3. $\frac{4}{15}$,   4. $\frac{1}{8}$
5. $\frac{2}{15}$,   6. $\frac{5}{6}$,   7. 3,   8. $\frac{1}{2}$
9. $2\frac{2}{3}$,   10. $\frac{5}{6}$,   11. $\frac{11}{45}$,   12. $\frac{9}{56}$,
13. $1\frac{1}{9}$,   14. $10\frac{1}{2}$,   15. $\frac{13}{16}$,   16. $\frac{4}{5}$
17. $2\frac{1}{2}$,   18. 6,   19. 14,   20. $1\frac{3}{7}$

B.
1. 3,   2. 4,   3. 8,   4. $1\frac{5}{6}$
5. $3\frac{1}{4}$,   6. 62,   7. $1\frac{1}{21}$,   8. $\frac{1}{3}$
9. 69,   10. 6,   11. $35\frac{3}{4}$,   12. $1\frac{3}{11}$
13. 74,   14. $7\frac{9}{10}$,   15. $9\frac{7}{8}$,   16. $46\frac{2}{3}$
17. $10\frac{3}{8}$,   18. $13\frac{13}{30}$,   19. $21\frac{1}{3}$,   20. 6

C.
1. 1530   2. 92   3. $\frac{7}{8}$ sq miles   4. $12
5. 110 km   6. $222\frac{3}{4}$ miles   7. $394   8. $46
9. $322\frac{1}{2}$ 10. $200

Problem Set 3: Page 67

A. 1. $1\frac{2}{3}$, 2. $1\frac{3}{4}$, 3. 9, 4. $\frac{1}{12}$
    5. $\frac{5}{16}$, 6. $\frac{4}{9}$, 7. 24, 8. $\frac{7}{16}$
    9. $\frac{8}{13}$, 10. $7\frac{1}{2}$, 11. 1, 12. $1\frac{1}{2}$
    13. $\frac{5}{28}$, 14. $1\frac{4}{5}$, 15. $2\frac{2}{5}$, 16. $\frac{1}{12}$

B. 1. 9, 2. $7\frac{1}{3}$, 3. 4, 4. $\frac{3}{4}$
    5. $1\frac{1}{3}$, 6. 6, 7. $1\frac{3}{4}$, 8. $2\frac{4}{7}$
    9. $1\frac{1}{5}$, 10. $8\frac{1}{3}$, 11. $1\frac{1}{4}$, 12. $1\frac{2}{3}$
    13. $\frac{6}{7}$, 14. $5\frac{1}{4}$, 15. $\frac{4}{5}$, 16. $5\frac{1}{3}$

C. 1. 16 2. $\frac{3}{8}$, 3. $\frac{1}{9}$ 4. $3\frac{1}{9}$,
    5. 18 6. 25 7. $\frac{6}{7}$, 8. 6
    9. $1\frac{1}{4}$ 10. $\frac{6}{29}$, 11. $17\frac{1}{2}$, 12. 17

D. 1. 34 mph 2. $1\frac{3}{8}$ 3. 57 4. 40
    5. $10\frac{8}{13}$ 6. 26 7. 26 8. 40

Problem Set 4: Page 81

A. 1. $1\frac{2}{5}$, 2. $1\frac{1}{3}$, 3. 1, 4. $\frac{2}{3}$
    5. $\frac{1}{4}$, 6. $\frac{1}{2}$, 7. $\frac{5}{12}$, 8. $\frac{3}{8}$
    9. $1\frac{1}{8}$, 10. $1\frac{1}{8}$, 11. $\frac{3}{4}$, 12. $\frac{1}{4}$
    13. $1\frac{3}{8}$, 14. $1\frac{2}{3}$, 15. $1\frac{4}{7}$ 16. $\frac{5}{8}$,
    17. $1\frac{1}{4}$, 18. $\frac{9}{20}$

B. 1. $1\frac{5}{8}$, 2. $\frac{1}{8}$, 3. $1\frac{11}{36}$, 4. $\frac{19}{36}$
    5. $\frac{29}{48}$, 6. $\frac{29}{35}$, 7. $\frac{53}{192}$, 8. $\frac{215}{216}$
    9. $\frac{13}{48}$, 10. $\frac{4}{35}$, 11. $\frac{5}{96}$, 12. $\frac{83}{216}$
    13. $1\frac{5}{8}$, 14. $4\frac{1}{36}$, 15. $5\frac{17}{48}$, 16. $4\frac{51}{56}$
    17. $\frac{3}{4}$, 18. $\frac{3}{8}$, 19. $1\frac{23}{48}$, 20. $\frac{51}{70}$

C. 1. $1\frac{2}{3}$, 2. $1\frac{13}{16}$, 3. $5\frac{1}{4}$, 4. $1\frac{4}{5}$
    5. $20\frac{3}{4}$ 6. $1\frac{89}{120}$, 7. $\frac{33}{40}$ 8. $1\frac{17}{60}$
    9. $3\frac{1}{12}$ 10. $2\frac{13}{40}$ 11. $5\frac{11}{48}$ 12. $1\frac{9}{10}$

D. 1. $36\frac{2}{3}$ hrs.; 2. $77\frac{3}{4}$ 3. $7\frac{7}{8}$ 4. (a) $28\frac{2}{5}$
       $238\frac{1}{3}$ or   (b) $19\frac{3}{4}$
       $238.33
    5. $414\frac{7}{8}$ 6. $12\frac{3}{8}$ 7. $2\frac{3}{16}$ 8. $8\frac{5}{8}$ lb
    9. $\frac{5}{12}$ lb 10. $53\frac{1}{8}$

Self-Test 2: Page 83

1. $1\frac{15}{16}$ 2. $3\frac{4}{11}$ 3. $\frac{15}{40}$ 4. $\frac{13}{17}$
5. $1\frac{19}{60}$ 6. $42\frac{7}{24}$ 7. $2\frac{7}{12}$ 8. 5
9. $\frac{26}{95}$ 10. (a) $31\frac{1}{5}$ gal; (b) $18\frac{1}{2}$ mpg

---

UNIT 3: DECIMAL NUMBERS

Problem Set 1: Page 93

A. 1. 0.8 2. 1.6 3. 1.5 4. 0.6 5. 0.9
    6. 1.2 7. 1.6 8. 0.7 9. 1.7 10. 0.1
    11. 0.7 12. 3.3 13. 0.5 14. 2.3 15. 1.8
    16. 1.4 17. 3.3 18. 5.2 19. 1.9 20. 9.9
    21. 4.8 22. 1.4 23. 9.0 24. 18.1 25. 5.5
    26. 1.4 27. 17.5 28. 1.6 29. 6.7 30. 0.6
    31. 3.6 32. 1.3 33. 1.8 34. 2.6 35. 2.6
    36. 0.4

B. 1. 21.01   2. $15.02   3. 1.617   4. 27.19

1. 21.01	2. $15.02	3. 1.617	4. 27.19
5. $30.60	6. 6.486	7. 78.17	8. $151.11
9. 5.916	10. 828.60	11. 16.2019	12. 1031.28
13. 63.7313	14. 238.24	15. 128.3685	16. 45.195
17. $27.59	18. 70.871	19. 108.37	20. 19.37
21. $15.36	22. 51.34	23. 1.04	24. 3.86
25. 42.33	26. 6.63	27. 6.52	28. 6.42
29. $36.18	30. 22.016	31. 2.897	32. $17.65
33. 6.96	34. 0.3759		

C.

1. 151.461	2. 602.654	3. 95.888	4. 91.15
5. 316.765	6. 14.67755	7. 16.0425	8. 35.4933
9. 19.011	10. 3.34974		

D. 1. (a) $8.58 (b) $2.15 (c) $4.02 (d) $3.53 (e) $37.02 (f) $10.40
2. $415.35   3. $46.84   4. $1.84   5. $225   6. $7469.37
7.

Month	A	B	C	D	Monthly Totals
January	$13,846.29	$12,465.23	$23,156.18	$ 9,476.18	$ 58,943.88
February	13,756.82	13,567.30	25,203.25	10,386.22	62,913.59
March	14,050.26	13,406.27	24,860.29	9,956.28	62,273.10
Quarter Total	$41,653.37	$39,438.80	$73,219.72	$29,818.68	$184,130.57

8. $189.06   9. 4,687.8   10. $723.62   11. 0.9086   12. $11.79

Box: Page 104

1. $2\frac{5}{6}$ 2. 84.5   3. $663.45

Problem Set 2: Page 107

A.

1. 0.00001	2. 0.1	3. 21.5	4. 0.06
5. 0.008	6. 0.014	7. 0.09	8. 0.72
9. 0.84	10. 0.009	11. 0.00006	12. 4.20
13. 1.4743	14. 2.18225	15. 0.5022	16. 0.03
17. .024	18. 2.16	19. 0.01476	20. 3.6225
21. 1.44	22. 80.35	23. 0.00117	24. 287.5
25. .03265	26. 0.0009255	27. 1223.6	

B.

1. 1300	2. 12.6	3. 0.045	4. 13
5. 126	6. 450	7. 60	8. 2000
9. 10,000	10. .037	11. 11.2	12. 6.6
13. 1900	14. 3256.25	15. 0.11	

C.

1. 3.33	2. 1.43	3. 0.83	4. 0.09	5. 10.53
6. 0.67	7. 37.50	8. 0.89	9. 0.12	10. 73.40
11. 2.62	12. 0.12	13. 33.86	14. 474.44	15. 5.00

1. 33.3	2. 1.1	3. 0.2	4. 6.7	5. 0.3
6. 0.1	7. 11.1	8. 285.7	9. 0.2	10. 16.0

1. 14.286	2. 0.023	3. 0.225	4. 65.000	5. 3.462
6. 13.640	7. 3.815	8. 2.999	9. 571.429	10. 1109.001

D.

1. $121.55	2. $126	3. $22.74	4. $194.74
5. $11,218.48	6. 18.5	7. $988, $288	8. 2653.615

9. $4179.24   10. (a) $2.67 (b) $8.64 (c) $3.04 (d) $35.80
11. $26.35   12. 39.1   13. $208.50
14. $77.28   15. $716.40 (15 × 12 × $3.98)

Problem Set 3: Page 115

A. 1. .50   2. .33   3. .67   4. .25   5. .50   6. .75
    7. .20   8. .40   9. .60   10. .80   11. .17   12. .83
    13. .14   14. .29   15. .43   16. .13   17. .38   18. .63
    19. .88   20. .10   21. .20   22. .30   23. .08   24. .17
    25. .25   26. .42   27. .58   28. .92   29. .06   30. .19
    31. .31   32. .44   33. .56   34. .69   35. .81   36. .94

B. 1. $\frac{3}{10}$ 2. $\frac{3}{4}$ 3. $\frac{11}{25}$ 4. $\frac{4}{5}$ 5. $\frac{3}{5}$ 6. $\frac{1}{40}$
    7. $\frac{2}{5}$ 8. $1\frac{3}{10}$ 9. $2\frac{1}{4}$ 10. $2\frac{1}{20}$ 11. $3\frac{4}{25}$ 12. $1\frac{1}{8}$
    13. $3\frac{11}{50}$ 14. $2\frac{1}{25}$ 15. $\frac{3}{40}$ 16. $10\frac{7}{8}$ 17. $\frac{7}{10000}$ 18. $\frac{3}{2500}$
    19. $\frac{17}{50}$ 20. $11\frac{21}{2000}$ 21. $6\frac{1}{500}$ 22. $4\frac{23}{200}$ 23. $\frac{7}{20}$ 24. $\frac{191}{200}$

C. 1. $10.56   2. $7.33   3. $18.06   4. $33.22
    5. $0.43   6. .278   7. $12.18   8. $186.83
    9. $5.19   10. $19.16 11. 91.5

Self-Test 3: Page 117

1. 36.0105   2. 15.0608   3. 142.086   4. 2.36
5. $19.56   6. $\frac{14}{25}$ 7. 1.3125   8. $163.67
9. $315.57   10. $1.59

UNIT 4: PERCENT

Problem Set 1: Page 129

A. 1. 40%   2. 10%   3. 95%   4. 3%   5. 30%
    6. 1.5%   7. 60%   8. 1%   9. 120%   10. 456%
    11. 225%   12. 775% 13. .3%   14. 300%   15. 80%
    16. 550%   17. 400% 18. 604% 19. 1000% 20. 33.5%

B. 1. 20%   2. 75%   3. 70%   4. 35%   5. 150%
    6. 25%   7. 10%   8. 50%   9. $37\frac{1}{2}$%   10. 60%
    11. 175%   12. 220% 13. 180% 14. 90%   15. $33\frac{1}{3}$%
    16. $216\frac{2}{3}$%   17. $66\frac{2}{3}$%   18. $68\frac{3}{4}$%   19. $191\frac{2}{3}$%   20. 330%

C. 1. .07   2. .03   3. .56   4. .15   5. .01
    6. .075   7. .90   8. 2.00   9. .003   10. .0007
    11. .0025 12. 1.50 13. .015 14. .063   15. .005
    16. .1225 17. 1.255 18. .667 19. .305   20. .085

D. 1. $\frac{1}{10}$ 2. $\frac{13}{20}$ 3. $\frac{1}{2}$ 4. $\frac{1}{5}$ 5. $\frac{1}{4}$
    6. $\frac{2}{25}$ 7. $\frac{9}{10}$ 8. $1\frac{7}{20}$ 9. $\frac{3}{100}$ 10. $\frac{3}{25}$
    11. $\frac{1}{200}$ 12. $\frac{3}{10000}$ 13. $\frac{9}{200}$ 14. $2\frac{1}{5}$ 15. $\frac{3}{200}$
    16. $\frac{1}{3}$ 17. $\frac{31}{400}$ 18. $\frac{13}{200}$ 19. $\frac{1}{6}$ 20. $\frac{1}{32}$

Problem Set 2: Page 143

A. 1. 80%   2. 20%   B. 1. 225   2. 7755
    3. $9   4. 64%   3. 25%   4. 18
    5. 15   6. 107   5. $.1974 or $.20   6. 180%
    7. 100%   8. 54   7. 160   8. $2.72
    9. 21   10. 60   9. 150%   10. $12\frac{1}{2}$%
    11. $12\frac{1}{2}$%   12. 150   11. $26\frac{2}{3}$%   12. 4.5

	13.	59	14.	80			13.	2%	14.	$6\frac{2}{3}\%$
	15.	$66\frac{2}{3}\%$	16.	500%			15.	17.5	16.	25%
	17.	100	18.	52%			17.	400%	18.	427
	19.	500	20.	$21.25			19.	22.1	20.	1600

C. 1. 32.5% (rounded)   2. $14,900   3. $805.60   4. 46¢
    5. $7.47   6. $43.22   7. 15%   8. $4.50; $4.77
    9. $688.50   10. $1196.25 11. $4598.10 12. $45,860.40
   13. $903.40   14. 4.33%   15. 92.5%   16. 5.5%
   17. 27   18. $7020   19. $3.30; $13.20 20. $3.40

Self-Test 4: Page 147

 1. 31.25%   2. 127%   3. 0.023   4. $\frac{7}{20}$ 5. 18%
 6. $482.56 7. 270   8. $60,800 9. 5%   10. 250,000

UNIT 5: MATHEMATICS OF BUYING AND SELLING

Problem Set 1: Page 159

A. 1. 8.00, 192.00   2. 18.50, 351.50   3. 14.40, 105.60
    4. 149.50, 500.50   5. 15.13, 259.87   6. 35.88, 427.12
    7. 8.28, 129.67   8. 51.91, 524.91   9. 16.85, 181.42
   10. 104.07, 745.51

B. 1. 214.13   2. 128.25   3. 442.68   4. 607.20   5. 278.43
    6. 227.11   7. 259.34   8. 259.48   9. 245.85 10. 279.75

C. 1. 27.325%   2. 38.8%   3. 31.15%   4. 31.184%   5. 42.625%
    6. 37.1%   7. 30.0025% 8. 18.775% 9. 40.96%   10. 33.445%

D. 1. 31.6%, 164.16   2. 29.62%, 246.33   3. 16.21%, 418.95
    4. 38.8%, 459.00   5. 21.34%, 667.82   6. 32.68%, 333.23
    7. 18.811%, 208.45   8. 35.2%, 247.85   9. 49.6%, 213.45
   10. 55.375%, 70.12

Problem Set 2: Page 167

A. 1. 4.00, 196.00   2. 10.50, 339.50   3. 10.00, 490.00
    4. 5.00, 245.00   5. 0, 700.00   6. 0, 625.00
    7. 14.25, 460.75   8. 16.84, 825.16   9. 22.47, 726.53
   10. 2.56, 125.44   11. 9.02, 441.98   12. 14.84, 727.16
   13. 8.50, 841.50   14. 11.20, 548.80   15. 0, 125.00
   16. 1.05, 33.95   17. 4.70, 230.30   18. 0, 972
   19. 14.40, 705.60   20. 19.05, 615.95

B. 1. 9.00, 291.00   2. 9.00, 441.00   3. 8.00, 392.00
    4. 15.00, 485.00   5. 8.25, 266.75   6. 0, 236.00
    7. 25.29, 817.66   8. 10.53, 252.66   9. 9.73, 476.64
   10. 5.05, 121.13   11. 7.18, 232.24   12. 8.37, 409.99
   13. 5.07, 248.18   14. 13.66, 441.63   15. 12.65, 619.80
   16. 16.32, 799.89   17. 2.53, 123.85   18. 0, 29.56
   19. 2.19, 70.77   20. 12.91, 632.35

Problem Set 3: Page 175

A. 1. 8975.00   2. 12,994.17   3. 1451.00

B. 1. (a) 8835   (b) 9190   (c) 8050
    2. (a) 12,960   (b) 13,015.78 (c) 12,802

3. (a) 1424.96 (b) 1476 (c) 1252
4. (a) 27,746.56 (b) 22,101.50 (c) 31,552.00

Problem Set 4: Page 189

A.
1. 172.11 2. 21.45 3. 43.76 4. 76.18 5. 99.26
6. 9.79 7. 36.42 8. 814.25 9. 23.14 10. 294.03
11. 15.79 12. 342.57 13. 37.44 14. 74.83 15. 41.43
16. 313.10 17. 57.69 18. 365.25 19. 14.27 20. 206.97

B.
1. 3.75, 28.75 2. 42.50, 292.50 3. 170.00, 214.20
4. 50.00, 67.50 5. 441.00, 47% 6. 108.75, 15%
7. 400.00, 76.00 8. 175.00, 73.50 9. 237.57, 292.21
10. 47.50, 285.00 11. 633.75, 30% 12. 178.20, 26.73
13. 164.22, 193.78 14. 138.42, 36% 15. 208.68, 52.17
16. 50.00, 206.26 17. 36.66, 189.41 18. 285.40, 56%
19. 27.50, 44% 20. 12.98, 42.48 21. 186.50, 160.39
22. 845.60, 1454.43 23. 256.80, 89.88 24. 17.50, 3.15
25. 695.40, 278.16

C.
1. 231.97, 57.99 2. 46.33, 10.17 3. 210.74, 250.88
4. 113.16, 152.92 5. 120.89, 23% 6. 171.99, 30%
7. 12.69, 42.30 8. 45.15, 122.03 9. 32.71, 52.76
10. 567.35, 378.23 11. 62.70, 24% 12. 6.54, 32.70
13. 129.00, 27% 14. 149.40, 249.00 15. 14.06, 4.44
16. 100.08, 142.97 17. 185.41, 104.29 18. 514.51, 46%
19. 72.15, 20.35 20. 176.85, 982.50 21. 852.80, 40%
22. 198.44, 360.80 23. 98.31, 655.40 24. 92.33, 263.80
25. 193.53, 322.55

Problem Set 5: Page 197

A.
1. 49.95, 199.80 2. 19.64, 95.86 3. 8.75, 64.15
4. 89.00, 267.00 5. 3.59, 16.36 6. 45.37, 84.27
7. 35.80, 53.70 8. 1.14, 21.56 9. 197.00, 659.50
10. 1097.24, 1868.26 11. 111.68, 186.14 12. 144.94, 308.01
13. 4.16, 14.34 14. 5.09, 24.83 15. 35.67, 202.13
16. 226.77, 529.12 17. 25.27, 46.94 18. 33.93, 55.37
19. 140.72, 115.13 20. 54.69, 140.63

B.
1. 182.50, 27% 2. 145.29, 13% 3. 246.80, 20%
4. 127.50, 15% 5. 107.00, 40% 6. 163.21, 24%
7. 375.64, 25% 8. 275.00, 36% 9. 274.08, 32%
10. 15.76, 20%

C.
1. 0.87, 15.37 2. 1.50, 76.50 3. 4.69, 192.25
4. 12.48, 262.13 5. 50.21, 886.97 6. 3.77, 129.52
7. 3.41, 38.39 8. 13.59, 466.75 9. 6.12, 128.60
10. 1.96, 34.68 11. 0.50, 10.06 12. 29.42, 764.84

Self-Test 5: Page 199

1. (a) 33.4% (b) 183.15
2. (a) 237.80 (b) 56.50 (c) 826.11 (d) 131.73 (e) 22.71 (f) 551.16
3. (a) 23,502.50 (b) 24,990 (c) 20,700 4. 24,090
5. (a) 297.42, 20% (b) 175.50, 231.66
   (c) 14.75, 1.18 (d) 40.50, 190.50

6. (a) 149.97, 40%   (b) 15.17, 3.33
   (c) 227.63, 875.50 (d) 199.96, 249.95
7. 6205.64
8. 251.72

UNIT 6: PAYROLL

Problem Set 1: Page 213

A. 1. $1250   2. $625   3. $1381.25   4. $257.50
   5. $195.25   6. $516.25   7. $352.50   8. $168
   9. $765   10. $458.75   11. $752.50   12. $275.50

B. 1. $1462.50; $731.25; $675; $337.50   2. $1215.50; $607.75; $561; $280.50
   3. $1280.50; $640.25; $591; $295.50   4. $767; $383.50; $354; $177
   5. $949; $474.50; $438; $219   6. $669.50; $334.75; $309; $154.50
   7. $1007.50; $503.75; $465; $232.50   8. $513.50; $256.75; $237; $118.50
   9. $1527.50; $763.75; $705; $352.50   10. $786.50; $393.25; $363; $181.50
   11. $1560; $780; $720; $360   12. $2340; $1170; $1080; $540
   13. $1541.67; $770.83; $711.54; $355.77
   14. $1875; $937.50; $865.38; $432.69
   15. $2362.50; $1181.25; $1090.38; $545.19

C. 1. 36, $113.04   2. 30, $88.50   TOTAL: $947.40
   3. 40, $138   4. 39, $122.85
   5. 31, $115.32   6. 38, $136.04
   7. 30, $97.50   8. 35, $136.15

D.

	Name	M	T	W	T	F	Total Hours	Regular Hours	Regular Rate	O.T. Hours	O.T. Rate	Regular Pay	O.T. Pay	Gross Pay
1.	Allgood, Harrell	9	8	10	8	9	44	40	$3.24	4	$4.86	$ 129.60	$ 19.44	$ 149.04
2.	Bertone, Marcus	9	9	9	9	8	44	40	2.96	4	4.44	118.40	17.76	136.16
3.	Edwards, Lawrence	8	8	8	8	8	40	40	3.46	0		138.40	0	138.40
4.	Gilmore, Marylina	9	8	8	9	10	44	40	3.50	4	5.25	140.00	21.00	161.00
5.	Hawkins, Rita	9	8	8	9	9	43	40	3.88	3	5.82	155.20	17.46	172.66
6.	Kimball, M. A.	8	7	8	8	6	37	37	3.76	0		139.12	0	139.12
7.	Parke, Robert	9	8	9	10	7	43	40	3.24	3	4.86	129.60	14.58	144.18
8.	Whitley, Debra	9	8	10	10	10	47	40	3.96	7	5.94	158.40	41.58	199.98
	TOTALS											1108.72	131.82	1240.54

E.

Name	M	T	W	T	F	Total Hours	Regular Hours	Regular Rate	O.T. Hours	O.T. Rate	Regular Pay	O.T. Pay	Gross Pay
1. Canaan, Rob-bye	8	7	6	5	4	30	30	$5.92	0	—	177.60	0	177.60
2. Culpepper, Lori	8	7	9	9	9	42	40	6.35	2	9.525	254.00	19.05	273.05
3. Moore, Charlene	9	8	9	8	9	43	40	5.80	3	8.70	232.00	26.10	258.10
4. Schein, Don	9	10	9	7	9	44	40	6.15	4	9.225	246.00	36.90	282.90
5. Schein, Pat	9	10	10	9	10	48	40	6.45	8	9.675	258.00	77.40	335.40
Totals											1167.60	159.45	1327.05

Problem Set 2: Page 221

A.

	Name	Pieces Completed M T W T F					Total Pieces	Rate	Gross Pay
1.	Anderson, Robert	65	72	71	68	66	342	$0.52	$ 177.84
2.	Burns, Thomas	62	71	65	62	64	324	0.49	158.76
3.	Dutton, Charles	75	79	75	76	81	386	0.47	181.42
4.	Jones, Roberta	76	74	75	75	76	376	0.51	191.76
5.	Perry, Jacqulyn	42	45	45	46	43	221	0.75	165.75
6.	Soliz, Elizabeth	82	83	84	84	81	414	0.48	198.72
7.	Tuttle, Wesley	95	96	92	93	91	467	0.46	214.82
8.	Williams, Velma	25	26	26	25	24	126	1.25	157.50
	TOTAL								$1446.57

B.

	Name	Bearings per Day M T W T F					Total	Gross Wages
1.	Adeleke, Joel	65	62	68	69	61	325	$ 131.00
2.	Hutte, Patsy	68	70	72	71	74	355	144.20
3.	Jones, Steve	62	64	63	62	64	315	126.60
4.	King, Richard	55	52	51	59	60	277	110.80
5.	Minor, Mark	69	70	75	76	74	364	148.16
6.	Myatt, Patti	82	81	85	87	82	417	172.33
7.	Reimler, Edward	75	76	77	77	76	381	155.64
8.	Tarrant, Betty	85	79	76	84	88	412	169.88
	TOTAL							$1158.61

C.

	Name	Total Sales	Gross Pay
1.	Biggs, Bobby	$ 25,700	$ 771.00
2.	Cao, Hung	42,500	1275.00
3.	Cummings, Sandra	35,500	1065.00
4.	Diglovanni, John	28,900	867.00
5.	Jenkins, Dorothy	42,300	1269.00
6.	Morgan, Geraldine	39,950	1198.50
7.	Pack, Kathy	24,200	726.00
8.	Walsh, Thomas	41,500	1245.00
	TOTAL	$280,550	$8416.50

D.

	Name	Total Sales	Gross Pay
1.	Brooks, Mendel	$ 4,241.76	$ 270.55
2.	Hampton, Charles	2,975.46	178.53
3.	Johnson, Leonard	3,152.34	190.66
4.	Kauble, Cecelia	3,851.85	239.63
5.	Martin, Jim	3,247.98	197.36
6.	Norton, Thomas	4,289.55	274.61
7.	Sitter, David	2,457.26	147.44
8.	Whibbey, Betty	4,557.86	297.42
	TOTAL	$28,774.06	$1796.20

E.

	Name	Salary	Sales	Commission	Gross Pay
1.	Caldwell, Rita	$ 150	$ 547.98	$13.70	$ 163.70
2.	Crownover, Vernon	145	425.56	10.64	155.64
3.	Devilbiss, Virgil	150	398.26	9.96	159.96
4.	Gaines, Russell	165	502.23	12.56	177.56
5.	Houston, Larry	160	655.93	16.40	176.40
6.	Morrell, Pamela	155	602.57	15.06	170.06
7.	Platt, Bill	250	125.57	3.14	253.14
8.	Weiner, Michael	170	509.98	12.75	182.75
	TOTAL	$1345	$3768.08	$94.21	$1439.21

Problem Set 3: Page 239

A.

	Name	Marital Status	No. of Exemptions	Gross Pay	Federal Income Tax	FICA	Other Deductions	Total Deductions	Net Pay
1.	Anderson, Lynda	M	2	$214.29	15.60	14.36	25.76	55.72	158.57
2.	Cusack, Mary	S	1	256.65	31.90	17.20	18.95	68.05	188.60
3.	Harris, Thomas	M	4	289.19	20.60	19.38	19.42	59.40	229.79
4.	Hendricks, Ray	M	3	198.57	10.90	13.30	23.57	47.77	150.80
5.	Lytle, Leslie	S	1	152.95	14.80	10.25	15.16	40.21	112.74
6.	McCormick, George	S	1	165.41	16.30	11.08	21.95	49.33	116.08
7.	Morgan, William	M	3	217.89	13.30	14.60	32.56	60.46	157.43
8.	Trivitt, Antoinette	S	2	187.42	16.40	12.56	25.18	54.14	132.28
	TOTALS			1682.37	139.80	112.73	182.55	435.08	1247.29

B.

Name	Marital Status	No. of Exemptions	Gross Pay	Federal Income Tax	FICA	Other Deductions	Total Deductions	Net Pay
1. Churchman, Stephen	S	1	516.42	62.27	34.60	42.50	139.37	377.05
2. Franklin, Gary	M	4	495.52	27.47	33.20	36.95	97.62	397.90
3. Hall, Margaret	M	2	455.26	32.63	30.50	35.14	98.27	356.99
4. Laws, Michael	S	1	314.57	28.85	21.08	29.36	79.29	235.28
5. Miller, Coy	S	2	298.42	20.18	19.99	25.42	65.59	232.83
6. Stevenson, Anthony	M	5	456.89	17.83	30.61	38.95	87.39	369.50
7. Woodard, Mike	M	3	567.49	43.23	38.02	45.56	126.81	440.68
8. Young, Judy	S	1	485.42	56.38	32.52	43.21	132.11	353.31
TOTAL			3589.99	288.84	240.52	297.09	826.45	2763.54

C.

Name	M T W T F	Total Hours	Regular Hours	Regular Rate	O.T. Hours	O.T. Rate	Regular Pay	O.T. Pay	Total Pay	Marital Status	No. of Exemptions	Federal Income Tax	FICA	Other Deductions	Total Deductions	Net Pay
1. Atkinson, Stephen	8 9 9 8 9	43	40	3.15	3	4.725	126.00	14.18	140.18	M	5	0	9.39	23.15	32.54	107.64
2. Brasier, Scott	8 9 9 10 9	45	40	3.18	5	4.77	127.20	23.85	151.05	S	1	14.80	10.12	22.95	47.87	103.18
3. Collins, Melinda	8 8 8 8 8	40	40	2.98	0		119.20	0	119.20	S	1	9.10	7.99	19.86	36.95	82.25
4. Keith, Thomas	8 7 7 8 7	37	37	2.86	0		105.82	0	105.82	M	2	2.70	7.09	18.49	28.28	77.54
5. Norman, Gary	8 9 10 10 10	47	40	3.28	7	4.92	131.20	34.44	165.64	M	4	5.00	11.10	25.62	41.72	123.92
6. Smith, Vanessa	9 10 10 10 8	47	40	3.45	7	5.175	138.00	36.23	174.23	S	2	14.90	11.67	24.86	51.43	122.80
7. Wade, Douglas	9 9 9 9 8	44	40	3.50	4	5.25	140.00	21.00	161.00	M	6	0.40	10.79	23.14	34.33	126.67
8. White, Ian	9 8 8 8 8	41	40	3.26	1	4.89	130.40	4.89	135.29	M	3	4.00	9.06	26.87	39.93	95.36
TOTALS							1017.82	134.59	1152.41			50.90	77.21	184.94	313.05	839.36

D.

Name	Salary	Sales	Commission	Total Pay	Marital Status	No. of Exemptions	Federal Income Tax	FICA	Other Deductions	Total Deductions	Net Pay
1. Adams, Jerry	325.00	3,825.36	95.63	420.63	M	3	0	28.18	42.96	71.14	349.49
2. Carr, Tim	350.00	4,156.75	103.92	453.92	S	1	31.29	30.41	39.51	101.21	352.71
3. Curtis, Cheryl	330.00	3,526.33	88.16	418.16	S	1	26.14	28.02	35.50	89.66	328.50
4. Fleming, Jimmie	300.00	2,856.89	71.42	371.42	S	2	10.53	24.89	29.98	65.40	306.02
5. Hise, Leslie	337.50	2,546.71	63.67	401.17	M	4	0	26.88	26.75	53.63	347.54
6. Landsberger, Dale	340.00	3,922.52	98.06	438.06	S	1	28.91	29.35	36.72	94.98	343.08
7. Staggs, Patty	800.00	4,526.52	113.16	913.16	M	1	77.07	61.18	53.86	192.11	721.05
8. Staggs, Virgil	800.00	2,851.27	71.28	871.28	M	1	70.55	58.38	52.49	181.42	689.86
TOTALS	3582.50	28,212.35	705.30	4287.80			244.49	287.29	317.77	849.55	3438.25

A.

Pay Period	Gross Pay	Federal Income Tax	FICA	Other Deductions	Total Deductions	Net Pay
Jan.	427.86	7.34	28.67	25.57	61.58	366.28
Feb.	435.95	8.31	29.21	26.89	64.41	371.54
March	431.56	7.79	28.91	26.14	62.84	368.72
First Quarter	1295.37	23.44	86.79	78.60	188.83	1106.54

B.

Week Ending	Gross Pay	Federal Income Tax	FICA	Other Deductions	Total Deductions	Net Pay
1/6	178.50	17.80	11.96	15.40	45.16	133.34
1/13	180.00	19.30	12.06	16.72	48.08	131.92
1/20	180.00	19.30	12.06	16.72	48.08	131.92
1/27	179.00	17.80	11.99	16.55	46.34	132.66
2/3	195.75	20.80	13.12	17.14	51.06	144.69
2/10	210.50	22.40	14.10	18.57	55.07	155.43
2/17	205.00	22.40	13.74	18.13	54.27	150.73
2/24	195.00	20.80	13.07	17.09	50.96	144.04
3/3	180.00	19.30	12.06	16.72	48.08	131.92
3/10	178.25	17.80	11.94	15.36	45.10	133.15
3/17	170.50	17.80	11.42	14.56	43.78	126.72
3/24	195.00	20.80	13.07	17.09	50.96	144.04
3/31	198.75	20.80	13.32	17.53	51.65	147.10
Quarter Totals	2446.25	257.10	163.91	217.58	638.59	1807.66

C.

Employee	Social Security No.	Quarterly Earnings	Federal Tax Withheld	FICA Withheld
Ashmore, Harvey	423-58-1096	1,954.37	145.29	130.94
Buckner, Billy	564-48-2037	2,043.89	152.53	136.94
Davis, Leslie	209-47-8943	2,516.47	205.27	168.60
Hatcher, Deborah	325-86-8132	2,056.82	153.48	137.81
Humphries, Tina	558-31-8942	2,956.42	245.37	198.08
Jones, Judy	356-68-3529	1,836,25	142.35	123.03
Kinchion, John	321-54-4687	2,156.83	155.27	144.51
McDonald, Mark	456-31-5101	3,215.67	289.36	215.45
Wallis, Charlotte	105-25-4015	2,259.21	195.47	151.37
Whitworth, Brenda	538-86-3925	1,997.43	146.38	133.83
TOTALS		22,993.36	1,830.77	1540.56

| 1. $564.50 | 2. $34.57 | 3. $238.92 | 4. $303.66 |

5.

Name	M T W T F	Total Hours	Regular Hours	Regular Rate	O.T. Hours	O.T. Rate	Regular Pay	O.T. Pay	Gross Pay
Andree, Denise	8 8 7 8 7	38	38	4.14	0		157.32	0	157.32
Cook, Chris	8 9 9 9 9	44	40	4.25	4	6.375	170.00	25.50	195.50
Gatlin, Dennis	8 9 7 9 10	43	40	3.86	3	5.79	154.40	17.37	171.77
Hernandez, Nasario	9 9 10 9 9	46	40	4.15	6	6.225	166.00	37.35	203.35
Walker, Robert	9 8 8 8 8	41	40	4.02	1	6.03	160.80	6.03	166.83
TOTALS							808.52	86.25	894.77

6.   FIT = $33.80; FICA = $17.92

7.

Pay Period	Gross Pay	Federal Income Tax	FICA	Other Deductions	Total Deductions	Net Pay
Jan.	657.50	24.90	44.05	34.96	103.91	553.59
Feb.	721.56	32.59	48.34	38.37	119.30	602.26
March	694.28	29.31	46.52	36.92	112.75	581.53
First Quarter	2073.34	86.80	138.91	110.25	335.96	1737.38

8.

Employee	Social Security No.	Quarterly Earnings	Federal Income Tax	FICA
DeLong, Pamela	445-66-1023	2,857.86	313.96	191.48
Griggs, Lesa	443-64-7270	3,062.57	328.76	205.19
Holleman, William	054-36-0250	2,525.78	288.45	169.23
McManus, Michael	526-11-3538	2,896.42	298.76	194.06
Swyden, Randy	548-32-4156	3,158.59	326.42	211.63
TOTALS		14,501.22	1556.35	971.59

UNIT 7: SIMPLE INTEREST

Problem Set 1: Page 255

A.	1. 97	2. 73	3. 69	4. 79	5. 310
	6. 151	7. 31	8. 275	9. 105	10. 52
	11. 74	12. 40	13. 98	14. 68	15. 156
	16. 157	17. 261	18. 112	19. 102	20. 89
	21. 131	22. 100	23. 29	24. 185	25. 67
	26. 18	27. 191	28. 133	29. 78	30. 291

B.	1. 32	2. 68	3. 68	4. 95	5. 94
	6. 99	7. 79	8. 23	9. 57	10. 58
	11. 46	12. 87	13. 62	14. 278	15. 161
	16. 290	17. 47	18. 37	19. 175	20. 117
	21. 322	22. 211	23. 60	24. 79	25. 214
	26. 101	27. 17	28. 18	29. 115	30. 305

C.	1. 66	2. 151	3. 149	4. 210	5. 172
	6. 114	7. 136	8. 75	9. 74	10. 252
	11. 52	12. 79	13. 152	14. 103	15. 50
	16. 99	17. 112	18. 126	19. 69	20. 232
	21. 39	22. 83	23. 314	24. 56	25. 84
	26. 131	27. 204	28. 253	29. 253	30. 41

Problem Set 2: Page 267

A.
1. 219 days; $85.50; $835.50
2. 73 days; $4.80; $204.80
3. 146 days; $26.00; $526.00
4. 73 days; $26.40; $851.40
5. 302 days; $66.11; $536.11
6. 48 days; $20.24; $875.24
7. 87 days; $9.68; $334.68
8. 196 days; $28.39; $478.39
9. 81 days; $13.32; $342.32
10. 200 days; $49.25; $905.25
11. 98 days; $24.17; $499.17
12. 304 days; $53.63; $640.63
13. 252 days; $92.95; $842.95
14. 241 days; $125.44; $1395.44
15. 83 days; $90.96; $2590.96
16. 39 days; $24.31; $1324.31
17. 179 days; $38.25; $688.25
18. 235 days; $108.45; $983.45
19. 29 days; $82.29; $6282.89
20. 81 days; $379.55; $10,179.55
21. 226 days; $755.80; $8405.80
22. 280 days; $770.77; $6970.77
23. 109 days; $392.37; $8892.37
24. 186 days; $471.38; $7746.38
25. 65 days; $200.19; $8550.19

B.
1. 40 days; $18.00; $918.00
2. 102 days; $8.50; $258.50
3. 45 days; $9.50; $409.50
4. 135 days; $40.50; $640.50
5. 129 days; $34.04; $534.04
6. 62 days; $13.26; $713.26
7. 79 days; $14.25; $544.25
8. 71 days; $15.64; $690.64
9. 157 days; $25.51; $350.51
10. 154 days; $36.36; $886.36
11. 84 days; $12.45; $439.45
12. 233 days; $101.45; $937.45
13. 249 days; $93.38; $843.38
14. 237 days; $125.41; $1395.41
15. 82 days; $91.11; $2591.11
16. 39 days; $24.65; $1324.65
17. 178 days; $38.75; $688.57
18. 232 days; $108.55; $983.55
19. 29 days; $83.66; $6283.66
20. 80 days; $381.11; $10,181.11
21. 223 days; $758.20; $8408.20
22. 276 days; $772.42; $6972.42
23. 108 days; $395.25; $8895.25
24. 186 days; $479.24; $7754.24
25. 65 days; $203.53; $8553.53

C.
1. 40 days; $8.00; $608.00
2. 72 days; $19.00; $519.00
3. 45 days; $5.63; $255.63
4. 135 days; $13.50; $313.50
5. 85 days; $21.25; $521.25
6. 230 days; $71.56; $771.56
7. 98 days; $23.89; $473.89
8. 249 days; $30.48; $405.48
9. 66 days; $16.64; $841.64
10. 65 days; $16.47; $496.47
11. 51 days; $6.62; $262.62
12. 51 days; $7.76; $445.76
13. 252 days; $94.50; $844.50
14. 241 days; $127.53; $1397.53

15. 83 days; $92.22; $2592.22
16. 39 days; $24.65; $1324.65
17. 179 days; $38.78; $688.78
18. 235 days; $37.90; $912.90
19. 29 days; $83.66; $6283.66
20. 81 days; $1333.89; $11,133.89
21. 226 days; $768.40; $8418.40
22. 280 days; $520.54; $6720.54
23. 109 days; $398.91; $8898.91
24. 186 days; $479.24; $7754.24
25. 65 days; $203.53; $8553.53

Box: Page 273

1. $278.57
2. 1397.40
3. 12.1%
4. 9.2%
5. 158
6. 91

Problem Set 3: Page 279

A.
1. 500
2. 1500
3. 3000
4. 1200
5. 1200
6. 2000
7. 675
8. 450
9. 1000
10. 900
11. 700
12. 600
13. 400
14. 600
15. 700
16. 5600
17. 2500
18. 8000
19. 6400
20. 6800
21. 650
22. 3600

B.
1. 12%
2. 11%
3. 14%
4. 9%
5. 14%
6. 9%
7. 6%
8. 10%
9. 6%
10. 28%
11. 14%
12. 18%
13. 12.5%
14. 9.5%
15. 8.5%
16. 15%
17. 17%
18. 19%
19. 12%
20. 26%
21. 16%
22. 18½%

C.
1. 45
2. 45
3. 120
4. 36
5. 72
6. 60
7. 30
8. 60
9. 36
10. 40
11. 36
12. 24
13. 90
14. 180
15. 36
16. 90
17. 30
18. 72
19. 60
20. 36
21. 60
22. 72

D.
1. 250
2. 400
3. 9%
4. 38.6%
5. 146
6. 90
7. 800
8. 180
9. 6%
10. 11%
11. 146
12. 900
13. 400
14. 500
15. 12%
16. 9%
17. 219
18. 72
19. 2500
20. 240
21. 19%
22. $5840

Self-Test 7: Page 285

1. 114; 112
2. 17.51; 767.51
3. 4.50; 504.50
4. 6.25; 456.25
5. 700
6. 9.5%
7. 90

UNIT 8: BANK DISCOUNT, COMPOUND INTEREST, AND PRESENT VALUE

Box: Page 294

A.
1. 9.60; 890.40
2. 19.25; 730.75
3. 13.20; 786.80
4. 24.44; 1475.56
5. 14.06; 735.94
6. 18.23; 1231.77
7. 20.05; 854.95
8. 21.83; 948.17
9. 23.92; 1206.08
10. 27.88; 1722.12

B.
1. 927.45; 9.58; 917.87
2. 2510.45; 64.43; 2446.02
3. 842.93; 13.91; 829.02
4. 1569.00; 25.56; 1543.44
5. 776.69; 14.56; 762.13
6. 1313.19; 19.15; 1294.04
7. 901.69; 20.66; 881.03
8. 995.87; 22.41; 973.46
9. 1275.92; 24.81; 1251.11
10. 1801.04; 28.69; 1772.35

Problem Set 1: Page 295

A.
1. 3.75; 496.25
2. 18; 1182
3. 7.50; 742.50
4. 0.67; 199.33
5. 3.75; 246.25
6. 3.83; 571.17
7. 3; 797
8. 10; 990
9. 4.86; 695.14

10.	5.06; 669.94	11.	13.33; 1486.67	12.	2.11; 122.89
13.	1.91; 248.09	14.	11.46; 888.54	15.	14.51; 1085.49
16.	17.50; 1482.50	17.	21.67; 1228.83	18.	7.31; 772.69

B.
1.	918; 11.48; 906.52	2.	255.63; 2.56; 253.07
3.	579.31; 2.41; 576.90	4.	1122; 4.21; 1117.79
5.	202.25; 0.67; 201.58	6.	1218; 9.14; 1208.86
7.	127.92; 0.64; 127.28	8.	261.25; 6.53; 254.72
9.	757.50; 5.68; 751.82	10.	1560; 39; 1521
11.	510; 7.27; 502.73	12.	682.59; 6.37; 676.22
13.	1041.67; 33.66; 1008.01	14.	725.38; 21.91; 703.47
15.	812.33; 6.91; 805.42	16.	1287.50; 13.52; 1273.98
17.	1368.25; 13.23; 1355.02	18.	918.75; 18.50; 900.25

Problem Set 2: Page 303

A.
1.	796.92	2.	1122.41	3.	425.61	4.	1752.90	5.	250.16
6.	2228.92	7.	764.01	8.	2063.17	9.	2402.37	10.	478.25
11.	303.49	12.	2539.47	13.	1127.65	14.	873.55	15.	1680.44
16.	3151.73	17.	1544.66	18.	643.40				

B.
1.	371.49	2.	1690.74	3.	1453.36	4.	164.39	5.	919.48
6.	3210.89	7.	2505.90	8.	764.90	9.	553.75	10.	203.88
11.	989.35	12.	510.60	13.	1594.86	14.	2113.65	15.	508.35
16.	1207.22	17.	790.58	18.	948.81	19.	1618.50	20.	1997.48

Box: Page 307

1.	(a)	$1191.02	(b)	$1195.62	(c)	$1196.68	(d)	$1197.20
2.	(a)	$680.58	(b)	$672.97	(c)	$671.21	(d)	$670.35

Problem Set 3: Page 309

A.
1.	287.66	2.	354.68	3.	160.17	4.	782.84	5.	149.81
6.	450.75	7.	746.07	8.	776.29	9.	231.19	10.	332.05
11.	647.08	12.	144.93	13.	1137.98	14.	358.34	15.	652.41
16.	439.90	17.	487.97	18.	485.60				

B.
1.	443.59	2.	512.92	3.	363.83	4.	665.59	5.	991.68
6.	185.34	7.	3937.83	8.	631.34	9.	519.73	10.	362.73
11.	630.05	12.	667.23	13.	556.84	14.	8416.56	15.	191.80
16.	630.17	17.	587.29	18.	232.88	19.	518.82	20.	685.99

Self-Test 8: Page 311

1.	1.50; 198.50	2.	206; 1.55; 204.45	3.	535.96
4.	817.19	5.	148.09		

UNIT 9: CONSUMER MATH

Problem Set 1: Page 331

A.

1.

No. 217	$	315	89

March 2 19 84

TO *Southland Mortgage Co.*

FOR _____

BALANCE	637	25
DEPOSITS		
TOTAL	637	25
AMOUNT THIS CHECK	315	89
BALANCE	321	36

JOE BROKE
1234 ELM STREET
ANYWHERE, STATE 00000

$\frac{39-81}{1234}$

217

March 2 19 84

PAY TO THE
ORDER OF *Southland Mortgage Company* $ 315.89

Three Hundred Fifteen and $\frac{89}{100}$ _____ DOLLARS

LAST NATIONAL BANK
ADDRESS
ANYWHERE, STATE 00000

SPECIMEN

MEMO _____ *Joe M. Broke*

⑈ 1234 ⑈ 081 ⑈ 190519 ⑈

2.

No. 218	$	56	25

March 4 19 84

TO *South Gas Company*

FOR _____

BALANCE	321	36
DEPOSITS		
TOTAL	321	36
AMOUNT THIS CHECK	56	25
BALANCE	265	11

JOE BROKE
1234 ELM STREET
ANYWHERE, STATE 00000

$\frac{39-1}{1234}$

218

March 4 19 84

PAY TO THE
ORDER OF *South Gas Company* $ 56.25

Fifty - Six and $\frac{25}{100}$ _____ DOLLARS

LAST NATIONAL BANK
ADDRESS
ANYWHERE, STATE 00000

SPECIMEN

MEMO _____ *Joe M. Broke*

⑈ 1234 ⑈ 081 ⑈ 190519 ⑈

3.

No. 219	$	15	82

March 10 19 84

TO *Bell Telephone Co*

FOR _____

BALANCE	265	11
DEPOSITS	432	50
TOTAL	697	61
AMOUNT THIS CHECK	15	82
BALANCE	681	77

JOE BROKE
1234 ELM STREET
ANYWHERE, STATE 00000

$\frac{39-81}{1234}$

219

March 10 19 84

PAY TO THE
ORDER OF *Bell Telephone Company* $ 15.82

Fifteen and $\frac{82}{100}$ _____ DOLLARS

LAST NATIONAL BANK
ADDRESS
ANYWHERE, STATE 00000

SPECIMEN

MEMO _____ *Joe M. Broke*

⑈ 1234 ⑈ 081 ⑈ 190519 ⑈

4.

No. 220 $ *123.36*
March 19 19 *84*
TO *Slow Foods*
FOR _____

BALANCE	681	79
DEPOSITS		
TOTAL	681	79
AMOUNT THIS CHECK	123	36
BALANCE	558	43

JOE BROKE
1234 ELM STREET
ANYWHERE, STATE 00000

$\frac{39\text{-}81}{1234}$ 220

March 19 19 *84*

PAY TO THE
ORDER OF *Slow Foods* $ *123.36*

One Hundred Twenty Three and $\frac{36}{100}$ _____ DOLLARS

LAST NATIONAL BANK
ADDRESS   SPECIMEN
ANYWHERE, STATE 00000

MEMO _____   *Joe M. Broke*

I: 1234 "081 I: 190519 "

5.

No. 221 $ *136.14*
March 25 19 *84*
TO *MM Finance Co.*
FOR _____

BALANCE	558	43
DEPOSITS		
TOTAL	558	43
AMOUNT THIS CHECK	136	14
BALANCE	422	29

JOE BROKE
1234 ELM STREET
ANYWHERE, STATE 00000

$\frac{39\text{-}81}{1234}$ 221

March 25 19 *84*

PAY TO THE
ORDER OF *Money Magic Finance Co.* $ *136.14*

One Hundred Thirty-Six and $\frac{14}{100}$ _____ DOLLARS

LAST NATIONAL BANK
ADDRESS   SPECIMEN
ANYWHERE, STATE 00000

MEMO _____   *Joe M. Broke*

I: 1234 "081 I: 190519 "

B.

BANK RECONCILIATION FORM

Outstanding Checks	
Number	Amount
219	15 | 82
221	136 | 14
Total	151 | 96

1. Compare the cancelled checks with your records.
2. List any outstanding checks.
3. Total the outstanding checks.
4. Enter the bank balance here: $ *570.10*
5. Add any deposits not on the summary.   +$ *0*

    +$ _____

6. Total $ *570.10*
7. Enter outstanding check total and subtract.   -$ *151.96*

Corrected Bank Balance $ *418.14*

8. Enter checkbook balance. $ *422.29*
9. Enter any service charge and subtract.   -$ *4.15*

Corrected Checkbook Balance $ *418.14*

Corrected Checkbook Balance and Corrected Bank Balance must be equal.

C.

Check No.	Date	Check Issued to or Deposit	Amount of Check		✔	Amount of Deposit	Balance 641	87
517	4/2	Front Apts.	317	55			324	32
518	4/2	Electric Company	45	17			279	15
519	4/5	Gas Company	52	51			226	64
	4/7	Deposit				275 58	502	28
520	4/10	Quick Foods	117	43			384	74
521	4/12	Money Magic Co.	319	86			164	88
522	4/15	Rita's Boutique	48	92			115	96
523	4/20	Wilson Company	89	14			26	82
524	4/21	Lou's Surplus	17	95			8	87
	4/25	Deposit				215 16	224	03

D.

BANK RECONCILIATION FORM

Outstanding Checks		
Number	Amount	
521	219	86
523	89	14
Total	309	00

1. Compare the cancelled checks with your records.
2. List any outstanding checks.
3. Total the outstanding checks.
4. Enter the bank balance here: $ 312.45
5. Add any deposits not on the summary. +$ 215.16

   +$ _____

6. Total $ 527.61
7. Enter outstanding check total and subtract. -$ 309.00

Corrected Bank Balance $ 218.61

8. Enter checkbook balance. $ 224.03
9. Enter any service charge and subtract. -$ 5.42

Corrected Checkbook Balance $ 218.61

Corrected Checkbook Balance and Corrected Bank Balance must be equal.

Problem Set 2: Page 343

A.
1. 532, 32
2. 500, 50
3. 82, 7
4. 169, 19
5. 290, 40
6. 440, 40
7. 410, 60
8. 482, 64
9. 151, 25
10. 88, 9
11. 940, 45
12. 1010, 60
13. 1402, 202
14. 822, 72
15. 870; 120
16. 561, 36
17. 2962, 612
18. 7922; 2144
19. 989, 144
20. 662, 92

B.
1. 480; 30; 18.5%
2. 203; 3; 16%
3. 1408; 158; 23.5%
4. 641; 41; 18.9%
5. 176; 1; 11.9%
6. 825; 25; 20.6%
7. 1090; 165; 23.5%
8. 885.20; 113.20; 19.0%
9. 1030; 45; 21.0%
10. 171.50; 15.50; 18.5%
11. 238; 1; 7%
12. 193; 4; 17.9%
13. 1614.40; 264.40; 23.1%
14. 264.50; 14.50; 24.9%
15. 547.00; 22.00; 24.2%
16. 781.40; 31.40; 20.9%
17. 3046.50; 471.50; 25.3%
18. 4653.00; 903.00; 25.7%
19. 911.00; 91.00; 24.2%
20. 389.10; 19.10; 22.2%

C.
1. 700; 42; 11.1%
2. 950; 133; 13.4%
3. 115; 8.05; 12.9%
4. 500; 32.50; 12%
5. 4400; 704; 15.4%
6. 500; 42.50; 15.7%
7. 400; 45; 14.2%
8. 6050; 1270.50; 13.6%
9. 5225; 836; 15.4%
10. 3347; 853.49; 16.5%
11. 246; 19.68; 14.8%
12. 428; 32.10; 13.8%
13. 5150; 927; 17.3%
14. 8125; 2071.88; 16.5%
15. 7075; 1680.31; 18.4%
16. 6300; 1456.88; 17.9%
17. 2800; 315; 14.2%
18. 6150; 1014.75; 15.8%
19. 7750; 2576.88; 18.6%
20. 5695; 1494.94; 17.0%

Problem Set 3: Page 353

A.
1. $\frac{1}{2}$%
2. $\frac{2}{3}$%
3. 12%
4. 24%
5. 18%
6. $\frac{3}{4}$%
7. 20%
8. 30%
9. $\frac{5}{6}$%
10. 36%
11. 22%
12. 16%
13. $1\frac{1}{2}$%
14. $1\frac{3}{4}$%
15. 14%
16. $1\frac{2}{3}$%
17. 28%
18. 26%
19. $1\frac{5}{6}$%
20. $1\frac{7}{12}$%

B.
1. 351.51, 5.27
2. 100.32, 1.50
3. 849.50, 14.87
4. 0, 0
5. 102.89, 1.29
6. 191.70, 2.88
7. 324.91, 4.87
8. 167.18, 2.09
9. 362.77; 3.63
10. 440.16, 6.60
11. 469.78; 8.22
12. 311.25; 5.71
13. 802.14; 12.03
14. 84.32; 1.26
15. 233.42; 4.08
16. 47.23; 0.79
17. 167.25; 2.93
18. 930.13; 15.50
19. 801.79; 12.03
20. 550.32; 9.17

C.
1. 635, 9.19
2. 595, 8.69
3. 725, 10.31
4. 902, 12.53
5. 1150, 15.63
6. 805, 11.31
7. 405, 6.08
8. 590, 8.63
9. 847, 11.84
10. 702, 10.03
11. 630, 9.13
12. 490, 7.35
13. 470, 7.05
14. 695, 9.94
15. 2130, 27.88
16. 396, 5.94
17. 795, 11.19
18. 841, 11.76
19. 696, 9.95
20. 645, 9.31

D.
1. 508; 7.62; 571.34
2. 116; 1.74; 144.24
3. 745; 13.04; 890.91
4. 400; 5.00; 464.65
5. 330; 4.95; 613.20
6. 130; 1.63; 144.61
7. 165; 2.89; 260.64
8. 847; 12.71; 948.83
9. 102; 1.53; 117.59
10. 182; 3.19; 466.76
11. 261; 4.57; 308.52
12. 730; 13.38; 766.88
13. 322; 4.83; 343.78
14. 761; 12.68; 806.44
15. 722; 12.64; 860.82
16. 253; 4.43; 292.69
17. 399; 7.32; 432.61
18. 501; 8.35; 527.54
19. 510; 7.65; 654.37
20. 75; 1.13; 151.32

Problem Set 4: Page 367

A. 1. 2.50; 375.00   2. 5.25; 1050.00   3. 4.80; 240.00
    4. 0.90; 22.50   5. 17.00; 1190.00   6. 9.60; 816.00
    7. 22.50; 6750.00   8. 8.75; 850.00   9. 5.95; 1041.25
  10. 9.60; 2400.00

B. 1. $47,050   2. $2381.25   3. $9337.50   4. $7180
    5. $22,537.50   6. $21,312.50   7. $925   8. $19,162.50
    9. $3158.75   10. $7500

C. 1. 80; 40   2. 87; 43.50   3. 86; 43
    4. 92; 46   5. 76; 38   6. 443.75; 221.88
    7. 425; 212.50   8. 92.50; 46.25   9. 93; 46.50
  10. 94; 47

D. 1. 82; 1022.50; 8.02%   2. 86; 988.75; 8.70%   3. 92; 995; 9.25%
    4. 73; 1010; 7.23%   5. 88.75; 1005; 8.83%   6. 412.50; 5125; 8.05%
    7. 92.50; 997.50; 9.27%   8. 460; 4912.50; 9.36%   9. 95; 971.25; 9.78%
  10. 87.50; 993.75; 8.81%

Self-Test: Page 369

1.

Check No.	Date	Check Issued to or Deposit	Amount of Check		✓	Amount of Deposit		Balance	
								655	29
942	8/1	Badview Apts.	325	50				329	79
943	8/4	Electric Co.	58	92				270	87
944	8/7	Acme Finance Co.	182	75				88	12
	8/10	Deposit				255	86	343	98
945	8/12	Rita's Boutique	53	18				290	80
946	8/20	Wards	32	95				257	85
947	8/20	Fast Foods	82	17				175	68
948	8/26	Gas Company	45	89				129	79
	8/27	Deposit				114	08	243	87
949	8/29	Lou's Surplus	29	86				214	01

2.

BANK RECONCILIATION FORM

Outstanding Checks		
Number	Amount	
945	53	18
947	82	17
949	29	86
Total	165	21

1. Compare the cancelled checks with your records.
2. List any outstanding checks.
3. Total the outstanding checks.
4. Enter the bank balance here: $ 261.97
5. Add any deposits not on the summary. +$ 114.08

+$ _____

6. Total $ 376.05
7. Enter outstanding check total and subtract. −$ 165.21

Corrected Bank Balance $ 210.84

8. Enter checkbook balance. $ 214.01
9. Enter any service charge and subtract. −$ 3.17

Corrected Checkbook Balance $ 210.84

Corrected Checkbook Balance and Corrected Bank Balance must be equal.

3. (a) 10.60 (b) 20.4%
4. (a) 690.63 (b) 12.6%
5. 20%
6. (a) 357.03 (b) 5.36 (c) 444.56
7. 165
8. 4455
9. 8.88%

UNIT 10: REAL ESTATE MATHEMATICS

Problem Set 1: Page 387

A. 1. 74 yds 2. 2.25 miles 3. 1.5 miles 4. 16.5 ft
   5. 108 in. 6. 144 in. 7. 15,840 ft 8. 3520 yd
   9. 5 yds 10. 17 ft 11. 12 sq. yd 12. 1.5 acres
   13. 3 acres 14. 126 sq. ft 15. 87,120 sq. ft 16. 16,940 sq. yd
   17. 459 cu. ft 18. 3.5 cu. ft 19. 5184 cu. in. 20. 5 cu. yd

B. 1. 11,480 sq. ft 2. 18,690 sq. ft 3. 13,775 sq. ft 4. 1316 sq. yd
   5. 8550 sq. ft 6. 1696 sq. yd 7. 58,425 sq. ft 8. 43,560 sq. ft
   9. 11,340 sq. ft 10. 15,660 sq. ft 11. 10,625 sq. ft 12. 1776 sq. yd
   13. 11,223 sq. ft 14. 65,772 sq. ft 15. 13,677 sq. ft 16. 15,876 sq. ft

C. 1. 90 sq. ft   2. 49.5 sq. ft   3. 100 sq. ft   4. 8550 sq. ft
    5. 15 sq. yd   6. 910 sq. ft   7. 149.5 sq. ft   8. 1638 sq. yd
    9. 6096 sq. ft   10. 108 sq. ft   11. 5842 sq. ft   12. 108 sq. ft
    13. 450 sq. ft   14. 22,137.5 sq. ft   15. 720 sq. ft

D. 1. 2100 sq. ft   2. 1520 sq. ft   3. 1925 cu. ft   4. 2024 sq. ft
    5. 4525 sq. ft   6. 2500 sq. ft

E. 1. 480 cu. ft   2. 6 cu. yd   3. 1155 cu. ft   4. $59,850
    5. $31.50   6. 1380 sq. ft   7. $27   8. $66,487.50
    9. 2900 sq. ft   10. $105,750   11. $65,000   12. $49.50
    13. 1450 sq. ft   14. 810 cu. ft   15. 2 cu. yd.

Problem Set 2: Page 395

A. 1. $262.50, $11.92, $29,988.08   2. $521.38, $21.91, $58,178.09
    3. $427.88, $19.43, $48,880.57   4. $416.67, $22.12, $49,977.88
    5. $243.87, $37.75, $32,478.39   6. $603.17, $69.51, $76,120.41
    7. $96.14, $118.65, $11,713.85   8. $50.96, $152.90, $7259.27
    9. $179.04, $45.75, $23,826.52   10. $323.22, $133.04, $45,498.38
    11. $270.95, $111.78, $36,014.56   12. $204.63, $48.13, $22,794.82
    13. $619.79, $15.23, $59,484.77   14. $674.38, $14.23, $62,235.77
    15. $917.13, $12.31, $75,889.69   16. $1080.45, $9.97, $82,310.03
    17. $682.85, $15.13, $61,827.84   18. $551.93, $0.57, $51,946.25

B. 1. 2, 2%, $1150   2. 4, 4%, $1400   3. 4, 4%, $2090
    4. 4, 4%, $1680   5. 6, 6%, $3780   6. 6, 6%, $2370
    7. 4, 4%, $3120   8. 2, 2%, $770   9. 6, 6%, $2700
    10. 6, 6%, $3360   11. 8, 8%, $3964   12. 4, 4%, $1576
    13. 4, 4%, $2512   14. 6, 6%, $4512   15. 2, 2%, $1052
    16. 2, 2%, $978   17. 2, 2%, $1277   18. 6, 6%, $3306

Problem Set 3: Page 403

A. 1. 15; $210.52   2. 2; $24.79   3. 26; $404.44   4. 4; $65.53
    5. 0; $0   6. 1; $10.54   7. 19; $152.13   8. 17; $245.50
    9. 9; $52.82   10. 28; $217.03   11. 12; $237.05   12. 2; $48.36
    13. 10; $166.50   14. 27; $353.11   15. 20; $340.51   16. 18; $316.07
    17. 4; $93.04   18. 28; $583.16

B. 1. 30; $27.25; buyer   2. 289; $366.27; seller
    3. 296; $669.24; seller   4. 258; $900.26; buyer
    5. 331; $194.96; buyer   6. 319; $830.73; seller
    7. 170; $754.81; seller   8. 282; $581.91; buyer
    9. 188; $180.56; seller   10. 104; $125.72; buyer

C. 1. 271; $283.04   2. 163; $115.46   3. 638; $330.81   4. 819; $562.68
    5. 293; $366.25   6. 968; $604   7. 104; $100.82   8. 801; $565.15
    9. 339; $283.44   10. 312; $273.58

Problem Set 4: Page 409

A. 1. $33,600; $796.32   2. $14,750; $299.43   3. $10,800; $369.36
    4. $10,215; $394.30   5. $29,250; $535.28   6. $27,300; $952.77
    7. $73,800; $690.77   8. $33,250; $1376.55   9. $17,640; $560.95
    10. $53,950; $532.49   11. $18,016; $967.46   12. $40,150; $1549.79
    13. $21,600; $203.47   14. $36,725; $2074.96   15. $20,575; $1779.74
    16. $38,320; $1134.27   17. $16,425; $139.94   18. $24,750; $226.96

B. 1. $14,980; $434.42   2. $36,595; $768.50   3. $11,000; $209
    4. $22,200; $754.80   5. $32,800; $1262.80   6. $18,830; $442.51
    7. $35,750; $643.50   8. $57,000; $1311   9. $15,750; $389.81
   10. $51,150; $1496.14   11. $13,500; $438.75   12. $51,300; $1102.95
   13. $18,375; $496.13   14. $18,080; $343.52   15. $13,225; $244.66
   16. $41,275; $1310.48   17. $37,740; $688.76   18. $40,425; $1192.54

Self-Test: Page 411

1. 12,325 sq. ft   2. 1530 sq. ft
3. (a) 108,900 sq. ft; (b) 2.5 acres   4. $56,875
5. $6275   6. 10 cu. yd
7. (a) $187.96; (b) $62.04;   8. (a) 6; (b) $2910
    (c) $26,473.96   10. $262.71 to the seller
9. $127.80   12. $988.38
11. $323.81

UNIT 11: INSURANCE

Problem Set 1: Page 429

A. 1. $136.90; $253.27; $369.63   2. $72.50; $134.13; $195.75
    3. $130.50; $241.43; $352.35   4. $119.70; $221.45; $323.19
    5. $243.60; $450.66; $657.72   6. $163.35; $302.20; $441.05
    7. $183.30; $339.11; $494.91   8. $123.50; $228.48; $333.45
    9. $129.60; $239.76; $349.92   10. $130.20; $240.87; $351.54
   11. $650; $1202.50; $1755   12. $237.80; $439.93; $642.06
   13. $167.40; $309.69; $451.98   14. $338.20; $625.67; $913.14
   15. $178.60; $330.41; $482.22   16. $315; $582.75; $850.50

B. 1. $45.77   2. $73.32   3. $27.97   4. $42.53
    5. $35.56   6. $85.18   7. $23.14   8. $53.35
    9. $86.12   10. $85.40   11. $111.93   12. $138.38

C. 1. 99; $122.05; $327.95   2. 71; $112.50; $262.50
    3. 158; $108.22; $141.78   4. 59; $74.68; $387.32
    5. 163; $300.85, $246.15   6. 182; $127.20; $84.80
    7. 153; $83.00; $115.00   8. 95; $135.79; $231.21
    9. 192; $255.15; $149.85   10. 178; $193.12; $202.88
   11. 87; $197.88; $384.12   12. 189; $468.10; $286.90

D. 1. $9,000; $12,000; $15,000   2. $7777.78; $11,666.67; $15,555.56
    3. $20,000; $30,000; $25,000   4. $27,500; $19,642.86; $7857.14
    5. $6285.71; $9428.57; $6285.71   6. $15,000; $20,000; $20,000
    7. $1276.60; $1595.74; $2127.66   8. $3920; $5600; $4480
    9. $33,428.57; $30,642.86;   10. $5000; $10,000; $5000
       $13,928.57
   11. $15,000; $20,000; $15,000   12. $20,000; $25,000; $30,000

E. 1. $10,702.05   2. $15,937.50   3. $5000   4. $27,884.62
    5. $1578.95   6. $13,671.88   7. $12,000   8. $30,000
    9. $24,615.38   10. $18,445.12   11. $47,894.74 12. $12,000

Problem Set 2: Page 439

A. 1. $369   2. $338   3. $462   4. $685
    5. $467   6. $660   7. $424   8. $431
    9. $331   10. $355   11. $654   12. $353
   13. $425   14. $437   15. $331

B. 1. (a) $19,000; (b) $400   2. (a) 23,200; (b) $2,000
    3. (a) $52,500; (b) 0   4. (a) $34,750; (b) $13,000
    5. (a) $250; (b) $1328   6. (a) $50; (b) $2793
    7. (a) $25,000; (b) $36,500 8. (a) $250; (b) $3967

Problem Set 3: Page 447

A. 1. $160.90   2. $74.10   3. $118.75   4. $609.20
    5. $1247.10   6. $729.25   7. $430.40   8. $648.45
    9. $264.50   10. $364.20   11. $1043   12. $433.60
   13. $387.36   14. $640.25   15. $208.40   16. $280.50
   17. $565   18. $875.10   19. $1998.90 20. $1210.44

B. 1. $110; $245; 3 yr, 159d   2. $820; $1980; 12 yr, 341d
    3. 0; 0; 0   4. $2580; $5820; 17 yr, 118d
    5. $525; $1225; 2 yr, 232d   6. $200; $475; 1 yr, 50d
    7. $3280; $7920; 12 yr, 341d   8. $160; $380; 1 yr, 50d
    9. $6450; $14,550; 17 yr, 118d 10. $3015; $6180; 21 yr, 47d
   11. 0; 0; 0   12. $375; $875; 2 yr, 232d
   13. $1230; $2970; 12 yr, 341d   14. $176; $392; 3 yr, 159d
   15. $10,050; $20,600; 21 yr, 47d 16. $280; $665; 1 yr, 50d
   17. $5025; $10,300; 21 yr, 47d   18. $4515; $10,185; 17 yr, 118d
   19. $270; $630; 2 yr, 232d   20. $1148; $2772; 12 yr, 341d

---

Self-Test 11: Page 449

1. (a) $263.20; (b) $486.92; $710.64
2. $73.53
3. (a) $193.18; (b) $258.82
4. (a) $237.15; (b) $289.85
5. A $7,500; B $12,500; C $10,000
6. A $10,000; B $15,000
7. $18,181.82
8. $330
9. $635
10. (a) $30,000; (b) $2325
11. (a) $250; (b) $1564
12. $149.50
13. $268.95
14. (a) $1032; (b) $2328; (c) 17 yr, 118 d

UNIT 12: DEPRECIATION

Problem Set 1: Page 463

A. 1.

Year	Annual Depreciation	Accumulated Depreciation	Book Value
0	—	—	$6,000
1	1,080	$1,080	4,920
2	1,080	2,160	3,840
3	1,080	3,240	2,760
4	1,080	4,320	1,680
5	1,080	5,400	600

2.

Year	Annual Depreciation	Accumulated Depreciation	Book Value
0	—	—	$650
1	150	$150	500
2	150	300	350
3	150	450	200
4	150	600	50

3.

Year	Annual Depreciation	Accumulated Depreciation	Book Value
0	—	—	$8,400
1	1,800	$1,800	6,600
2	1,800	3,600	4,800
3	1,800	5,400	3,000
4	1,800	7,200	1,200

4.

Year	Annual Depreciation	Accumulated Depreciation	Book Value
0	—	—	800,000
1	31,200	31,200	768,800
2	31,200	62,400	737,600
3	31,200	93,600	706,400
4	31,200	124,800	675,200

5.

Year	Annual Depreciation	Accumulated Depreciation	Book Value
0	—	—	240,000
1	18,480	18,480	221,520
2	18,480	36,960	203,040
3	18,480	55,440	184,560
4	18,480	73,920	166,080
5	18,480	92,400	147,600
6	18,480	110,880	129,120
7	18,480	129,360	110,640
8	18,480	147,840	92,160
9	18,480	166,320	73,680
10	18,480	184,800	55,200

B.   1.

Year	Units Produced	Annual Depreciation	Accumulated Depreciation	Book Value
—	—	—	—	$4,200
1980	2,500	500	500	3,700
1981	3,800	760	1,260	2,940
1982	4,900	980	2,240	1,960
1983	3,900	780	3,020	1,180
1984	2,900	580	3,600	600

2.

Year	Miles	Annual Depreciation	Accumulated Depreciation	Book Value
—	—	—	—	8,400.00
1980	26,410	2,165.62	2,165.62	6,234.38
1981	28,550	2,341.10	4,506.72	3,893.28
1982	24,190	1,983.58	6,490.30	1,909.70
1983	20,850	1,709.70	8,200.00	200.00

3.

Year	Units Produced	Annual Depreciation	Accumulated Depreciation	Book Value
—	—	—	—	5,000.00
1980	15,000	937.50	937.50	4,062.50
1981	17,000	1,062.50	2,000.00	3,000.00
1982	19,000	1,187.50	3,187.50	1,812.50
1983	20,000	1,250.00	4,437.50	562.50
1984	6,000	375.00	4,812.50	187.50
1985	3,000	187.50	5,000.00	0

4.

Year	Units Produced	Annual Depreciation	Accumulated Depreciation	Book Value
—	—	—	—	6000.00
1980	1576	788.00	788.00	5212.00
1981	1652	826.00	1614.00	4386.00
1982	1856	928.00	2542.00	3458.00
1983	1885	942.50	3484.50	2515.50
1984	1842	921.00	4405.50	1594.50
1985	1989	994.50	5400.00	600.00

5.

Year	Balloons Inflated	Annual Depreciation	Accumulated Depreciation	Book Value
—	—	—	—	650.00
1980	403	100.75	100.75	549.25
1981	506	126.50	227.25	422.75
1982	553	138.25	365.50	284.50
1983	515	128.75	494.25	155.75
1984	423	105.75	600.00	50.00

Problem Set 2: Page 477

A.   1.

Year	Annual Depreciation	Accumulated Depreciation	Book Value
0	—	—	6000.00
1	2400.00	2400.00	3600.00
2	1440.00	3840.00	2160.00
3	864.00	4704.00	1296.00
4	518.40	5222.40	777.60
5	177.60	5400.00	600.00

2.

Year	Annual Depreciation	Accumulated Depreciation	Book Value
0	—	—	650.00
1	325.00	325.00	325.00
2	162.50	487.50	162.50
3	81.25	568.75	81.25
4	31.25	600.00	50.00

3.

Year	Annual Depreciation	Accumulated Depreciation	Book Value
0	—	—	8400
1	4200	4200	4200
2	2100	6300	2100
3	900	7200	1200
4	0	7200	1200

4.

Year	Annual Depreciation	Accumulated Depreciation	Book Value
0	—	—	800,000.00
1	48,000.00	48,000.00	752,000.00
2	45,120.00	93,120.00	706,880.00
3	42,412.80	135,532.80	664,467.20
4	39,868.03	175,400.83	624,599.17

5.

Year	Annual Depreciation	Accumulated Depreciation	Book Value
0	—	—	240,000.00
1	48,000.00	48,000.00	192,000.00
2	38,400.00	86,400.00	153,600.00
3	30,720.00	117,120.00	122,880.00
4	24,576.00	141,696.00	98,304.00
5	19,660.80	161,356.80	78,643.20
6	15,728.64	177,085.44	62,914.56
7	7,714.56	184,800.00	55,200.00
8	0	184,800.00	55,200.00
9	0	184,800.00	55,200.00
10	0	184,800.00	55,200.00

B. 1.

Year	Annual Depreciation	Accumulated Depreciation	Book Value
0	—	—	6000
1	1800	1800	4200
2	1440	3240	2760
3	1080	4320	1680
4	720	5040	960
5	360	5400	600

2.

Year	Annual Depreciation	Accumulated Depreciation	Book Value
0	—	—	650
1	240	240	410
2	180	420	230
3	120	540	110
4	60	600	50

3.

Year	Annual Depreciation	Accumulated Depreciation	Book Value
0	—	—	8400
1	2880	2880	5520
2	2160	5040	3360
3	1440	6480	1920
4	720	7200	1200

4.

Year	Annual Depreciation	Accumulated Depreciation	Book Value
0	—	—	800,000
1	60,000	60,000	740.00
2	57,600	117,600	682,400
3	55,200	172,800	627,200
4	52,800	225,600	574,400

5.

Year	Annual Depreciation	Accumulated Depreciation	Book Value
0	—	—	240,000
1	33,600	33,600	206,400
2	30,240	63,840	176,160
3	26,880	90,720	149,280
4	23,520	114,240	125,760
5	20,160	134,400	105,600
6	16,800	151,200	88,800
7	13,440	164,640	75,360
8	10,080	174,720	65,280
9	6,720	181,440	58,560
10	3,360	184,800	55,200

Problem Set 3: Page 491

A. 1.

Year	Annual Depreciation	Accumulated Depreciation	Book Value
			6000
1980	270	270	5730
1981	1080	1350	4650
1982	1080	2430	3570
1983	1080	3510	2490
1984	1080	4590	1410
1985	810	5400	600

2.

Year	Annual Depreciation	Accumulated Depreciation	Book Value
			650
1980	125	125	525
1981	150	275	375
1982	150	425	225
1983	150	575	75
1984	25	600	50

3.

Year	Annual Depreciation	Accumulated Depreciation	Book Value
			8400
1980	300	300	8100
1981	1800	2100	6300
1982	1800	3900	4500
1983	1800	5700	2700
1984	1500	7200	1200

4.

Year	Annual Depreciation	Accumulated Depreciation	Book Value
			800,000
1980	18,200	18,200	781,800
1981	31,200	49,400	750,600
1982	31,200	80,600	719,400
1983	31,200	111,800	688,200

5.

Year	Annual Depreciation	Accumulated Depreciation	Book Value
			240,000
1980	12,320	12,320	227,680
1981	18,480	30,800	209,200
1982	18,480	49,280	190,720
1983	18,480	67,760	172,240
1984	18,480	86,240	153,760
1985	18,480	104,720	135,280
1986	18,480	123,200	116,800
1987	18,480	141,680	98,320
1988	18,480	160,160	79,840
1989	18,480	178;640	61,360
1990	6,160	184,800	55,200

B. 1.

Year	Annual Depreciation	Accumulated Depreciation	Book Value
			6000.00
1980	600.00	600.00	5400.00
1981	2160.00	2760.00	3240.00
1982	1296.00	4056.00	1944.00
1983	777.60	4833.60	1166.40
1984	466.56	5300.16	699.84
1985	99.84	5400.00	600.00

2.

Year	Annual Depreciation	Accumulated Depreciation	Book Value
			650.00
1980	270.83	270.83	379.17
1981	189.59	460.42	189.58
1982	94.79	555.21	94.79
1983	44.79	600.00	50.00
1984	0	600.00	50.00

3.

Year	Annual Depreciation	Accumulated Depreciation	Book Value
			8400
1980	700	700	7700
1981	3850	4550	3850
1982	1925	6475	1925
1983	725	7200	1200
1984	0	7200	1200

4.

Year	Annual Depreciation	Accumulated Depreciation	Book Value
			800,000.00
1980	28,000.00	28,000.00	772,000.00
1981	46,320.00	74,320.00	725,680.00
1982	43,540.80	117,860.80	682,139.20
1983	40,928.35	158,789.15	641,210.85

5.

Year	Annual Depreciation	Accumulated Depreciation	Book Value
			240,000.00
1980	32,000.00	32,000.00	208,000.00
1981	41,600.00	73,600.00	166,400.00
1982	33,280.00	106,880.00	133,120.00
1983	26,624.00	133,504.00	106,496.00
1984	21,299.20	154,803.20	85,196.80
1985	17,039.36	171,842.56	68,157.44
1986	12,957.44	184,800.00	55,200.00
1987	0	184,800.00	55,200.00
1988	0	184,800.00	55,200.00
1989	0	184,800.00	55,200.00
1990	0	184,800.00	55,200.00

C. 1.

Year	Annual Depreciation	Accumulated Depreciation	Book Value
			6000
1980	450	450	5550
1981	1710	2160	3840
1982	1350	3510	2490
1983	990	4500	1500
1984	630	5130	870
1985	270	5400	600

2.

Year	Annual Depreciation	Accumulated Depreciation	Book Value
			650
1980	200	200	450
1981	190	390	260
1982	130	520	130
1983	70	590	60
1984	10	600	50

3.

Year	Annual Depreciation	Accumulated Depreciation	Book Value
			8400
1980	480	480	7920
1981	2760	3240	5160
1982	2040	5280	3120
1983	1320	6600	1800
1984	600	7200	1200

4.

Year	Annual Depreciation	Accumulated Depreciation	Book Value
			800,000
1980	35,000	35,000	765,000
1981	58,600	93,600	706,400
1982	56,200	149,800	650,200
1983	53,800	203,600	596,400

5.

Year	Annual Depreciation	Accumulated Depreciation	Book Value
			240,000
1980	22,400	22,400	217,600
1981	31,360	53,760	186,240
1982	28,000	81,760	158,240
1983	24,640	106,400	133,600
1984	21,280	127,680	112,320
1985	17,920	145,600	94,400
1986	14,560	160,160	79,840
1987	11,200	171,360	68,640
1988	7,840	179,200	60,800
1989	4,480	183,680	56,320
1990	1,120	184,800	55,200

Problem Set 4: Page 499

1.

Year	Depreciation	Accumulated Depreciation	Book Value
—	—	—	$17,800
1	$4,450	$4,450	13,350
2	6,764	11,214	6,586
3	6,586	17,800	0

2.

Year	Depreciation	Accumulated Depreciation	Book Value
—	—	—	$32,900
1	$4,935	$4,935	27,965
2	7,238	12,173	20,727
3	6,909	19,082	13,818
4	6,909	25,991	6,909
5	6,909	32,900	0

3.

Year	Depreciation	Accumulated Depreciation	Book Value
—	—	—	$2850.00
1	$427.50	$427.50	2422.50
2	627.00	1054.50	1795.50
3	598.50	1653.00	1197.00
4	598.50	2251.50	598.50
5	598.50	2850.00	0

4.

Year	Depreciation	Accumulated Depreciation	Book Value
—	—	—	$14,600
1	$3650	$3,650	10,950
2	5548	9,198	5,402
3	5402	14,600	0

5.

Year	Depreciation	Accumulated Depreciation	Book Value
—	—	—	$58,500
1	$4680	$4,680	53,820
2	8190	12,870	45,630
3	7020	19,890	38,610
4	5850	25,740	32,760
5	5850	31,590	26,910
6	5850	37,440	21,060
7	5265	42,705	15,795
8	5265	47,970	10,530
9	5265	53,235	5,265
10	5265	58,500	0

6.

Year	Depreciation	Accumulated Depreciation	Book Value
—	—	—	$8250.00
1	$2062.50	$2062.50	6187.50
2	3135.00	5197.50	3052.50
3	3052.50	8250.00	0

Self-Test 12A: Page 501

1.

Year	Annual Depreciation	Accumulated Depreciation	Book Value
0	—	—	$6000
1	1080	1080	4920
2	1080	2160	3840
3	1080	3240	2760
4	1080	4320	1680
5	1080	5400	600

2.

Year	Annual Depreciation	Accumulated Depreciation	Book Value
0	—	—	$6000.00
1	$2400.00	$2400.00	3600.00
2	1440.00	3840.00	2160.00
3	864.00	4704.00	1296.00
4	518.40	5222.40	777.60
5	177.60	5400.00	600.00

3.

Year	Annual Depreciation	Accumulated Depreciation	Book Value
0	—	—	$6000
1	$1800	$1800	4200
2	1440	3240	2760
3	1080	4320	1680
4	720	5040	960
5	360	5400	600

4.

Year	Units	Annual Depreciation	Accumulated Depreciation	Book Value
—	—	—	—	$9700.00
1980	12,560	$2336.16	$2336.16	7363.84
1981	17,890	3327.54	5663.70	4036.30
1982	13,120	2440.32	8104.02	1595.98
1983	6,430	1195.98	9300.00	400.00

5.

Year	Depreciation	Accumulated Depreciation	Book Value
—	—	—	$13,200
1	$3300	$3,300	9,900
2	5016	8,316	4,884
3	4884	13,200	0

Self-Test 12B: Page 503

1.

Year	Annual Depreciation	Accumulated Depreciation	Book Value
			4200
1980	300	300	3900
1981	900	1200	3000
1982	900	2100	2100
1983	900	3000	1200
1984	600	3600	600

2.

Year	Annual Depreciation	Accumulated Depreciation	Book Value
			4200
1980	700	700	3500
1981	1750	2450	1750
1982	875	3325	875
1983	275	3600	600
1984	0	3600	600

3.

Year	Annual Depreciation	Accumulated Depreciation	Book Value
			4200
1980	480	480	3720
1981	1320	1800	2400
1982	960	2760	1440
1983	600	3360	840
1984	240	3600	600

4.

Year	Units Produced	Annual Depreciation	Accumulated Depreciation	Book Value
				9700.00
1980	12,560	2336.16	2336.16	7363.84
1981	17,890	3327.54	5663.70	4036.30
1982	13,120	2440.32	8104.02	1595.98
1983	6,430	1195.98	9300.00	400.00

5.

Year	Depreciation	Accumulated Depreciation	Book Value
—	—	—	$13,200
1	$3300	$3,300	9,900
2	5016	8,316	4,884
3	4884	13,200	0

Problem Set 1: Page 525

A.

1.

PEKARA MANUFACTURING COMPANY COMPARATIVE BALANCE SHEETS DECEMBER 31, 1983 AND 1984			Increase or Decrease	
	1983	1984	Amount	Percent
Assets				
Current Assets				
Cash	$ 27,000	$ 30,000	$ 3,000	11.1%
Merchandise inventory	78,000	85,000	7,000	9.0%
Accounts receivable	75,000	76,000	1,000	1.3%
Total current assets	$180,000	$191,000	$11,000	6.1%
Plant and Equipment				
Land	$ 20,000	$ 20,000	$ 0	0%
Buildings	72,000	72,000	0	0%
Equipment	85,000	92,000	7,000	8.2%
Total plant and equipment	$177,000	$184,000	$ 7,000	4.0%
Total assets	$357,000	$357,000	$18,000	5.0%
Liabilities and Capital				
Current Liabilities				
Accounts payable	$ 72,000	$ 75,000	$ 3,000	4.2%
Wages payable	25,000	27,000	2,000	8.0%
Total current liabilities	$ 97,000	$102,000	$ 5,000	5.2%
Long-Term Liabilities				
Mortgage payable	80,000	78,000	2,000*	2.5%*
Total liabilities	$177,000	$180,000	$ 3,000	1.7%
Pekara, Capital	180,000	195,000	15,000	8.3%
Total liabilities and capital	$357,000	$375,000	$18,000	5.0%

2.

BRAKEBILL LAB SUPPLY, INC.
COMPARATIVE BALANCE SHEETS
JUNE 30, 1983 AND 1984

	1983	1984	Increase or Decrease	
			Amount	Percent
Assets				
Current Assets				
Cash	$ 5,000	$ 6,000	$1,000	20.0%
Merchandise inventory	16,000	20,000	4,000	25.0%
Accounts receivable	9,000	9,500	500	5.6%
Total current assets	$30,000	$35,500	$5,500	18.3%
Plant and Equipment				
Land	$ 4,000	$ 4,000	$ 0	0%
Buildings	15,000	16,000	1,000	6.7%
Equipment	16,000	18,000	2,000	12.5%
Total plant and equipment	$35,000	$38,000	$3,000	8.6%
Total assets	$65,000	$73,500	$8,500	13.1%
Liabilities and Capital				
Current Liabilities				
Accounts payable	$ 7,000	$ 8,500	$1,500	21.4%
Wages payable	5,000	5,500	500	10.0%
Total current liabilities	$12,000	$14,000	$2,000	16.7%
Long-Term Liabilities				
Mortgage payable	13,000	12,500	500*	3.8%*
Total liabilities	$25,000	$26,500	$1,500	6.0%
Brakebill, Capital	40,000	47,000	7,000	17.5%
Total liabilities and capital	$65,000	$73,500	$8,500	13.1%

B.

1.

PEKARA MANUFACTURING COMPANY COMPARATIVE BALANCE SHEETS DECEMBER 31, 1983 AND 1984	1983		1984	
	Amount	Percent	Amount	Percent
Assets				
Current Assets				
Cash	$ 27,000	7.6%	$ 30,000	8.0%
Merchandise inventory	78,000	21.8%	85,000	22.7%
Accounts receivable	75,000	21.0%	76,000	20.3%
Total current assets	$180,000	50.4%	$191,000	50.9%
Plant and Equipment				
Land	$ 20,000	5.6%	$ 20,000	5.3%
Buildings	72,000	20.2%	72,000	19.2%
Equipment	85,000	23.8%	92,000	24.5%
Total plant and equipment	$177,000	49.6%	$184,000	49.1%
Total assets	$357,000	100.0%	$357,000	100.0%
Liabilities and Capital				
Current Liabilities				
Accounts payable	$ 72,000	20.2%	$ 75,000	20.2%
Wages payable	25,000	7.0%	27,000	7.2%
Total current liabilities	$ 97,000	27.2%	$102,000	27.2%
Long-Term Liabilities				
Mortgage payable	80,000	22.4%	78,000	20.8%
Total liabilities	$177,000	49.6%	$180,000	48.0%
Pekara, Capital	180,000	50.4%	195,000	52.0%
Total liabilities and capital	$357,000	100.0%	$375,000	100.0%

2.

BRAKEBILL LAB SUPPLY, INC. COMPARATIVE BALANCE SHEETS JUNE 30, 1983 AND 1984				
	1983		1984	
	Amount	Percent	Amount	Percent
Assets				
Current Assets				
Cash	$ 5,000	7.7%	$ 6,000	8.2%
Merchandise inventory	16,000	24.6%	20,000	27.2%
Accounts receivable	9,000	13.8%	9,500	12.9%
Total current assets	$30,000	46.2%	$35,500	48.3%
Plant and Equipment				
Land	$ 4,000	6.2%	$ 4,000	5.4%
Buildings	15,000	23.1%	16,000	21.8%
Equipment	16,000	24.6%	18,000	24.5%
Total plant and equipment	$35,000	53.8%	$38,000	51.7%
Total assets	$65,000	100.0%	$73,500	100.0%
Liabilities and Capital				
Current Liabilities				
Accounts payable	$ 7,000	10.8%	$ 8,500	11.6%
Wages payable	5,000	7.7%	5,500	7.5%
Total current liabilities	$12,000	18.5%	$14,000	19.0%
Long-Term Liabilities				
Mortgage payable	13,000	20.0%	12,500	17.0%
Total liabilities	$25,000	38.5%	$26,500	36.1%
Brakebill, Capital	40,000	61.5%	47,000	63.9%
Total liabilities and capital	$65,000	100.0%	$73,500	100.0%

Problem Set 2: Page 539

A.

1.

TALL COMPANY COMPARATIVE INCOME STATEMENTS FOR THE YEARS ENDED DECEMBER 31, 1983 AND 1984				
			Increase or Decrease	
	1983	1984	Amount	Percent
Income				
Sales	$95,000	$110,000	$15,000	15.8%
Less sales returns	2,000	2,500	500	25.0%
Net sales	$93,000	$107,500	$14,500	15.6%
Cost of goods sold:				
Inventory, January 1	$13,000	$ 12,000	$ 1,000*	7.7%*
Purchases	62,000	71,000	9,000	14.5%
Goods available for sale	$75,000	$ 83,000	$ 8,000	10.7%
Less inventory, December 31	12,000	14,000	2,000	16.7%
Cost of goods sold	$63,000	$ 69,000	$ 6,000	9.5%
Gross profit	$30,000	$ 38,500	$ 8,500	28.3%
Operating expenses:				
Selling	$12,000	$ 15,000	$ 3,000	25.0%
Rent	3,000	3,500	500	16.7%
Utilities	500	1,000	500	100.0%
Miscellaneous	500	1,500	1,000	200.0%
Total operating expenses	$16,000	$ 21,000	$ 5,000	31.3%
Net Income	$14,000	$ 17,500	$ 3,500	25.0%

2.

			Increase or Decrease	
	1983	1984	Amount	Percent
Income:				
Sales	$460,000	$515,000	$55,000	12.0%
Less sales returns	3,000	4,000	1,000	33.3%
Net sales	$457,000	$511,000	$54,000	11.8%
Cost of goods sold:				
Inventory, January 1	$ 39,000	$ 42,000	$ 3,000	7.7%
Purchases	359,000	402,000	43,000	12.0%
Goods available for sale	$398,000	$444,000	$46,000	11.6%
Less inventory, December 31	42,000	41,000	1,000*	2.4%*
Cost of goods sold	$356,000	$403,000	$47,000	13.2%
Gross profit	$101,000	$108,000	$ 7,000	6.9%
Operating expenses:				
Selling	$ 32,000	$ 35,000	$ 3,000	9.4%
Rent	15,000	16,000	1,000	6.7%
Utilities	7,000	7,500	500	7.1%
Miscellaneous	11,000	10,500	500*	4.5%*
Total operating expenses	$ 65,000	$ 69,000	$ 4,000	6.2%
Net Income	$ 36,000	$ 39,000	$ 3,000	8.3%

E. LESLIE, INC.
COMPARATIVE INCOME STATEMENTS
FOR THE YEARS ENDED DECEMBER 31, 1983 AND 1984

B.

1.

TALL COMPANY COMPARATIVE INCOME STATEMENTS FOR THE YEARS ENDED DECEMBER 31, 1983 AND 1984	1983		1984	
	Amount	Percent	Amount	Percent
Income:				
Sales	$95,000	102.2%	$110,000	102.3%
Less sales returns	2,000	2.2%	2,500	2.3%
Net sales	$93,000	100.0%	$107,500	100.0%
Cost of goods sold:				
Inventory, January 1	$13,000	14.0%	$ 12,000	11.2%
Purchases	62,000	66.7%	71,000	66.0%
Goods available for sale	$75,000	80.6%	$ 83,000	77.2%
Less inventory, December 31	12,000	12.9%	14,000	13.0%
Cost of goods sold	$63,000	67.7%	$ 69,000	64.2%
Gross profit	$30,000	32.3%	$ 38,500	35.8%
Operating expenses:				
Selling	$12,000	12.9%	$ 15,000	14.0%
Rent	3,000	3.2%	3,500	3.3%
Utilities	500	0.5%	1,000	0.9%
Miscellaneous	500	0.5%	1,500	1.4%
Total operating expenses	$16,000	17.2%	$ 21,000	19.5%
Net Income	$14,000	15.1%	$ 17,500	16.3%

2.

E. LESLIE, INC. COMPARATIVE INCOME STATEMENTS FOR THE YEARS ENDED DECEMBER 31, 1983 AND 1984				
	1983		**1984**	
	Amount	Percent	Amount	Percent
Income:				
Sales	$460,000	100.7%	$515,000	100.8%
Less sales returns	3,000	0.7%	4,000	0.8%
Net sales	$457,000	100.0%	$511,000	100.0%
Cost of goods sold:				
Inventory, January 1	$ 39,000	8.5%	$ 42,000	8.2%
Purchases	359,000	78.6%	402,000	78.7%
Goods available for sale	$398,000	87.1%	$444,000	86.9%
Less inventory, December 31	42,000	9.2%	41,000	8.0%
Cost of goods sold	$356,000	77.9%	$403,000	78.9%
Gross profit	$101,000	22.1%	$108,000	21.1%
Operating expenses:				
Selling	$ 32,000	7.0%	$ 35,000	6.8%
Rent	15,000	3.3%	16,000	3.1%
Utilities	7,000	1.5%	7,500	1.5%
Miscellaneous	11,000	2.4%	10,500	2.1%
Total operating expenses	$ 65,000	14.2%	$ 69,000	13.5%
Net Income	$ 36,000	7.9%	$ 39,000	7.6%

Problem Set 3: Page 547

A.　1.　2.3　　　　2.　2.1　　　　3.　1.9　　　　4.　2.0
　　5.　2.4　　　　6.　1.8　　　　7.　2.7　　　　8.　2.1

B.　1.　1.2　　　　2.　0.9　　　　3.　1.1　　　　4.　1.5
　　5.　1.2　　　　6.　0.9　　　　7.　1.7　　　　8.　1.3

C.　1.　8%　　　　2.　20%　　　　3.　15%　　　　4.　5%
　　5.　8.5%　　　6.　24%　　　　7.　9.3%　　　　8.　12.3%

D.　1.　13,000; 7.8　　2.　21,500; 7.6　　3.　47,500; 13.4　　4.　28,250; 8.2
　　5.　19,000; 13.7　　6.　54,250; 6.8　　7.　7750; 24.8　　8.　18,000; 11.7

Self-Test: Page 549

1.

<table>
<tr><td colspan="5">J. NICHOLS, INC.
COMPARATIVE BALANCE SHEETS
DECEMBER 31, 1983 AND 1984</td></tr>
<tr><td></td><td colspan="2">1983</td><td colspan="2">1984</td></tr>
<tr><td></td><td>Amount</td><td>Percent</td><td>Amount</td><td>Percent</td></tr>
<tr><td>**Assets**</td><td></td><td></td><td></td><td></td></tr>
<tr><td>*Current Assets*</td><td></td><td></td><td></td><td></td></tr>
<tr><td>Cash</td><td>$ 12,000</td><td>10.7%</td><td>$ 13,000</td><td>10.4%</td></tr>
<tr><td>Merchandise inventory</td><td>45,000</td><td>40.2%</td><td>48,000</td><td>38.2%</td></tr>
<tr><td>Accounts receivable</td><td>30,000</td><td>26.8%</td><td>35,000</td><td>27.9%</td></tr>
<tr><td>Total current assets</td><td>$ 87,000</td><td>77.7%</td><td>$ 96,000</td><td>76.5%</td></tr>
<tr><td>*Plant and Equipment*</td><td></td><td></td><td></td><td></td></tr>
<tr><td>Land</td><td>$ 8,000</td><td>7.1%</td><td>$ 8,000</td><td>6.4%</td></tr>
<tr><td>Buildings</td><td>12,000</td><td>10.7%</td><td>17,000</td><td>13.5%</td></tr>
<tr><td>Equipment</td><td>5,000</td><td>4.5%</td><td>4,500</td><td>3.6%</td></tr>
<tr><td>Total plant and equipment</td><td>$ 25,000</td><td>22.3%</td><td>$ 29,500</td><td>23.5%</td></tr>
<tr><td>Total assets</td><td>$112,000</td><td>100.0%</td><td>$125,500</td><td>100.0%</td></tr>
<tr><td>**Liabilities and Capital**</td><td></td><td></td><td></td><td></td></tr>
<tr><td>*Current Liabilities*</td><td></td><td></td><td></td><td></td></tr>
<tr><td>Accounts payable</td><td>$ 33,000</td><td>29.5%</td><td>$ 34,500</td><td>27.5%</td></tr>
<tr><td>Wages payable</td><td>5,000</td><td>4.5%</td><td>5,500</td><td>4.4%</td></tr>
<tr><td>Total current liabilities</td><td>$ 38,000</td><td>33.9%</td><td>$ 40,000</td><td>31.9%</td></tr>
<tr><td>*Long-Term Liabilities*</td><td></td><td></td><td></td><td></td></tr>
<tr><td>Mortgage payable</td><td>15,000</td><td>13.4%</td><td>18,000</td><td>14.3%</td></tr>
<tr><td>Total liabilities</td><td>$ 53,000</td><td>47.3%</td><td>$ 58,000</td><td>46.2%</td></tr>
<tr><td>J. Nichols, Capital</td><td>59,000</td><td>52.7%</td><td>67,500</td><td>53.8%</td></tr>
<tr><td>Total liabilities and capital</td><td>$112,000</td><td>100.0%</td><td>$125,500</td><td>100.0%</td></tr>
</table>

2.

<table>
<tr><td colspan="6">S. KAMM, INC.
COMPARATIVE INCOME STATEMENTS
FOR THE YEARS ENDED DECEMBER 31, 1983 AND 1984</td></tr>
<tr><td rowspan="2"></td><td rowspan="2">1983</td><td rowspan="2">1984</td><td colspan="2">Increase or Decrease</td></tr>
<tr><td>Amount</td><td>Percent</td></tr>
<tr><td>*Income:*
Sales
Less sales returns</td><td>$79,000
4,000</td><td>$85,000
3,500</td><td>$6,000
500*</td><td>7.6%
12.5%*</td></tr>
<tr><td>Net sales</td><td>$75,000</td><td>$81,500</td><td>$6,500</td><td>8.7%</td></tr>
<tr><td>*Cost of goods sold:*
Inventory, January 1
Purchases</td><td>$24,000
52,000</td><td>$27,000
55,000</td><td>$3,000
3,000</td><td>12.5%*
5.8%</td></tr>
<tr><td>Goods available for sale
Less inventory, December 31</td><td>$76,000
27,000</td><td>$82,000
30,000</td><td>$6,000
3,000</td><td>7.9%
11.1%</td></tr>
<tr><td>Cost of goods sold</td><td>$49,000</td><td>$52,000</td><td>$3,000</td><td>6.1%</td></tr>
<tr><td>*Gross profit*
Operating expenses:
Selling
Rent
Utilities
Miscellaneous</td><td>$26,000

$ 9,000
4,000
1,500
500</td><td>$29,500

$10,000
4,500
1,600
400</td><td>$3,500

$1,000
500
100
100*</td><td>13.5%

11.1%
12.5%
6.7%
20.0%*</td></tr>
<tr><td>Total operating expenses</td><td>$15,000</td><td>$16,500</td><td>$1,500</td><td>10.0%</td></tr>
<tr><td>*Net Income*</td><td>$11,000</td><td>$13,000</td><td>$2,000</td><td>18.2%</td></tr>
</table>

3. 2.4   4. 1.2   5. 1.8   6. 15%

THE ELECTRONIC CALCULATOR

Exercises A-1: Page 554

8. 0.7   9. 6.5   10. 3.37   11. 4.25   12. 1.3333333
14. 0.16   15. 3021   16. 2.1

Exercises A-2: Page 554

1. $4.20   2. $0.40 or 40¢   3. $0.02 or 2¢
4. $11.74 rounded   5. $431.75 rounded   6. $3,786,506.10
7. $0.39 or 39¢   8. $6.01   9. $29,214
10. 4.1   11. 189.0   12. 3278.6
13. 0.2   14. 4.3   15. 5.7
16. 0.49   17. 3.70   18. 137.52
19. 12.48   20. 6.85   21. 4.01

Exercises A-3: Page 556

1. 3976	2. 97.28	3. 1110
4. 258,741	5. 28.8	6. 1.3672131
7. 1678.55	8. 61,351.369	9. 0.036
10. 0.0364422	11. $48.24	12. $16.391304
13. 8.25	14. $58.50	15. $2.20
16. $0.05 or 5¢	17. $1181.25	

Exercises A-4: Page 560

1. 121	2. 53.89	3. 230.17
4. 16.5	5. 814	6. 1.895
7. $4.92	8. $7.63	9. $0.29
10. 1,202.96	11. 4.298346	12. 64.6416
13. 0.0376534	14. 2.1331578	15. 86.970697
16. 1.2265625	17. 9.4333333	18. 1.4607329
19. 12.625	20. 8.7456	21. 25.416666
22. $688.16	23. 18.3	

Exercises A-5: Page 562

1. 4 R5	2. 14 R3	3. 11 R7
4. 24 R6	5. 8 R15	6. 1 R42
7. 3 R2	8. 4 R766	9. 4 R3353
10. 103 R5	11. 29 R25	12. 12 R15
13. 4 days, 16 hours	14. 30 feet, 10 inches	15. 3 hours, 32 minutes

METRIC SYSTEM

Exercises A-6: Page 565

1. 4720 dg	2. 784.2 cℓ	3. 372 dkm
4. 370 mg	5. 5.287 kℓ	6. 4230 mm
7. 84,000 g	8. 0.728 dkℓ	9. 0.458 km
10. 7500 cg	11. 3.42 dkm	12. 820 mg
13. 4.3 hℓ	14. 3.5 km	15. 320 g
16. 4200 mm	17. 10 km	18. 850 mℓ
19. 0.85 ℓ	20. 2.5 kg	

Exercises A-7: Page 566

1. unlikely	2. likely	3. unlikely	4. unlikely
5. unlikely	6. likely	7. unlikely	8. unlikely
9. likely	10. unlikely		

Exercises A-8: Page 568

1. 1.97 cm	2. 2.86 lb	3. 11.34 ℓ
4. 78.8 in.	5. 4.54 kg	6. 3.105 mi
7. 5.676 ℓ	8. 2.27 kg	9. 6.21 mi
10. 5.28 gal	11. 3.94 in.	12. 3.152 in.
13. 16,000 m	14. 141.5 g	15. 110 lb
16. 7.56 ℓ	17. 4.24 qt	18. 68.1 kg
19. 88.55 km	20. 4.57 m	

Index